The Culture of Flushing

The Nature | History | Society series is devoted to the publication of high-quality scholarship in environmental history and allied fields. Its broad compass is signalled by its title: nature because it takes the natural world seriously; history because it aims to foster work that has temporal depth; and society because its essential concern is with the interface between nature and society, broadly conceived. The series is avowedly interdisciplinary and is open to the work of anthropologists, ecologists, historians, geographers, literary scholars, political scientists, sociologists, and others whose interests resonate with its mandate. It offers a timely outlet for lively, innovative, and well-written work on the interaction of people and nature through time in North America.

Other books in the series include *States of Nature: Conserving Canada's Wildlife in the Twentieth Century*, by Tina Loo, and *Shaped by the West Wind: Nature and History in Georgian Bay*, by Claire Elizabeth Campbell.

General Editor: Graeme Wynn, University of British Columbia

The Culture of Flushing

A Social and Legal History of Sewage

JAMIE BENIDICKSON

FOREWORD BY GRAEME WYNN

UBC Press • Vancouver • Toronto

© UBC Press 2007

All rights reserved. No part of this publication may be reproduced, stored in a retrieval system, or transmitted, in any form or by any means, without prior written permission of the publisher, or, in Canada, in the case of photocopying or other reprographic copying, a licence from Access Copyright (Canadian Copyright Licensing Agency), www.accesscopyright.ca.

15 14 13 12 11 10 09 08   5 4 3 2

Printed in Canada on ancient-forest-free paper (100% post-consumer recycled) that is processed chlorine- and acid-free, with vegetable-based inks.

Library and Archives Canada Cataloguing in Publication Data

Benidickson, Jamie
   The culture of flushing: a social and legal history of sewage / Jamie Benidickson.

(Nature, history, society, ISSN 1713-6687)
Includes bibliographical references and index.
ISBN: 978-0-7748-1291-7 (bound); 978-0-7748-1292-4 (pbk.)

   1. Sewage – Social aspects – History. 2. Sewage – Law and legislation – History. I. Title. II. Series.

TD730.B45 2006   363.72'8   C2006-905360-X

Canadä

UBC Press gratefully acknowledges the financial support for our publishing program of the Government of Canada through the Book Publishing Industry Development Program (BPIDP), and of the Canada Council for the Arts, and the British Columbia Arts Council.

This book has been published with the help of a grant from the Canadian Federation for the Humanities and Social Sciences, through the Aid to Scholarly Publications Programme, using funds provided by the Social Sciences and Humanities Research Council of Canada.

UBC Press
The University of British Columbia
2029 West Mall
Vancouver, BC V6T 1Z2
604-822-5959 / Fax: 604-822-6083
www.ubcpress.ca

Contents

Foreword: Risk and Responsibility in a Waste-Full World / vii
Graeme Wynn

Acknowledgments / xxiii

Introduction: The Culture of Flushing / 3

1. The Advantage of a Flow of Water / 11
2. Navigating Aquatic Priorities / 31
3. A Source of Civic Pride / 57
4. The Water Closet Revolution / 78
5. Municipal Evacuation / 98
6. Learning to Live Downstream / 128
7. The Bacterial Assault on Local Government / 154
8. The Dilutionary Impulse at Chicago / 183
9. Separating Water from the Waterways / 213
10. Streams Are Nature's Sewers / 244
11. Riparian Resurrection / 267
12. Governing Water / 291

Conclusion: Water Quality and the Future of Flushing / 322

Notes / 333

Suggested Reading / 391

Illustration Credits / 395

Index / 397

Foreword by Graeme Wynn

RISK AND RESPONSIBILITY IN A WASTE-FULL WORLD

GRADE-SCHOOL HUMOUR is not something that one normally associates with scholarly books from university presses, but in the final pages of this wide-ranging and thought-provoking study, lawyer-historian Jamie Benidickson carries readers back to the schoolyard with a story about a six-year-old caller asking the hosts of a Calgary phone-in radio show why Tigger put his head in the toilet. While listeners with fond memories of Christopher Robin and Eeyore and the other characters who wandered A.A. Milne's Hundred Acre Wood may have racked their memories for an answer, the anxious folks in the studio wondered what was coming next. The answer was predictably simple, if not quite what most people expected: "He was looking for Pooh."

Like most successful wordplays, this double entendre, delightfully calculated to produce uncontrollable giggles among kindergarten children, works on several levels. It is both innocuous and arresting. People don't normally look in the toilet for anything, least of all what they might most expect to find there. In the First World, at least, the toilet is a thoroughly taken-for-granted part of the household sanitation system. It is designed to hide – in some countries it is still known by the earlier, less euphemistic name of water *closet* – and to carry away, by the flushing action of water, human excreta. The room in which it is housed is a sanctuary in which ubiquitous, but now intensely private, actions can be performed covertly, and the device itself provides seemingly quick, quiet, and effective disposal of the evidence of these necessary deeds. Together, the terms "toilet," "WC," and "throne room" address and reinforce cultural taboos – in "polite" Western societies no right-thinking persons urinate in public (and only the wretchedly sick put their heads in the toilet).

At the same time, and with typical recourse to economies of scale and engineering solutions, toilets connected to sewage systems solve a local problem by moving it elsewhere, to be dealt with "properly." There is hardly room for surprise at advertisements insisting that the wash-down pedestal offers (more euphemistic wordplay) "the best seat in the house" and comfort beyond expectation. Those who live in relative affluence, in the urban centres of the developed world, have become so accustomed to the convenience of flushing, to the miles of subterranean pipes that carry away their waste, and to the promise that vaguely perceived problems of pollution are being taken care of, that few give much thought to the novelty, the rarity, and the consequences of all of this. Out of sight, out of mind.

Such blind amnesia is dangerous. In Vancouver and Toronto, in Chicago, Hamburg, Manchester, and Paris it is easy to forget that more than two and a half billion of the world's people live without basic sanitation facilities. According to statistics compiled by the United Nations, almost two in every three residents of sub-Saharan Africa and South Asia suffer this plight; in Afghanistan and Ethiopia, fewer than one in ten have access to adequate sanitation. Poverty, suffering, and death are the corollaries of such statistics; in most parts of southern Africa and south Asia, disease is endemic, and infant and child mortality rates are extremely high. According to UNICEF, "highly infectious, excreta-related diseases such as cholera still affect whole communities in developing countries. Diarrhoea, which is spread easily in an environment of poor hygiene and inadequate sanitation, kills about 2.2 million people each year, most of them children under five."[1]

It is also easy to forget, in the hyperclean environments of modern suburban dwellings where counters and basins are sponged with anti-bacterial wipes to "stop the spread of germs," just how recently and vigorously the scourge of disease plagued both rural dwellers and those who lived in major cities. Nineteenth-century newspapers from small towns across North America carried dozens of poignant accounts of individuals and families cut down by various afflictions. Let a single entry, from Nova Scotia's *Eastern Chronicle* of 15 February 1853, stand for them all: "DIED: Of dysentery at Mount Thom, Pictou County, on the 15th ult, Abigail, aged 5 months; on the 16th Alexander Hugh, aged 11 years; on the 19th Daniel Smith, aged 7 years; on the 28th Elizabeth Sarah, aged 13 years; and on the 30th Mary Welsh, aged 14 years, all children of Mr. Abiel Brown of that place; also on the 28th, Nancy, his wife, aged 46 years."

In July 1849, *The Times* of London published an equally heartrending but perhaps even more disturbing document in the form of a letter from a resident of London's Soho Square. It read in part, "we are . . . as it may be living in [a] wilderness, so far as the rest of London knows anything of us, or as the rich and great people care about. We live in much dirt and filth. We ain't got no priviz, no dust bins, no drains, no water splies and no drain or suer in the hole place . . . The stench of a Gully-hole is disgustin. We, all of us, suffer and numbers are ill, and if Colera comes Lord Help Us . . . we are living like pigs and it ain't fair we should be ill treted."[2]

The Cholera did come, and rather famously as things turned out. In 1854, "there were upwards of five hundred fatal attacks of cholera in ten days" in St. James Parish, Soho. John Snow, a physician who lived and worked in the area, and who had earlier begun to understand the spread of the disease as a consequence of both contagion (person-to-person transmission) and contaminated water supplies, began a detailed investigation of the tragedy. By interviewing survivors and mapping the incidence of death in what he called a "diagram of the topography of the outbreak," he concluded that those who died had either consumed water from a pump in Broad Street or had a close association with someone who did so. Work on a slightly earlier cholera outbreak in South London had already led Snow to the view that death rates varied between areas supplied by different water companies. They were generally higher in districts served by the Southwark and Vauxhall Water Company, which drew its supplies from the River Thames, near Battersea, and lower where the Lambeth Company provided water from an intake further upstream.[3]

These were monumental conclusions, even though they were resisted by several of Snow's colleagues who clung to older miasmatic theories of disease transmission. Cholera was the "critical illness" of the mid-nineteenth century, the key challenge to understanding in contemporary medical science, and debate about its causes forced both public officials and society at large to reconsider long-established attitudes toward waste and the environment.[4] Linking differences in mortality rates to the locations of water company intakes pointed to the increasing pollution of the Thames as it flowed through London. The river at Battersea was more befouled than at Putney, and compared to that at Rotherhithe, downstream, even the water of Battersea might seem relatively clean. Along the winding course of the Thames through London, sewer outfalls and street drains, as well as tributary streams that were themselves little more than sewers, poured the detritus and excreta of crowded humanity (and of the horses and dogs and cows and chickens and pigs upon which people depended) into the river.

The Thames was not unique. By the mid-nineteenth century, dozens of rivers that began as "pure pellucid babbling brook[s]" in the uplands were transformed into "weary forlorn and morally bedraggled huss[ies]" by their passage through large industrial cities (see page 14 herein). In Manchester, the Irwell and the Irk suffered such ignominious fates. Frederick Engels described the latter in *The Condition of the Working Class in England:* it was, he wrote in the mid-nineteenth century,

> a narrow, coal-black, foul-smelling stream, full of *débris* and refuse . . . In dry weather, a long string of the most disgusting, blackish-green, slime pools are left standing on this [right] bank, from the depths of which bubbles of miasmatic gas constantly arise and give forth a stench unendurable . . . the stream itself is checked every few paces by high weirs, behind which slime and refuse accumulate and rot in thick masses. Above the bridge are tanneries, bonemills and gasworks, from which all drains and refuse find their way into the Irk which receives further the contents of all the neighbouring sewers and privies.

A few years later, the Irwell was characterized, equally graphically, as a "hapless river" that "loses caste as it gets among the mills and the print works. There are myriads of dirty things given it to wash, and whole wagon-loads of poisons from dye-houses and bleach-yards thrown into it to carry away; steam-boilers discharge into it their seething contents, and drains and sewers their fetid impurities; till at length it rolls on . . . considerably less a river than a flood of liquid manure, in which all life dies, whether animal or vegetable."[5]

These were classic cases of what economists have called the tragedy of the commons. Tradition, and the law, had conferred upon owners of riverbank properties the right to use the water flowing by. No one owned the stream, but all held an equal claim to its benefits. In time it became clear that this was not much of a right at all if the flow had to continue unreduced and unobstructed, and thus the test of legal entitlement became "reasonable use" – use that did not injure other claimants, upstream or down. Much of this discussion turned on questions of impoundment or extraction: it was not reasonable to stop the natural flow of the river entirely; it was not unreasonable to allow a riparian landowner to drink from the stream – and so on. But "reasonableness" is something defined in and by particular contexts; its meaning is never entirely stable. One discouraged late twentieth-century commentator suggested that the very idea of reasonableness approached "the ultimate in amorphousness" (see page 29 herein).

And what of contamination, rather than obstruction and withdrawal? Was it reasonable to pollute? What was pollution anyway? Changing circumstances and shifting currents of understanding made these slippery questions indeed. Rivers carried things away. Silt, from banks broken down by cows that came to drink, moved downstream and was deposited elsewhere; a bucket of slops thrown into the current would disperse and (appear to) disappear; so too a pail of urine. When people were few and discharges were limited, rivers seemed capable of absorbing and removing items thrown into them without suffering irreparable harm. They possessed a certain assimilative capacity that allowed them to serve as nature's sewers. Moreover, because all rivers run eventually to the sea and the sea is enormous, it is not difficult to regard the assimilative capacity of the hydrological system as infinite. Thus the engineers' mantra, "the solution to pollution is dilution" – practiced (if not formally expressed) since the construction, two and a half millennia ago, of Rome's Cloaca Maxima (or great sewer) to drain wastewater from the low-lying areas of the city into the Tiber River and thence the Mediterranean Sea. So long as diseases were attributed to the miasmas (or gases) emanating from putrefying organic matter, removing such material from human proximity made sense.

Well before Engels looked down on the Irk from Ducie Bridge and John Snow investigated the South London water companies, however, the flaws inherent in applying the doctrine of reasonable use to a common property resource and depending upon the assimilative capacity of drainage networks to deal with waste were becoming evident. As human populations increased and industrial uses expanded, the destructive and toxic loads imposed on rivers rose almost exponentially. In the 1820s, well over a hundred sewers poured their offensive output into the Thames between Chelsea and London Bridge. By 1850, some 250 tons of fecal matter entered the river each day. To human waste from overtaxed cesspits and privies, abattoirs added their refuse, tanners the noxious elements they used to cure hides, chemical works various kinds of acids and caustic sodas, cloth manufacturers an assortment of bleaches, dyes, and clays, and so on. By one estimate, most of the two million animals sold at Smithfield market each year in the mid-nineteenth century were slaughtered nearby, in cellars and sheds from which blood and offal seeped into public sewers.[6]

Individually, these discharges would have been bad enough; collectively, they were simply overwhelming. Viewing the unsightly, evil-smelling accumulations of decomposing material along the riverbanks, residents of almost any rapidly expanding nineteenth-century city might have concluded that their waste-disposal system was malfunctioning. Yet so long as one riparian landowner felt entitled to use the river as a drain, others would probably act in concert because the costs of doing otherwise would be borne by them alone. Through the eighteenth and early nineteenth centuries, the law was of little help on such issues because it was generally interpreted and made to serve practical ends and to encourage rather than restrict economic activity.[7]

Once the connection between contaminated water and the deadly, dreaded scourge of cholera became apparent, however, the need for action became clear. Sanitation quickly came to head the agenda of late-nineteenth-century urban reform. By century's end, in North America, physical and moral hygiene had been thoroughly conflated, and sanitation was entangled with questions of social improvement. Young women were instructed in "domestic science," which encompassed the various important arts of housekeeping and emphasized hygiene standards. Settlement-house volunteers were advised that they would find immorality, dirt, disease, unthrift, and all the faults that produced poverty and misery among those with whom they worked. But for three or four decades after 1850, water was the major focus of concern. Politicians debated and engineers schemed, municipal officials agitated, and the public sought means to improve water supplies for urban dwellers as they endeavoured to deal more effectively with the challenges of handling cities' sewage, or waste.

This was no easy task. In Hamilton, Ontario, the proportion of homes served by sewers increased from 10 to 50 percent in the last quarter of the nineteenth century, but the pipes ran to the bay, and well before 1900 shorelines near outfalls were covered with what members of the local Board of Health described as "an accumulation of the most disgusting and filthy matter." In an effort to address the situation, earth was shovelled atop the foul-smelling morass, but this was not a sufficient solution. The Board of Health recommended that the city create a buffer around the site by acquiring the adjoining property and that it raise "the pay of the men working at the outlet."[8] Recognizing that ice for the refrigeration of food was harvested from the increasingly polluted harbour, the Board of Health also sought to monitor ice quality and to restrict ice cutting to areas distant from sewer outlets.

In Vancouver, where urban expansion came late and well after the need for clean water and effective waste disposal were understood, the sheer pace of growth and the costs of infrastructure development stretched the city's capacity to cope with demands. Good water was piped into the city from the North Shore mountains, but sewage disposal posed more difficulties. Expert but expensive plans were compromised, combined wastewater and sewage systems sometimes disgorged human waste onto local beaches (instead of into the Fraser River or the Gulf of Georgia where it was "designed" to go), and thousands of houses were built without connections to the sewage system. When septic tanks failed, pollution contaminated lakes and streams within the city, even into the 1950s.[9]

In Toronto, where efforts to establish a reliable domestic water supply system dated from the mid-nineteenth century, the challenges were also significant. The Toronto Gas, Light & Water Company, established in 1843, distributed water in wooden pipes to those who could afford its service, but in 1858 fewer than one in ten Toronto dwellings had running water. Lake Ontario, one of the largest bodies of fresh water in the world, seemed an obvious source from which to satisfy the city's needs. Some learned residents described the lake water as "of most extra-ordinary purity," but others demurred, noting that the city of forty thousand disgorged its sewage into the harbour, that the Don River hardly ran clean, and that the shipping which plied the lakes also contributed pollution to their waters. The situation did not improve with time. In the 1890s, "all the sewage of 175,000 people" was being discharged "in its crude state into [the] tideless and practically stagnant harbour" (see page 217 herein). Still, the costs of more effective wastewater disposal deferred action. Toronto's first wastewater treatment plant, at Ashbridges Bay, was not built until 1910.

In the meantime, the city attempted to step over the problem. After establishing a public waterworks service in 1872, it built a filtration basin on Toronto Island. Water was drawn into this from the lake, rather than the harbour, and pumped through a pipe under the bay to an open-storage reservoir on high ground. Yet water quality remained suspect, and residents spoke of the "drinkable sewage" that flowed from their taps. The next step entailed construction of a wooden intake pipe extending a kilometre into Lake Ontario. When this broke in the 1890s, polluted harbour water entered the supply system and produced a typhoid epidemic. More engineering works followed, including the development of a steel intake pipe, the building of a brick-lined tunnel from the island to the mainland, and the construction of a slow-sand filtration plant, the largest of its kind in the world, in 1912.[10]

Two decades later, reflecting both the expanding city's voracious demand for clean water and the twentieth century's confidence in engineering solutions, work began on the R.C. Harris Filtration Plant at the eastern edge of Toronto. Subsequently expanded to double its original capacity, it remains the city's largest such facility (capable of producing 950 million litres of potable water a day) and supplies the domestic needs of almost half of Toronto's residents. The plant draws water from Lake Ontario through two intakes located approximately 2.5 kilometres out in the lake and cleans it through an elaborate series of processes including pre-chlorination (adding chlorine to the water before filtration), super-chlorination (adding very high doses of chlorine), and de-chlorination (adding chemicals including sulphur dioxide and ammonium to remove excess chlorine and to stabilize the remainder, which helps keep the water safe between the filtration plant and the homes, schools, and industries to which it is directed).

As improbable as it may seem, Torontonians take a great deal of pride in this place. The city notes that it was one of the first in North America to treat its water with chlorine and that it led Canada in the pre-, super-, and de-chlorination trifecta. Dozens of magazines and websites have described the Harris Plant as "a cool place" to visit, have characterized it as "quietly dramatic" and "magical," and have even suggested that it is, "arguably," the city's "most inspiring and elegant landmark." This certainly owes something to the Art Deco grandeur of the enormous structure housing the filtration works, and its use as a setting for several movies and TV programs. Yet there is more to it than this. Until recently, there were tours of the plant where, it was noted, "the city's water is tested every four hours." These were so popular that visitors were advised to "call ahead" to ensure a place on the tour of their choice. Even the Toronto Green Community, the Toronto Field Naturalists, and the Sierra Club were enjoined in November 1998 to "Come and see Toronto's urban ecosystem in action!" on a visit to the plant widely known as the "Palace of Purification."[11]

Some of those who admire the way in which function and form have been brought into harmony in the R.C. Harris Filtration Plant might also recall that its construction plays a prominent part in Michael Ondaatje's 1987 novel *In the Skin of a Lion*. As critics have noted, Ondaatje here reminds readers that "the monuments of the modern city bear traces of divergent social meanings and purposes." He also uses darkness "as a blindfold to shield characters from the truth and to provide them with protection from the burden and responsibility of knowing the truth." His purposes, critics would argue, are to honour the contributions of migrant workers to the making of Canada and to the destabilization of "a linear historical view of Canadian society as a top down initiative orchestrated by a predominantly Anglo ruling, or managerial class."[12]

By the same token, Ondaatje's focus on the construction of the filtration plant reminds us of the reasons for its existence. Would immigrant workers have endured eight-hour shifts tunnelling through "brown, slippery darkness" to realize the "mad scheme by Commissioner Harris to collect lake water 3,300 yards out in the lake" had Lake Ontario not been polluted?[13] Would the city have needed its waterworks had it not so enthusiastically embraced the culture of flushing?

When members of the Toronto Field Naturalists, Green Community, and Sierra Club were encouraged, a week or two after their visit to the filtration plant, to "get a look at what happens to the water that came from R.C. Harris!" at the Ashbridges Bay Sewage Treatment Plant, some among them might have wondered about chickens and eggs or carts and horses; those who knew their Ondaatje might have worried about blindfolds and truth. But all would have exercised themselves unnecessarily in doing so, because the fundamental significance of their linked visits surely lay in the fact that all things are connected, that Ashbridges Bay was simply another part of the "urban eco-system in action," and that the Palace of Purification was also a Monument to Flushing.[14]

Faced with evidence of the rising human and ecological costs of pollution in the 1970s, Canadian prime minister Pierre Trudeau borrowed a phrase from a then "well known and widely hailed . . . masterpiece of Indian oratory" that was understood to speak to "present day concerns." This was a speech attributed to the American Indian chief Seattle on the shores of Puget Sound in the 1850s, and Trudeau invoked it to ask "What is befalling the Earth?"[15] With *The Culture of Flushing*, Jamie Benidickson seeks to move us from puzzlement to understanding. In the process he has produced an uncommonly wide-ranging and boundary-crossing book. It is a reflection of the specialization of the modern-day research enterprise and widespread suspicion of overarching meta-narratives (which are seen to impose particular and limited views on complex multifaceted circumstances) that relatively few contemporary scholars choose the grand sweep and write on a national, continental, or global canvas. The proportion of those who do so may be higher among environmental historians than among scholars in other disciplines, perhaps because many environmental historians believe that their subject transcends familiar political boundaries, and perhaps because the field of environmental history seems inclined to produce and attend to large stories of environmental destruction, human hubris, and so on.

Yet *The Culture of Flushing* stands out, even in this company. Its focus is on the use of water bodies for refuse disposal and its analysis is rooted in the law, but it is written with the general reader rather than the technical specialist in mind. It engages questions of health, aesthetics, politics, engineering, public policy, and environmental citizenship. It is full of details, some of them arcane, drawn from an almost dizzying spectrum of times and places, contexts and settings. But Benidickson, who teaches environmental law, administrative law, and legal history in the University of Ottawa and who has written charmingly of the place of the canoe in Canadian life, as well as of recreation, resources, and Aboriginal rights in Northern Ontario, has a talent for summarizing complex bodies of law and regulation.[16] He can also turn a phrase. Consider his delicious summary of a dissembling Ontario politician's position on the degradation of the Spanish River: "This exercise in verbal gymnastics was balance-beam work of Olympic calibre."

In the end, *The Culture of Flushing* amounts to much more than the sum of its diffuse parts because Benidickson consistently finds order (and meaning) in the complex swirl of factors causing and shaping aquatic pollution. English, American, and Canadian experiences with growing quantities and types of waste jettisoned into water bodies (with varying degrees of forethought, treatment, and legality) are juxtaposed and engaged one with another to reveal a great deal about attitudes, assumptions, and values, both persistent and fleeting, that shaped environmental behaviour across a broad range of circumstances. By painting across this expansive canvas, Benidickson makes a significant contribution to environmental history. There are many legal texts that deal with water rights, but few combine legal approaches and a rich understanding of context to address the questions posed by sewage and waste disposal in what have sometimes been called "effluent societies." Concerned about environmental degradation, Benidickson leads his readers to face its causes and, in meeting the enemy, to discover, Pogo-like, that "he is us," both individually and collectively.

This is important because it is all too easy to read the story of wastewater and effluent treatment as a progressive one, with improvements following one another in more or less regular succession, and to assume that the culture of flushing has solved the problems of effluent disposal. Thus literature from Toronto Water, a division of the city government responsible for "all aspects of water production, transmission and distribution, wastewater collection and treatment, and storm water collection, transmission and treatment," notes the advances made since mid-century. In 1953, eighteen wastewater treatment plants served the thirteen municipalities of Toronto. Most of them were too "overloaded and ill-equipped" to cope with dramatic postwar population growth. With the advent of metropolitan government in 1954, small, inadequate treatment plants were replaced by larger, more efficient works; by 1960 four major wastewater treatment facilities served the entire city. They collected wastewater through an extensive underground sewer system that was refined and extended until, early in the twenty-first century, it encompassed 1,301 kilometres of combined sewers, 4,305 kilometres of storm sewers, 371 kilometres of watercourse, 76 wastewater pumping stations, 4,048 kilometres of sanitary sewers, and 358 kilometres of trunk sewers. At the wastewater treatment plants, solids, chemicals, and other undesirables are removed from the liquid that courses through this elaborate labyrinth. Between 72 and 97 percent of suspended solids, biological oxygen demand, and phosphorus are said to be eliminated before it is released into the natural water supply.[17]

This may seem admirable, but there is little cause for complacency. The waters of Lake Ontario are far from pure. Scientific studies conducted in the late 1980s found "measurable concentrations of hundreds of toxic pollutants" in the lake. Much of this contaminant load enters Lake Ontario through the Niagara River, to be sure, and the "impact zones" from local sources of pollution along the Toronto waterfront are relatively small. Rivers flowing into the lake are now reasonably clean, but sewage treatment plants (STPs) continue to release impurities. According to the authors of one of these reports, the "elimination of chlorobenzene loadings from the Toronto Main STP would be quite beneficial to the Lake Ontario ecosystem." Yet this may be difficult to achieve because chlorinated benzenes are widely used to disinfect urinals.[18]

Across Canada, cities run the gamut in the effectiveness of their sewage treatment.[19] In September 2004, the Sierra Legal Defense Fund ranked twenty-two urban centres on its National Sewage Report Card. Calgary led the way with an A+ for implementing 100 percent tertiary treatment and ultraviolet light disinfection. Despite having the toughest sewer-use by-law in the country, Toronto scored only B-, a grade reflecting its dependence on secondary treatment and the discharge of almost ten billion litres of untreated sewage and runoff. Montreal, which conducts primary sewage treatment only and dumps 3.6 billion litres of raw sewage into the St. Lawrence River each year, and Victoria, which performs preliminary screening only and discharges more than 34 billion litres of raw sewage into surrounding waters annually, were at the bottom of the class. Vancouver, Halifax, and St. John scored badly, with D grades. Due to the cost-saving construction of combined sewer and stormwater drains in the twentieth century, Vancouver is plagued by a large number of raw sewage overflows into stormwater outlets, and two of its STPs, on Iona Island at the mouth of the Fraser River and at Lions Gate on the North Shore of Burrard Inlet, provide only primary treatment. "It is simply shocking," concluded a representative of the Georgia Strait Association, that the "hundreds of toxic harmful chemicals such as heavy metals, persistent organic pollutants and PCBs," as well as the organic matter and micro-organisms in "the toxic soup we call sewage," are "regularly being dumped into our lakes, rivers and oceans."[20]

But what are the alternatives? Better treatment is expensive. The city of Vancouver is spending between $13 and $20 million a year in a fifty-year project to replace combined sewers with separate stormwater drains and sewage pipes. Upgrading the Iona plant to secondary treatment will cost $500 million. The net replacement value of the existing Greater Vancouver Regional District wastewater system, with all its flaws, is estimated at $13 billion. We might invest in some improvements (such as those being undertaken in Vancouver) and limit growth in hope that current or slightly reduced levels of pollution will not have disastrous deleterious long-term effects on aquatic ecosystems. We might hope for technological solutions – at one end of the spectrum, this might mean incorporating composting toilets and small-scale sewage treatment plants utilizing "living machine" technology (water-based sewage-digesting organisms) or solar aquatics (plant-based filtration) into buildings that are "off-the-grid"; at the other, it might entail the construction of separate waste systems for industrial users. We might change people's behaviours, and thus reduce both the quantity and the toxic content of water in the waste stream. Or we might continue as we are, and sacrifice rivers, lakes, seas, and perhaps even some oceans to heedless and wasteful consumption.

None of these possibilities is straightforward. The difficulties of implementing all but the last of them are many. The costs of each are considerable. In every case it is virtually impossible to apportion expenditures and benefits fully and accurately. Because the law relating to pollution generally places the burden of proof upon those whose interests are damaged – reflecting the nineteenth-century conviction that "industrial production will be benign unless demonstrated otherwise" – redress is generally possible only after harm has been done.[21] Moreover, the law typically requires plaintiffs to prove that the degradation stems from a particular, identifiable source, and it often places the costs of legal action, at least initially, on those less able to pay, the victims rather than the perpetrators of pollution.

In the end, the issues raised by the challenges of waste management resolve into problems of risk and responsibility.[22] Risks are typically defined by probabilities. Those who face them must deal with uncertainty: How much is too much? Can we "chance our arm" and bet "against the odds"? What are the costs of entering, and the rewards of success, in this game? Are returns assessed on the long or the short term? Who bears the costs of a misguided gamble? Where does liability rest when hurt and harm reach beyond those who initiate the action? Although regulations attempt to clarify and codify the basic "how much is too much" question of risk, by establishing admissible concentrations of toxic substances in parts per million, for example, the other concerns, bound up with issues of responsibility, accountability, and obligation are much slipperier. Even to raise such queries is to confront large questions about the kind of society in which we live and about how we might respond to the circumstances that they anticipate.

Pondering these issues early in the twenty-first century, it is difficult to proceed very far without engaging the work of sociologist Ulrich Beck. In two books published in the 1990s, *Risk Society* (1992) and *Ecological Politics in an Age of Risk* (1995), Beck sought to place questions of environmental degradation at the centre of a new conceptualization of modern society. Arguing that contemporary hazards, including pollution, have far more extensive spatial impacts than those characteristic of the nineteenth century, that they are often more toxic and potentially catastrophic, that their effects are cumulative and irreversible, and that their pathways through the environment are often essentially invisible to the population at large, he insists that long-standing hazard management strategies which depended (however creakingly) on the identification of polluters and the payment of reparation cannot hope to deal effectively with the challenges of the present and future. In this view, contemporary "risk societies" face an awful dilemma: "at the very time when threats and hazards seem to become more dangerous and more obvious, they simultaneously slip through the net of proofs, attributions and compensation that the legal and political systems attempt to capture them with."[23]

In response, Beck identifies three possible development paths for the twenty-first century. The first is to soldier on, in classic laissez-faire style, wedded to the nineteenth and early twentieth centuries' faith in progress and the conviction that human ingenuity and new technologies can save the earth from the ills inflicted upon it. The second seeks amelioration through more active participation of the state and its agencies in regulation and remediation of environmental damage. And the third imagines a situation in which development would be prohibited until those who advocated it demonstrated, to exacting standards, that their planned activities would not damage the environment.

Each of these options has sociological and political corollaries. The first broadly reflects the politics of the right, and, in the vision of Beck and his interpreters, probably also entails efforts to shore up "the disintegrating institutions of industrial society" (such as the nuclear family and segregated gender roles) in the face of challenges posed by the more reflexive form of modernity being spawned by new risks and hazards. The second, which has been characterized as "an ecological variant of the welfare state," envisages the centralization and bureaucratization of environmental governance and the location of decision making firmly in the purview of designated scientists and institutions. In political terms, this is the preferred model of social-democratic government. It holds the promise of bringing the environmental consequences of technological development under greater public scrutiny and control than would the first option. But, convinced that the world has changed and that effective governance of risk societies requires devolution rather than centralization, Beck regards this model as ill-adapted to the challenges of the present. The third alternative turns on the development of what has been termed differential politics, in which political power is exercised in different forms on different elements of the body politic. Ultimately, this approach, favoured by those on the political left and members of environmental movements, would open the hitherto essentially "private sphere of economic decision-making to political debate and control" by both the state and the "institutions of civil society."[24]

Lest much of this seem both abstract and remote, let us return once more to the Palace of Purification on the shores of Lake Ontario. This "Art Deco extravaganza," this "veritable Versailles of Public Works" is indisputably redolent of those years after the First World War when engineers, scientists, and other professionals were awarded the keys to the future and charged with the task of building a new society upon the ruins of the old order fractured by the conflict of 1914-18. The R.C Harris Filtration Plant, it has been well said, constituted "a declaration of faith – faith in the future and the power of technology."[25] But it was more than this. It enshrined the conviction that government could and should provide its citizens with clean water. It was necessitated, in a sense, by the unregulated despoliation of the environment as a growing population embraced the culture of flushing, and it was built, at enormous cost, for the public good. Further investment expanded and upgraded the facility in the 1950s, large sums were spent on restoration of the building in the last years of the century, and the quality of the water exiting the plant is still tested "every four hours." Governments today are much less willing to invest in the public good (and take pride in that investment) than they were half a century and more ago, however. The market, efficiency, and public-private partnerships are the ideological mantras of the new millennium. The recent hollowing out of the state in many jurisdictions seized by neoliberal (or is it "neoconservative," for it sometimes seems to matter not!) conviction has slashed public spending on environmental monitoring and regulation and undermined prospects for well-judged, scientifically based environmental management. The vacuum has yet to be filled. We are left, Beck might say, with the complacent remnants of a safety state, to face the ascendance of "organized irresponsibility" – a "concatenation of cultural and institutional mechanisms by which political and economic elites effectively mask the origins and consequences of the catastrophic risks and dangers of late industrialization."[26]

All of this is a long way removed from Tigger and Pooh and the bodily functions alluded to by a mischievous child in Calgary, but it is well to remember – as Jamie Benidickson demonstrates so skilfully in the pages that follow – that the practice of treating waterways as nature's sewers is both long-standing and fraught with danger. In the wake of the widely publicized tragedy at Walkerton, Ontario, where seven people died due to the contamination of groundwater by manure and a failure to properly monitor and maintain water quality, Canadians may need few reminders of the hazards of water pollution. But the crucial question, on which the history recounted here throws a good deal of light, is how we came to the particular conjuncture that Walkerton represents: where young and old die from drinking tap water in one of the world's most affluent societies.

Although the proximate reasons for the Walkerton tragedy are well known, and range from human negligence to the enormous quantities of animal waste discharged into the environment with the intensification of agriculture, its antecedent causes are far more complex and disquieting. They are rooted in the propensity, inherent in industrial society through the last two hundred years and more, to value economic activity more highly than environmental quality, and to privilege development over ecosystem health. Whether manifest in the disruption of habitat by logging, the extinction of species due to market hunting, or the pollution of air and water by discharge of all manner of waste into the environment, this tendency has exacted a rising toll. It has also, at least implicitly, framed a series of choices about the balance between private and public interest, between professional and public knowledge, between conciliatory and coercive management, between political and bioregional administration, and among local, regional, and national priorities.

In recounting how societies have grappled with these issues, with more or less effectiveness, through space and time, as the culture of flushing became increasingly normalized and insidious, Benidickson shines a light on the past. In doing so, he illustrates that events which we order chronologically, as traces of time's unerring linear, progressive (arrowlike) passage, might also be regarded as cyclical. At one scale, the problems of Walkerton look a lot like those in John Snow's London. History may not repeat itself, exactly, but Aldous Huxley was surely right when he said that "The art of living together without turning the city into a dunghill has [to be] repeatedly discovered." Our environmental victories, Benidickson and Huxley know – and all who read this book should recognize – "are in no sense definitive or secure" (see page 290 herein). By reminding us of this, Benidickson encourages us to think again about the choices we make, the risks we take, and the responsibilities we have, as we navigate our ways through what has become a conspicuously waste-full, and menacing, world.

Acknowledgments

THOSE OF US WHO WRITE ARE MUCH INDEBTED to our listeners or readers for their sympathetic curiosity, for their thoughtful criticism, and often for their patience. Colleagues and friends including Constance Backhouse, Philippe Crabbé, Toby Gelfand, Bill Kaplan, Gerry Killan, Tony Scott, Harry Swain, and John Wadland endured most of this manuscript at a point in the process where it was not very far along the road to presentability. I am grateful for their advice and for the fact that we were all still speaking in the aftermath. Others, such as Elizabeth Brubaker, John Evans, Robert Glennon, and Jon Schladweiler, provided much-appreciated guidance or reaction on individual chapters. Anonymous readers contributed greatly with the insights and suggestions that I can reward only with a sincere expression of thanks and the hope that it will reach them somehow. Late in the game, Graeme Wynn, the general editor of the Nature|History|Society series, of which I am honoured to be a part, provided extremely useful guidance for a final round of revision. Jane McWhinney and editorial staff led by Darcy Cullen at UBC Press contributed greatly to readability.

I am also appreciative in a different way to others whose reaction to the fact that I was exploring aspects of the social and legal history of sewage fell short of enthusiastic endorsement. There is nothing like a peremptory change of subject in cocktail conversation to test one's level of commitment to a research agenda.

Research assistants responded magnificently to inquiries ranging in precision from "I'm looking for stuff about sewers" to "Did Edwin Chadwick ever exchange any thoughts about sewage treatment with Baron Bramwell?" Thus, sincere thanks to Sean Bawden, Elaine Borg, Karen Chambers, Joseph Griffiths, Will Hinz, Peter Keen, Maria Kotsopoulous, Adam Kubelka, Carolyn Magwood, Bruce Newey, Christina Vechsler, and Bill Wade. At a decisive period in the success or failure of this effort, I benefitted from the extraordinary dedication of Melanie Mallet, a gifted assistant for whose contribution and partnership I am extremely grateful. If any of you have not yet received a copy of this volume, please get in touch.

As I was beginning to confront the scope of an inquiry that had not seemed entirely unmanageable when it got under way, a visitorship at the Faculty of Law, University of Calgary, provided an enormously valuable opportunity to impose some shape on the project. The Social Sciences and Humanities Research Council and the Law Foundation of Ontario subsequently provided much-appreciated financial support.

My daughter is reported to have informed her grade four class a number of years ago that her father was writing a book about the "S—" word. Yucko. But thanks anyway, Nicola.

The Culture of Flushing

Introduction

THE CULTURE OF FLUSHING

"THE CENTRAL QUESTION OF WATER LAW," according to the distinguished American legal scholar Joseph Sax, is "why, and how, water is different" from other commodities.[1] Water is, in some respects, ordinary property in the sense that one can buy and sell it, or lease and transfer it, in the course of using and profiting from it. On the other hand, Sax points out, water differs significantly from other forms of private property: "If I own a wristwatch, a bag of potatoes* or a house, I can use them or leave them unused. I can crush my watch under my heel, or live alone in a ten-bedroom mansion while others yearn for places to sleep."[2] Water, on the other hand, cannot be owned unless it is put to use.

What makes water different is an intriguing question. Is there, Sax asks, some unique physical quality that makes conventional private ownership of water unsuitable? Or have we endowed water with some special social value — some "publicness," as he puts it — that causes us to regard it in a different light? Certainly, evolving claims about the public interest in water have often exerted important constraints on private claims for water use. Is any such special public quality unique to water, or does water law represent "the cutting edge of a new view of private/public relations that will come more and more to the forefront?"[3] The fascination of the question is only heightened by forceful and articulate, if not widespread, insistence that public interests in water may actually be best safeguarded

---

* In the case of potatoes, time limits may apply.

through private means, and by the ascendancy of water issues on local, national, and international agendas around the world. Most readers, alas, will find my subject less engaging.

Not to put too fine a point on it, this volume's point of departure is your personal bodily functions, or rather functions that may have appeared to be personal until their integration through flushing with public waters and watercourses was legislated, engineered, and financed at some cost during the nineteenth and twentieth centuries. Viewed from a somewhat more technical perspective, a series of revolutionary initiatives, first associated with the water-borne removal of organic materials from burgeoning cities, then with the bacterial transmission of disease, and eventually involving chemical and mechanical means of purifying water supplies, now underpins vast networks of municipal infrastructure linking waste to water.

The practice of dumping or discharging waste into rivers, streams, lakes, and oceans is long-standing. It received official sanction in Europe and North America, however, only in the nineteenth century, when a peculiar combination of circumstances accompanying urban and industrial growth encouraged sewerage and water-borne waste removal. The introduction of water supply systems in municipalities across Europe and North America increased pressure to remove wastewater from sodden urban landscapes, particularly after the water closet replaced the privy pit and cesspool. A widespread contemporary belief that disease originated in the decay of organic materials suggested that great advances in public health might be achieved by using municipal water flow to flush household wastes into the waterways and to direct manure from the streets to the same destination. Running water, presumed to purify itself, was not considered to be at serious risk from sewerage. This cluster of ideas, emerging as the common law faced growing pressure from intense river usage and before the role of bacteria in disease transmission was understood, nurtured flushing on a grand scale.

Water became a "sink" by design.[4] Indeed, observers have been known to remark that "water is one of the most valuable media for the disposal of municipal, industrial and agricultural residuals."[5] All too frequently, it has been assumed that this is a primary purpose of water and waterways. It has even been argued on occasion that such usage enjoys the exalted legal status of a right, a central element of our perilous fantasy that the planet was created for human convenience. The evolution of these views, the practices on which they have rested, and some of the values that constrain and condone such perceptions of water, along with some of the consequences, are the subject of this book: the culture of flushing.

Domestic plumbing arrangements and municipal wastewater removal evolved in a context that only dimly appreciated the possible environmental implications of flushing and largely discounted the significance of any that were apparent. Water became a dumping ground because the choice was often made to do so. It certainly seemed convenient, sometimes even necessary. It served prevailing conceptions of public health. It appeared to be economical. Beyond this, flushing became acceptable as the legal system gradually accommodated the discharge of municipal and industrial wastes alongside such historic aquatic priorities as navigation, fisheries, and the property rights of riparian landowners. Even though we have repeatedly been made aware of adverse consequences, much of the intellectual framework supporting the culture of flushing remarkably endures.

The practicalities of household plumbing, water supply, and sewage management are ordinarily relegated to the trade school and engineering curriculums or to municipal planning offices. They emerge prominently in major media only when the loss of human life can be attributed to deficiencies in water supply or in waste and wastewater disposal practices. The experience of Walkerton, Ontario, is among the latest incidents to provoke sadness and outrage. The death of seven people and the illness of many hundreds of others could be traced to a water supply contaminated by the application of manure to a nearby field. A rigorous public inquiry pursued a wide range of plausible suspects, including human error, systems malfunction, government neglect, and agricultural practices.[6] The humanity of the judge presiding over the investigation, in combination with the creative settlement of a class action lawsuit for $50 million, helped to restore, at least to some degree, the badly shaken confidence of a severely damaged community. The rest of us can now sit back to await the next unanticipated tragedy of this kind, for in the absence of crisis, urban infrastructure is hardly a matter of dinnertime conversation. The human contribution to the water cycle, typically leading from water purification facilities to sewage treatment plants by way of tubs, sinks, and the stylish descendants of the water closet, has never seemed particularly riveting or worthy of attention. Why do the consequences of flushing not otherwise register on the public consciousness?

Twenty-first-century debate about water is now well under way. Water wars – a phenomenon expected to differ significantly from the water fights of my youth – are widely forecast. The focus of such conflicts is ordinarily presumed to be the availability of water, yet this vital concern is not unrelated to its quality, for variations in water quality may constrain water use or influence costs. If it should seem desirable to retain water quality, or to regain it where it has been lost, there may be some virtue in examining the social and legal processes that led to deterioration and in considering mechanisms designed to forestall that result.

What can be done to restore degraded waterways and groundwater in order to safeguard natural ecosystems and the integrity of water supplies is a question asked apprehensively today and with increasing frequency around the world. The dilemma presented by the need to sustain environmental quality alongside economic activity has become virtually all-pervasive in industrialized societies. Various ways of managing that reconciliation have been adopted. These have included legal control measures, economic incentives designed to influence the behaviour of water users, and technological initiatives whose adoption promises – always – more than it will ever deliver.

If we set aside the views of ocean desalinators and other quick fixers, the answers, offered from various quarters, advance a series of competing visions. Those who are disheartened by or distrustful of public water protection measures champion private rights as the instrument most likely to lead us back from the brink. This approach would rely upon individual property owners to defend their personal interests and in so doing to protect environmental quality. Others, who attribute deterioration in the quality or supply of water to disregard of protective standards, call for more vigorous enforcement of regulatory regimes. Those who associate elements of a water crisis with the absence or breakdown of markets urge the creation or restoration of such markets as a remedial strategy. If water were appropriately valued and priced accordingly, economic incentives would operate to ensure conservation and protection efforts.

Each of these alternatives imagines a supportive framework or context to contribute some order and structure for the processes and institutions that will make the world a better place, for that, one presumes, is the universal aim of reformers. Although that context is vastly wider than the scope ordinarily given to water law, it is worthwhile to reflect briefly on that narrower field as a preliminary means of dispelling the "magic answer" syndrome. For if you are already the possessor of one of the magic answers, you will have little inclination to turn, let alone to contemplate, the pages that follow, and I will have lost a valued reader. And, given my publisher's projection of the market for a volume such as this, I may have lost a not insignificant percentage of forecast royalties.

The physical qualities of water, as has often been observed, reduce its susceptibility to conventional norms of private ownership. Thus, though private entitlements have certainly been acknowledged in systems recognizing the effect of appropriation or embodying principles of allocation under licence, for example, the natural, public, and common or community features of water have almost always presented countervailing claims. The physical characteristics of water also determine opportunities for the economical utilization of water for a variety of interdependent purposes. These characteristics create, as noted from an economic perspective, externalities and dis-economies that must be taken into account if water resources are to be used efficiently. An example of a water or water-related service that cannot readily be subdivided into units for sale and purchase by individual consumers is flood control. Similarly, the availability of economies of scale has often resulted in water development activities such as hydro-electric power production being characterized as natural monopolies. Moreover, some water services, such as water as a general environmental amenity, for example, are not readily marketable. This line of analysis has led commentators to remark that the characteristics of water have implications for the design of institutional arrangements: "The design of water resources institutions must take into account these characteristics if water is to be developed and utilized so as to realize the most benefits practicable."[7]

It has often appeared desirable to develop institutions with reference to the distinctive features of the relevant subject matter and to some function intended to be accomplished. The presumption, of course, is that our understanding of the relevant features is reasonably reliable, and that those responsible for policy choices have appropriately identified the function or functions to be served. These, in the case of water, have ranged from navigation and fishing to water supply, irrigation, recreation, and power production, among other claims on the waterways. To be a little bit off in one respect or the other risks misconceiving the institutions, a factor quite apart from the impact on the design and operation of institutions of vast forces of inertia and self-interest. In the context of governing water, let's just say that we are still some distance down the learning curve. Indeed, it might well be argued that efforts genuinely directed toward the protection of water quality as a priority have been very rare. One needn't go that far, however, to point out major limitations.

Those in possession of relevant knowledge have not always been in effective communication with each other. Thus, scientific understanding of water (as this has evolved) has not always been aligned with legal principles, nor these with practices in engineering or public health. Many public officials have therefore struggled with limited success to identify the long-term interests of their communities and to pursue these effectively. Moreover, public officials struggle eternally over the proper allocation of responsibility for water quality between national, state/provincial, and local governments.

A series of important choices repeatedly arises. One of these involves the degree of reliance to be placed on professional opinion over the perceptions and preferences of the public, the former not infrequently divided, and the latter, though democratic, not necessarily informed. There has also been ongoing debate about the extent to which water supply and management are matters of a local nature and community based, rather than regional or national. Correspondingly, should the boundaries of administration, management, or even governance be civic and municipal/provincial, or more closely aligned with ecological boundaries and watersheds? The design of water and sewerage systems presents the further difficulty of choosing between public and private mechanisms, or some combination of these. In either case, consideration must be directed toward consumers' willingness to pay and the means of ensuring payments, whether through taxation, rate setting, or user charges. To promote compliance with whatever arrangements are adopted for water supply and sewerage, further consideration must be given to the style of enforcement, generally described as entailing conciliatory and coercive alternatives, each of which involves different implications from the perspective of cost, effectiveness, and so on.

Public interests, once consisting of rights to fishing and navigation, are now understood – at least by some – to encompass more than mere uses. Language has emerged to identify those interests – ecosystem integrity and function, biodiversity, sustainability. The last two concepts are still sufficiently new that my word processor has kindly underlined them in red, with the implication that I should find something else to talk about. Nonetheless, a good deal of thought is being given to what these concepts mean and to how they might be preserved or realized. The fundamental options have not changed all that significantly: markets, regulation, property rights, education. We may well wonder whether they are sufficiently adaptable to be up to the task.

Few of the policy makers, lawyers, and government officials who have espoused various models over the years have measured the effect of their preferred approaches against environmental standards. The tendency has been rather to evaluate regulation, property regimes, or markets from the perspective of institutional norms such as participation, democracy, efficiency, and, in the case of water, maximum benefits for multiple users. If one takes conservation, ecological integrity, or biodiversity as a frame of reference, what happens? Some of our institutions undermine or erode the foundations of life on earth less quickly and recklessly than others. It's not much of an accomplishment.

Nineteenth-century industrial development and accompanying urban growth significantly affected the use of waterways and their quality. Navigation, fishing, and mechanical power production at mill sites have always been prominent on that list of uses. Municipal water supply and waste removal (less romantic and less legendary) have attracted comparatively less attention, and orders of magnitude less than they deserve. The adoption and design of collective water supply and sewage systems were facilitated and shaped by developments in engineering, the natural sciences, public health, and law and municipal finance, among other professional specializations, which cumulatively if not cooperatively fostered, tolerated, and at times even encouraged a culture of flushing. The adoption of flushing found support in practices and institutions that reflected values, attitudes, and assumptions whose meandering course is a vital part of the story. As a use of water, waste removal presented its own peculiar demands, whose insidious ascendance and subsequent decline within the legal hierarchy in Britain, the United States, and Canada are this volume's central theme. There are, of course, important differences among and within the legal regimes of these jurisdictions, even though they generally share common law traditions. In relation to the practice of flushing, however, striking similarities are notable, whether these resulted from shared legal assumptions or the influence of other professionals whose trans-Atlantic careers and exchanges encouraged the continuing diffusion of learning throughout the public health and sanitary engineering communities.

A discussion of our varied and changing relationships with water over the past two centuries will help to illuminate several things. First, deeply rooted social values have shaped public attitudes to water. Those attitudes, in turn, have influenced choices about what to do and what not to do when threats to water quality have emerged. Choices, once made, have often been embodied in legal systems and institutions. On a continuing basis these too have been influential, insulating flushing from close scrutiny. The unfolding of this process can perhaps help to identify what needs to be overcome if we are to do better – better, that is, from the longer-term perspectives of sustainability, biodiversity, and ecological integrity.

With reference to the early nineteenth century, Chapters 1 and 2 introduce key aspects of the cluster of ideas and values that influenced decisions about the use of water. Included here is a description of early – and largely unsuccessful – attempts by navigation, fishing, and riparian interests to check or forestall a growing inclination to discharge waste materials into water. Chapters 3, 4, and 5 describe the introduction of municipal water supply systems by expanding urban communities; rapid growth in water consumption for a variety of purposes, including domestic flushing via the water closet; and the resultant need – influenced by the contemporary understanding of disease – to remove water and wastes from urban centres. Chapters 6 and 7 address the consequences of nineteenth-century sewage for downstream residents, for public health, and for the search for treatment procedures to reduce the volume of raw sewage and untreated industrial waste discharged. Many of the general themes are illustrated quite dramatically in the history of Chicago's water supply and sewage. As Chapter 8 shows, the Chicago experience generated considerable legal controversy and some particularly vigorous and imaginative defences of flushing before significant sewage treatment measures were finally implemented.

Recognition of the role of water-borne bacteria in the transmission of disease during the late nineteenth and early twentieth centuries encouraged a re-examination of flushing. Implementation of sewage treatment proceeded only slowly, especially as resources began to shift toward drinking-water treatment as a more direct means of safeguarding human health, a transition outlined in Chapter 9. Resistance to wastewater and sewage treatment was frequently encouraged by rudimentary calculations of costs and benefits that largely excluded the possible environmental impacts of flushing. The calculus penetrated legal thinking and contributed to the unwillingness of judges to check pollution. As the twentieth century advanced, some waterways were explicitly dedicated to waste removal, reflecting a powerful assumption – documented in Chapter 10 – that streams are nature's sewers. Early environmentalists, resolute anglers, and concerned policymakers responded to the widespread degradation of waterways with a range of strategies described in Chapter 11 as a riparian resurrection. As Chapter 12 indicates, any number of solutions to the challenges of deteriorating water quality were explored. Some approaches were purely technological, and some emphasized competing legal strategies; other means of governing water involved new institutional models, the participation of higher levels of government, new regulatory agencies, or efforts to enhance awareness of watersheds themselves as suitable units of organization. Late-twentieth-century water quality initiatives combining further waves of litigation and a series of statutory reforms must now be integrated alongside new technology with contemporary insights into the importance of biodiversity and sustainability.

I

The Advantage of a Flow of Water

ON THE EVE OF A PERIOD of urban and industrial transformation, three celebrated eighteenth-century figures interested themselves in water in remarkably influential ways. To set the stage, and to anticipate sometimes intricate interconnections between and among a variety of scientific and professional perspectives on water quality, it will be helpful to refer briefly to their views. The economist Adam Smith's assertion that "nothing is more useful than water" seems high praise indeed on the part of the author of *The Wealth of Nations,* until it is juxtaposed with his further observation that water "will purchase scarce any thing; scarce any thing can be had in exchange for it."[1] In fact, for Smith, these characteristics made water a public good, something of undeniable importance, though most difficult to govern or control. William Blackstone, Smith's contemporary and the renowned author of the mid-eighteenth-century *Commentaries on the Laws of England,*[2] firmly espoused the proposition that first occupancy was the governing legal principle in water rights: "If a stream be unoccupied, I may erect a mill thereon, and detain the water; yet not so as to injure my neighbour's prior mill, or his meadow: for he hath, by the first occupancy acquired a property in the current."[3] In other words, simply by appropriating water for use, an industrious entrepreneur seemingly could secure the legal entitlement to control at least some of Smith's public good.

At the very moment of Blackstone's invitation to the enterprising to assert their interest in Adam Smith's very useful and freely available waterways, Antoine-Laurent Lavoisier was engaged in early research into the actual composition of water. The French scientist had abandoned his own early legal training to pursue an almost infinite variety of disciplined scientific inquiries, which earned him a leading place among the founders of modern chemistry. A paper he presented to the French Academy of Sciences in 1768 on the analysis of water samples foreshadowed his admission to that celebrated body and initiated a very promising line of experimentation. Was water, as eighteenth-century commentators widely believed, one of the world's essential qualities like earth, air, and fire – vitally necessary, but otherwise quite unremarkable?[4] Lavoisier was soon able to establish that, contrary to prevailing opinion, water, when repeatedly distilled, would not be converted or "transmuted" to earth: as he commented, the notion "that a mass of water can, without addition, without loss of its substance, change itself into a mass of earth, is repugnant."[5] By subsequently analyzing water into its component parts, he showed that it was not an element.[6] In the words of Emmanuel Le Roy Ladurie, Lavoisier "gave a decisive push to the secularization and demystification of water." The transformation was profound: "Mystery gave way to science, religion to technology and salvation to health."[7] Chemistry became the leading water science of the age, eventually offering valuable insights to a growing body of engineers, municipal officials, and others concerned in practical ways with the supply and utilization of water.

Yet if Lavoisier's efforts might casually be characterized as directed toward understanding the physical nature of water, others – in the tradition of Adam Smith and William Blackstone – confronted seemingly more practical and immediate matters. Who was entitled to use water, and for what purposes? If disputes arose, including conflicts over water quality, how would they be resolved? What priorities would prevail?

We may begin to appreciate the nature of the water quality puzzle by considering the nineteenth-century foundations of water law. These are revealed in judicial and early legislative responses to the clashes, conflicts, and controversies that pitted individual users – often mill owners – against each other, or which saw broader community interests asserted against private claimants.

Embracing Corruption

As the nineteenth century opened, the Irwell was a busy stream descending from the Pennines to England's North Lancashire Plain. Here lay Manchester, already a textile centre of long standing, to whose development and prosperity the Irwell, despite its modest size, contributed significantly as a source of power for local mills, a source of water for a growing network of canals – and a means of conveying urban and industrial wastes away from a city whose population in 1801 exceeded eighty thousand.[8]

Eventually, two Irwell mill owners clashed over their respective entitlements to the stream's flow. Bealey, a whitster who established his mill and machinery in 1787, had been unable to operate his bleaching works for substantial periods of time due to shortages of water – indeed, its virtual absence from the riverbed. Attributing the lack of water to the 1791 enlargement of a sluiceway by his upstream neighbour, Shaw, he initiated legal action.[9] Shaw and his associates, whose own weir and sluiceways on the Irwell had been progressively expanded through much of the eighteenth century, boldly replied that their prior use meant they were "at liberty to appropriate to themselves as much more as they wanted, in the same manner as they had several times done before the plaintiff's works and sluice were erected and made."[10] To put Shaw's claim to the river a bit more bluntly, he got there first.

At that point, several judges rejected Shaw's claim to exclusive use of the river and to his total subordination of Bealey's interests. As Sir Simon Le Blanc summarized his view, "After the erection of works and the appropriation by the owner of the land of a certain quantity of the water flowing over it, if a proprietor of other land afterwards take what remains of the water before unappropriated, the first mentioned owner, however he might before such second appropriation have taken to himself so much more, cannot do so afterwards."[11] Essentially, what Le Blanc described was a first-come, first-served system with significant encouragement to help yourself to seconds before other guests arrived.

In concurring with his colleague, Lord Ellenborough, the chief justice, expressed a broader understanding of current law. In general, he observed, "Every man has a right to have the advantage of a flow of water on his own land without diminution or alteration." Here was a stronger presumption of equal entitlement: neighbouring water users along a waterway were limited in principle by the claims of others, even if those competing uses were not actually being exercised. It was possible, nonetheless, that prior occupation or actual use of the waterway might secure greater rights. An incentive therefore existed to take a very full measure indeed, and to do so sooner rather than later. Ellenborough spelled out the legal consequences for the downstream owner: "And though the stream be either diminished in quantity or even corrupted in quality, as by means of the exercise of certain trades, yet if the occupation of the party so taking or using it have existed for so long a time as may raise the presumption of a grant, the other party whose land is below must take the stream, subject to such adverse right."[12]

Although the actual case centred on the flow or volume of the Irwell, Ellenborough's formulation of the applicable principle of law notably addressed the condition or quality of the water, specifically the possibility he described as "corruption." The reference was somewhat unusual, for although navigation, fishing, and waterpower development at mill sites were matters of long-standing public interest, early nineteenth-century law had comparatively little to say about contamination. It was more noteworthy still that Ellenborough's understanding of the relevant principles seemed to accommodate elements of a right to pollute. The nature of such a right, its scope and limitations, and perhaps more fundamentally the basis on which the corruption of water might be justified, were left for another day.

However the principles of "corruption" might ultimately be explained, the Irwell was in practice transformed from the late eighteenth century. The use of the river for drinking water, and for such domestic purposes as washing, was abandoned as the tanneries, gasworks, and chemical and textile mills of Lancashire discharged their residues into its course, where it joined the raw sewage from Manchester's burgeoning population. By 1854 Manchester's city surveyor showed no hesitation in referring to the Irwell and other waterways in the vicinity as "nothing but main sewers."[13] A student of textile wastes later likened the Irwell, "a pure pellucid babbling brook" at its source at Irwell Springs, to a fair maid who reached the Manchester Ship Canal above Tafford Bridge as a "weary, forlorn and morally bedraggled hussy." Textile producers involved in bleaching, dyeing, calico printing, and finishing operations, having established their entitlement to the clean water of the Irwell and its tributaries, subsequently discharged effluent into the watercourse. "Thus the fair maiden does not travel very far before her hair is dyed blue, green, or some other unnatural colour; her eyes are burned by acids or caustic soda and her skin befouled by starch and china clay." The outpourings of fellmongers, tanners, tar distillers, paper makers, bone boilers, chemical manufacturers, and gasworks, attracted to power sites along the waterways, all took their toll.[14]

Frederick Engels, in documenting the condition of the working classes in England, also recorded conditions on the nearby Irk, a waterway entirely typical of the district. He too painted a dire picture, describing the Irk as "a narrow, coal black, foul-smelling stream full of debris and refuse, which it deposits on the shallower right bank. In dry weather, a long string of the most disgusting, blackish green, slime pools are left standing on this bank, from the depths of which bubbles of miasmatic gas constantly arise and give forth a stench unendurable even on the bridge forty or fifty feet above the surface of the stream." Because the Irk's flow was impeded by high weirs at short intervals, slime and refuse from tanneries, bone mills, and gasworks accumulated to "rot in thick masses," along with sewage and privy wastes discharged wantonly and without apparent objection.[15]

The Irk and the Irwell were extreme, though not unique, examples of the deterioration of nineteenth-century industrial waterways. Numerous such waterways had long been put to service for both industrial and municipal waste. As Sir Samuel Martin remarked in adjudicating another mid-nineteenth-century water dispute, the use of flowing water to operate machinery was "as old as the law." He considered the law to be well settled in favour of such uses, for "it is at once beneficial to the owner and to the Commonwealth." However beneficial mill operations may have been from the perspective of power production, they were also a source of refuse and waste materials. These were typically discharged, without much thought to the consequences, into waterways whose natural flow was increasingly subordinated to industrial priorities. Mills, explained Martin, were occasionally situated directly upon the stream bank, but more frequently a little distance away, in which case water was conveyed along an artificial cut, or goit, leading from the stream. Having turned the mill wheel, the flow was directed back to the stream via the tail goit. Martin estimated the value throughout England of such water rights to be in the range of "hundreds of thousands, perhaps millions."[16] As high as their value may have been, later generations would ultimately come to reflect upon their cost.

Water mills proliferated in North America as well, beginning with local grist- and sawmills, and then, in connection with the expanding lumber trade, moving into small-scale ironworks and textile industries. Along the tributaries of the northeastern river network, where tens of thousands of potential sites could be found, development was widespread and often intensive. Estimates and census records for the United States suggest that from the few thousand mills in existence even before the revolutionary break from England, numbers rose steadily to the tens of thousands by the mid-nineteenth century, reaching a peak of perhaps a hundred thousand. As late as 1919, the US Federal Census of Manufactures reported twenty thousand water mills still in use.[17]

At particularly favourable locations, clusters of mill sites operated together. As described by Louis C. Hunter in his monumental *History of Industrial Power in the United States,* by 1824 some twenty mills and factories operated at Vergennes, Vermont, around a series of three falls in

Otter Creek. A decade later at Seneca Falls, New York, where a sudden fall in the Seneca River provided opportunities for four dams, thirty manufacturing establishments supported a population of roughly three thousand. Sharp drops in the Black River en route to Lake Ontario from Watertown, New York, and in the Nashua River in the vicinity of Fitchburg, Massachusetts, similarly encouraged a concentration of industrial activity, harnessing waterpower and generating wastes.[18]

On both sides of the Atlantic, mills were generally encouraged by legislation that recognized their economic contribution; or they were so firmly established by the nineteenth century that their adverse implications for water quality were difficult to overcome. As Manchester sought to discharge its own wastes, the city perceived itself as a victim of existing mills and their associated infrastructure along regional waterways: "the damming up of . . . streams into a series of stagnant pools of filthy water, receiving the whole sewage of the town, is a monstrous evil."[19] Despite decrying the "evil" effects of mills, dams, and sewage in towns, mid-century observations widely took rivers for granted as a vehicle for the removal of municipal waste – further downstream.

To address related concerns in the United States, some statutes attached obligations to mill privileges to reflect a popular understanding of the emergence of disease from pollution. Michigan legislators, for example, required millpond owners to keep ponds clear of such "substances as may produce disease." A still more elaborate provision called for annual drainage to facilitate cleaning of the ponds "whose rank vegetable growths indicated the half-hidden impurities on which they are nourished [and when] emptied and the whole bottom exposed [revealing] from the faint and disgusting odor, what dangerous stuff has been fostered in the half-stagnant water." Yet such requirements were hardly universal, and their scientific foundations failed to endure. Some of these water quality provisions were subsequently eliminated, or limited in their application to existing mills; they were seldom, if ever, included in the legislation of more westerly American jurisdictions. The overwhelming tendency in mill legislation was therefore to treat water as a source of power, and to confine mill operators' obligations to compensating riparian neighbours for the value of submerged lands.[20] A Massachusetts health official summed up the perception of dispirited observers: "There will never be wanting advocates of any application of natural forces which may lead to individual or corporate profit, while considerations of public health are always less obvious."[21] After the adoption of municipal water-borne sewage systems, and following a good deal of learning in the aftermath, Marshall Ora Leighton warned in his 1902 report to the United States Geological Survey that millponds might become "grievous nuisances" by storing pollution from upstream municipalities.[22]

Less obvious environmental damage also occurred when waterways were engineered and manipulated in the interests of power production. Mills, storage reservoirs, and the operation of associated facilities have the effect of eliminating several features of their natural regimes. Established ranges and patterns of flow, water temperature, and sediment transport may be replaced with rapidly fluctuating flows that alter conditions significantly. Low flows may run higher and high flows are generally lower, eliminating processes of cleansing and nutrient deposition associated with the annual flood pulse.[23] As a consequence of alterations brought about by damming, ecosystems along waterways may experience severe transformations or destruction. Upstream, the terrestrial ecosystem submerged by millponds or reservoirs will disappear, along with the original river ecosystem now replaced by standing water. Downstream, ecosystem effects are particularly noticeable in relation to fish, especially species prevented by dams from reaching spawning grounds. But all organisms adapted to the natural cycle of river fluctuations are adversely affected when dams, channels, and diversions alter the flow.[24]

Potential controversy over water had become so prevalent by the mid-nineteenth century that contemplation of these seemingly arcane disputes was by no means confined to professional circles. In George Eliot's novel *The Mill on the Floss* (1860), a tragic clash between Mr. Tulliver of Dorlcote Mill and a certain Mr. Pivart, a newcomer to the area whose planned irrigation works jeopardize Tulliver's personal and financial circumstances, figures prominently. Tulliver's understanding that "water is water" is rooted in generations of undisturbed enjoyment of the flow of the Ripple, a tributary of the Floss: "It's plain enough what's the rights and the wrongs of water, if you look at it straightforward; for a river's a river, and if you've got a mill, you must have water to turn it." He also knows, of course, that "water's a very particular thing – you can't pick it up with a pitch fork. That's why it's been nuts to Old Harry [the Devil] and the lawyers." The old miller attributes Pivart's triumph in the litigation, and the ensuing precipitous decline in his own fortunes, to the deviousness of his opponent's attorney, who had bested his own counsel in the contest of law – a process he quite aptly sums up as "a sort of cock-fight."[25] Although Mr. Tulliver may have believed that the lawyers saw water-related litigation solely in terms of profit, in fact the rights and wrongs of water continued to bedevil the legal profession throughout the century.

Certainly there were more conflicts and more cases. Increasing and more intense usage not only multiplied the sources of friction among river users but complicated the resulting tensions as more than a handful of immediate neighbours were affected. "In consequence of so many water mills," wrote John Sutcliffe in his 1816 work on reservoirs and canals, "the country is never free from litigations and vexatious lawsuits respecting erecting, repairing or raising mill-weirs, by which the peace and harmony of neighbours and friends are often destroyed."[26] In 1833 the author of an American water law treatise noted that in the ten years since his first edition there had been more decisions about water rights than in the entire history of the common law up to that time.[27] As we shall see, later in the century water rights cases were to be even more prevalent and vigorously contested.[28]

A growth in the volume of disputes is often either the cause or the consequence of new principles working themselves out. Many who have reflected on the matter, generally in relation to problems of water supply and power production rather than water quality, are confident in concluding that in this transitional period the rules were indeed changing.[29] There is less agreement on how to explain the transformation. Some point to an interventionist thrust on the part of judges, others to the influence of rising costs and the increasing complexity of concluding agreements among river users, and some to a historic ebb and flow between the claims of occupancy and ownership as mediating mechanisms.

One important argument is simply that increasing industrialization brought about changes in the rules, both directly and indirectly. Manufacturing, especially of textiles, along New England's rivers accentuated concerns about efficiency in waterpower use. This objective in turn shaped the context within which water law evolved, and thereby to some degree shaped the law's content as well. The argument, elaborately presented in numerous historical and theoretical accounts, explains changes in late eighteenth- and nineteenth-century water law to a position that increasingly privileged industrial activity for its perceived contribution to the general well-being of the community.

Morton Horwitz, whose provocative *Transformation of American Law* stimulated much debate, describes the late-eighteenth-century premises of the common law as "fundamentally anti-developmental." Land, rather than being a productive asset whose value lay in its productivity or instrumental capacity, was understood to be "a private estate to be enjoyed for its own sake."[30] Conflicts over new uses had therefore customarily been resolved in favour of inactivity. Subsequently, however, means of reconciliation, or balancing principles emerged, oriented around the inherently judicial concept of reasonable use. These principles were invoked to facilitate productive activity, notwithstanding its impact on traditional and established interests. In the opinion of historian Theodore Steinberg, author of *Nature Incorporated,* crucial changes took place in the late 1700s "as cultural norms and legal rules evolved away from a vision of water consistent with long-standing, agricultural uses."[31] Thereafter, water law in New England developed in conformity with the evolving requirements of the region's industrial economy. By mid-century an instrumental approach to water use oriented toward maximization of economic growth predominated.[32]

The conception of law as "instrumental" has been explained as a view in which legal rules are essentially "tools devised to serve practical ends."[33] Carol Rose, an insightful legal theorist, views judicial decisions also as an attempt to encourage economic activity, although her account is rooted in difficulties that earlier doctrines presented to anyone trying to negotiate successful accommodations between rival water users. Growing numbers of mills and other industrial river uses rendered the task of negotiating contractual agreements to resolve the complexities of diversion, interruption of flows, pollution, and other issues insurmountable under the earlier doctrines of ancient use and prior occupancy. Given such obstacles to successful negotiations (costs of transacting, as they are sometimes described), Rose observes, courts in industrializing states sought alternative doctrines more likely to enhance the overall value of river usage. Resource economists Tony Scott and Georgina Coustalin have a similar understanding of judicial objectives. In an attempt to assess and explain the long-term evolution of water law, they suggest that we "look at the dynamics of change in water law through the forces of supply and demand." Courts and legislatures can be said to supply change over time in response to evolving demands for authorized river and water use.[34] Depending upon the attributes of any particular waterway, such as its levels, flows, and fluctuations, opportunities arise for a range of human uses. On the demand side, whereas navigation (or transport) and fishing once predominated, technological changes prompted additional demands for irrigation and power production. Other uses, notably municipal water supply, waste removal – and eventually wildlife habitat, recreation, and aesthetic appreciation – subsequently introduced new tensions to the process of allocation.

The values being maximized were, of course, values recognized by and important to contemporaries. The limits of their knowledge and aspirations thus circumscribed their understanding of river functions, eliminating other values from whatever calculus applied. In the early nineteenth century, for instance, those interested in the use of waterways had power for industrial purposes as a primary objective, even though, by attracting industrial development and urbanization to the shorelines, power production was accompanied by waste production and the use of water for purposes of disposal. The adverse consequences of textile and paper making, among other industrial activities whose waste management programs involved flushing, were rarely integrated into decision making. Indeed, they were often discounted, while waste removal – at times surreptitiously and on occasion through deliberation – secured an abiding place in the inventory of river functions.

For the Use and Comfort of Man

A sampling of nineteenth-century judicial opinion and commentary reveals some of the interaction and convergence between industrial usage of waterways and community values and public interests. Despite differences in the pace and style of urban and industrial development, similar challenges over water use and allocation eventually arose throughout Europe and North America. Lawyers, engineers, scientists, and public health officials, among the growing ranks of interested professionals on both sides of the Atlantic, sought guidance from each other, or wherever it might be found.

Justice Joseph Story, who practised law in Salem, Massachusetts, from 1810 to 1823, and James Kent, chancellor of the New York Court of Chancery from 1814 to 1823, were particularly influential American commentators whose widely respected judicial opinions and academic writings encouraged a more flexible approach to water rights than the traditional principles of natural flow and prior appropriation formally allowed.[35] In *Tyler v. Wilkinson*,[36] Story, then a circuit justice in Rhode Island, invoked general principles of water law as expressed in cases such as *Bealey v. Shaw* to underpin his own reformulation of the rights of shoreline owners in a manner more in keeping with principles derived from the French Civil Code.[37] He discerned "a very strong and controlling current of authority" supporting, as a point of departure, the understanding that "every proprietor upon each bank of a river is entitled to the land, covered with water, in front of his bank, to the middle thread of the stream." Ownership by "riparian proprietors" entailed the right to the use of the water flowing by "in its natural current, without diminution or obstruction." This did not entail ownership of the water itself "but a simple use of it, while it passes along." In consequence, "no proprietor has a right to use the water to the prejudice of another," whether upstream or down. This, Story explained, was "the necessary result of the perfect equality of right among all the proprietors of that, which is common to all." It was important to add – and Story did so with emphasis – that the doctrine did not require "that there can be no diminution whatsoever, and no obstruction or impediment whatsoever, by a riparian proprietor, in the use of the water as it flows; for that would be to deny any valuable use of it." The law therefore allowed "of that, which is common to all, a reasonable use," with the extent of that use governed or ultimately limited on the basis of injury to other proprietors.[38]

In a related elaboration of these ideas, Kent endorsed the understanding that in using water as an incident to land, a riparian might indeed interfere with its flow – detaining it in effect – as long as that proprietor did not "*unreasonably* detain" the water.[39] To be sure, the traditional language of natural flow, complete with suitable reference to principles from *Bealey v. Shaw*, was well represented in the writings of Story and Kent: "Every proprietor of lands on the banks of a river has naturally an equal right to the use of the water which flows in the stream adjacent to his lands, as it was wont to run . . . without diminution or alteration. No proprietor has a right to use the water, to the prejudice of other proprietors, above or below him, unless he has a prior right to divert it, or a title to some exclusive enjoyment. He has not property in the water itself, but a simple usufruct while it passes along."[40] Ultimately, however, this reformulation of water law was much less restrictive than the point of departure: "All that the law requires of the party, by or over whose land a stream passes, is, that he should use the water in a reasonable manner, and so as not to destroy or render useless, or materially diminish, or affect the application of the water by the proprietors above or below on the stream."[41]

In light of the fairly firm manner in which natural flow might have been applied – and the basis of enduring tradition that might have supported it – one wonders what justification underpinned the more flexible formulation of riparianism espoused by Chancellor Kent. He was remarkably frank in his explanation: "Streams of water are intended for the use and comfort of man; and it would be unreasonable, and contrary to the universal sense of mankind, to debar every riparian proprietor from the application of the water to domestic, agricultural, and manufacturing purposes, provided the use of it be made under the limitations which have been mentioned."[42]

Recognizing that the strictures of the rule might severely impede if not entirely prohibit use, the formulation carefully provided that "this rule must not be construed literally, for that would be to deny all valuable use of the water to the riparian proprietors. It must be subjected to the qualifications which have been mentioned, otherwise rivers and streams of water would become utterly useless, either for manufacturing or agricultural purposes."[43] Embedded in such sentiments was an important corollary: a virtual condemnation of unproductive resources – unproductive, that is, of manufacturing and material advancement. The suggestion that rivers and streams not employed for manufacturing or agricultural purposes were essentially being wasted became a widely accepted article of faith. The moral framework left little room for ecological functions, or even aesthetic appreciation. Lacking human intervention, waterways fell beyond the range of contemplation of those committed to the quest for improvement, however vital they might later appear.

At roughly the same time as the *Tyler v. Wilkinson* litigation was under way in the United States, a comparable controversy was unfolding in England. In *Wright v. Howard,* despite resistance to reasonable use as a mediating principle, the court clearly asserted, "Every proprietor has an equal right to use the water which flows in the stream, and consequently no proprietor can have the right to use the water to the prejudice of any other." Only with the consent of fellow riparians could a proprietor diminish the volume of water that would otherwise flow to downstream proprietors, or cause water to back up on the proprietors above.[44] If the judgment appeared to retain elements of natural flow language, the adoption of equal rights was certainly far removed from a regime in which prior use or occupancy, as previously endorsed in judgments such as *Bealey v. Shaw,* constituted the basis of entitlement to water use.

At mid-century, perhaps by way of critical comment on the principles emerging from the United States, another English court suggested that the effect of Chancellor Kent's formulation was that "in America a very liberal use of the water, for the purposes of irrigation, and for carrying on manufactures, has been allowed."[45] Yet shortly afterwards, and again with explicit reference to the American perspective, the "reasonable use" version of riparian doctrine, permitting what was described as "a very liberal use of water," had also been adopted in England. Claims of exclusivity, whether based on earlier notions of ancient rights or the concept of prior occupancy, had given way to the principle of correlative rights, a doctrine that may have facilitated water use at the cost of certainty. As expressed in another English decision, a proprietor's entitlement to enjoy "the benefit and advantage of the water flowing past his land" was neither absolute nor exclusive. It was, rather, "subject to the similar rights of all the proprietors of the banks on each side to the reasonable enjoyment of the same gift of Providence."[46]

Just as significant as the somewhat abstract statement of entitlement was discussion of the application of reasonable use to distinguish between lawful and unlawful uses. The circumstances of each case would be determinative. By way of example, it was thought highly improbable that it would be acceptable for the owner of an extensive tract of porous soil to irrigate thousands of acres on a continuous basis from an abutting stream with the consequence that the quantity of water remaining was seriously diminished. On the other hand, it was suggested, "one's common sense would be shocked" to imagine that a riparian owner "could not dip a watering-pot into the stream, in order to water his garden, or allow his family or his cattle to drink it." The distinction between reasonable and impermissible uses was "entirely a question of degree," and difficult, if not impossible, to define with precision. The court was confident nonetheless that resolving particular cases would be simple enough. Indeed, the immediate situation was one such example, for the defendant's irrigation was intermittent rather than continuous, occurred only when the river was full, produced no perceptible diminution of the water, and resulted in no damage to the operation of the mill whose owner had initiated the unsuccessful legal claim.[47]

In light of these developments, it is clear enough that Tulliver's fictionalized conflict with his upstream antagonist came at a poor time for his own prior use of a water mill to succeed handily against the newcomer's irrigation venture. However, mechanisms other than an uncertain judicial process might be employed to clarify water rights. One involved private arrangements devised between the water users themselves to govern the use of waterways. The other was legislative intervention.

The Regulated Waterway

In 1830 and 1831, spring freshets on the Scugog River within the central Canadian Trent-Kawartha watershed demolished much of the original stonework of a mill dam erected by William Purdy, releasing water that had accumulated behind the rudimentary structure. Renewed building efforts brought greater success, resulting in what John Langton, a contemporary observer, imagined might be "the largest mill dam in the world."[48] The construction transformed the Scugog from a meandering stream into a river, as it was widened and its current virtually stilled; an area roughly thirty miles in length and width was submerged.

The extensive flooding caused immediate dissatisfaction among landowners in the vicinity, including some holders of other valuable mill privileges. Nicol H. Baird, an engineer and surveyor in the employ of a regional canal, reported in 1835 that the "destructive scale" of Purdy's dam greatly exceeded the impact of similar dams with which he was familiar.[49] Baird had observed hay meadows covered to a depth of nine feet; drowned and decayed timber was widespread. The remnants of settlers' roofs close to the waterline even bore witness to a number of compulsory evacuations. Instead of advancing the interests of farmers, Purdy's operations had obliterated the lands of some and severely devalued the holdings of others.

Baird contended that Purdy's Mill, more efficiently operated, would have needed less water and could accordingly have alleviated the upstream hardship caused by the overflow.[50] To appease area residents, he recommended lowering the existing dam to about half its then-height of twelve feet and proposed other supplementary and remedial engineering, including a second dam and a lock to facilitate access to the more elaborate canal network of the surrounding district. Purdy's reluctance to accommodate the settlers' desires for such access to wider markets only added to their resentment. Armed settlers attacked the mill in 1838. Three years later, when an outbreak of malaria was popularly – if not entirely appropriately – attributed to the mill dam and the flooding it produced, the structure was again targeted by a desperate mob.[51] Rioters attacked Purdy's facilities on two separate occasions in December 1841. Shortly after Christmas, weakened by the assaults, the dam collapsed.[52]

Other rules and other circumstances might have altered the relationships between Purdy and the neighbours. Contemporary observers attributed the failure of local residents to take any legal action against mill owners to a state of poverty so severe as to exclude them from access to whatever lawful forms of redress were otherwise obtainable.[53] In addition, anticipating complications that might arise from stored water overflowing neighbouring lands, Purdy had originally requested that the interests of those holding lots adjacent to the mill site be specifically subordinated to flooding caused by the dam. Such a concession, had it been granted, would have afforded a large measure of immunization for Purdy's Mill against claims by any flooded and aggrieved fellow residents. A grant in the terms requested, though too generous to the builder, might have redirected the energies of the dissatisfied settlers toward monetary compensation rather than elimination of the dam. Had Purdy not also pre-empted fishing on the river by placing traps at the mill and then selling the captured fish at two pence each, he might have spared himself still further animosity.[54]

Although Purdy failed to obtain legal immunity for the effects of his appropriation of the Scugog, developers elsewhere had been more successful in securing such mechanisms. Official encouragement of mill dams was long-standing throughout Europe and North America. Arrangements not unlike William Purdy's request for immunity against aggrieved riparian neighbours had in fact been made in a number of jurisdictions. Mill legislation typically allowed a landowner to develop a water mill and mill dam on a non-navigable stream conditional upon payments to other riparians who experienced flooding as a consequence of the water being held back. The injured proprietors, in other words, were deprived – for a price – of their shorelines. They lost the right to preserve their properties through legal action to prohibit the flooding that would result from mill construction. Legal historian J. Willard Hurst has described Wisconsin's approach to mill dams, for example, as one that "showed general favor toward developing waterpower, by giving a blanket license to riparian owners to build dams on non-navigable streams." By way of qualification, and to afford a measure of protection to injured landowners, the dam builder's initiative was subject to an "absolute requirement that fair compensation be paid."[55]

Such legislative mechanisms embodied the understanding that mills, although in private hands, made an important public contribution to the well-being of the communities they served. Apart from navigational uses of waterways, whose claim to pre-eminence will shortly be considered, waterpower enjoyed "a special and favoured position in the eyes of government." This privileged status reflected a firmly entrenched assumption that mills – gristmills and sawmills in particular – were important elements of community infrastructure. An entitlement to special forms of aid and encouragement derived from their perceived public character and their benefit to the general population.[56] "The man who first erects a mill in a new country," reads one early-nineteenth-century court decision in Pennsylvania, "is considered as a public benefactor, and no subject ought to be treated with more tenderness, no possession more respected."[57] In Wisconsin, as Hurst explains, both general legislation and specially crafted authorizations were employed to effect the desired mix of promotion, reconciliation, and regulation. It was appropriate, Wisconsin jurists acknowledged, that mill dam construction be favoured by the legislature in a new country with limited capital and a sparse, widely scattered population. A law of this sort, having the effect of encouraging investment, enterprise, and associated conveniences, in the opinion of one court, "enhances the value of land, advances its settlement, and promotes general civilization."[58]

Massachusetts courts also construed mill acts sympathetically, protecting exclusive rights to water insofar as power sites had actually been appropriated for mill purposes.[59]

In their design and implementation, mill dam authorizations might address – either explicitly or by implication – the relationship between the benefits anticipated from construction on the one hand, and a complex range of impacts and implications for alternative stream uses on the other. For example, what consideration would be accorded neighbouring mill sites that might be subject to a reduced head (height of fall) from backflooding, or to reduced, interrupted, or unpredictable flows? What provisions, if any, would secure a measure of future flexibility in the event of technological advances, or unanticipated consequences? Legislators devoted sometimes more, sometimes less attention and reflection to such questions. What they did not address – or addressed only rarely – was the significance of mill dams for the quality and condition of the waterways.

Waterpowered manufacturing production, as noted by Hunter, involved one or more simple and repetitive operations: grinding, crushing, pounding, pressing, sawing, pumping, or blowing. Lamentably, flushing must be added to the inventory of activity concentrated along waterways where power was available. Certain mills – notably tanneries, works associated with textile manufacturing, and sawmills – discharged substantial quantities of waste into the waterways.[60] Having approved such enterprises – indeed, having endorsed them as facilities contributing to the public welfare in some respects by their very existence – many jurisdictions would later face difficulty when the consequences of their discharges, even including the destruction of traditional water supply sources, came to be more fully understood.

As well as legislative intervention, direct agreements reached between groups of proprietors might create a regime that allowed little room, if any, for residual or underlying public interests. Manufacturing works along the River Tonge in England's Lancashire district, for example, used settling ponds to remove elements of contamination from water drawn onto their premises before it could be put to use in the various businesses of the neighbourhood. But in time the accumulated residues had to be removed. By local practice, "generally on Saturday afternoon," these ponds were flushed out. During flushing time the water descending the stream was so foul as to be unusable for any purpose: "Of course every manufacturer is aware of what is going to happen, and is alive to the danger and shuts off the intake and lets the foul water pass," remarked a high court judge presiding over a challenge to these apparently quite conventional arrangements. Not until egregious abuse by one upstream manufacturer provoked a downstream bleaching works to object to the long-standing routine was this convenient conspiracy of contamination checked.[61]

In neighbouring Yorkshire, systematic sharing of the flow of the Bowling Beck at Bradford permitted worsted-spinners and dyeworks to coexist with a colliery and the town's common sewers. The arrangement depended upon cloughs, or doors, that were installed across the beck upstream from the worst of the waste inflow. With the beck closed off during daylight hours, comparatively clean water was held back while the lower waterway was essentially dedicated to effluent and mine water. Between ten o'clock in the evening and about six the following morning, the woollen mills opened the clough to collect a usable if not ideal water supply for their own purposes. Eventually, lower volumes of runoff and increased contamination from additional manufacturing concerns undermined the viability of this practical accommodation of private interests competing for water and waste-removal services.[62]

Although the challenges of negotiating could be significant, particularly where water supplies were limited, private parties were on occasion able to regulate the use of rivers – portions of them, at least – through formal agreements designed to govern the allocation of water. Six paper mill operators clustered around Newton Lower Falls on Massachusetts's Charles River, for instance, concluded an agreement in 1816 concerning the overall apportionment of the flow amongst themselves, consenting in addition to a schedule of alternating usage in periods of particularly low water.[63] It was not uncommon for industrial requirements to exceed the natural flow of many streams. This deficiency could be alleviated by holding water back for storage when mills were not operating and releasing it later as required.[64] Such practices might, of course, present a problem for downstream mills at the start of the week if their upstream counterparts had been accumulating pondage over Saturday night and Sunday in order to ensure their own supply of power. The pressures of necessity eventually brought mill owners on many streams together in joint programs for upstream storage and stream control that had little to do with any natural functions of the waterway.[65]

Regulation by private agreement was perhaps most readily accomplished in conjunction with the sale of land. Here, the deed itself recorded any understandings such as those related to priorities, or with respect to anticipated future requirements and responsibilities for the maintenance of shared facilities. These were precisely the circumstances that inspired Granby pioneer industrialist John Horner to specify certain conditions when transferring part of his holdings along Quebec's Yamaska River in 1835. Horner had previously sold a downstream gristmill to another local entrepreneur, but anticipated construction of a sawmill on the remaining part of his own lands. He therefore attempted through negotiations to ensure priority for the necessary waterpower when he sold a tannery site across the river to Harlow Miner. Thus, the deed subordinated the waterpower needs of Miner's tannery to the requirements of Horner's proposed sawmill and required Miner to keep a tight flume and maintain his end of an existing storage dam.

Alas, Horner's careful arrangements unravelled when yet another local industrialist, Francis Gilmour, acquired both the downstream gristmill and the site that Horner had expected to develop opposite Miner's tannery. Gilmour then proceeded to operate a sluiceway on the upper dam in a manner that ensured the successful operation of the original downstream gristmill. The effect was to draw down water that might otherwise have been retained to power Miner's tannery.

Relations between Miner and Gilmour deteriorated enough that this local colonial dispute, having worked its way through Canadian courts, eventually reached the Judicial Committee of the Privy Council in London. The distinguished panel of jurists included the Right Honourable Sir John Coleridge, a nephew of Samuel Taylor Coleridge (whose sentiments about water remain somewhat better known from "The Rime of the Ancient Mariner"). For their part, the judges made plain the importance of riparian ownership for the fate of the waterways. Initially, every riparian proprietor enjoyed a right to "what may be called the ordinary use of the water flowing past his land." This could involve, for example, "the reasonable use of the water for his domestic purposes and for his cattle, and this without regard to the effect which such use may have, in case of a deficiency, upon proprietors lower down the stream." In addition, riparian ownership entitled the shoreline proprietor to a further right to use water "for any purpose, or what may be deemed the extraordinary use of it, provided that he does not thereby interfere with the rights of other proprietors, either above or below him."[66]

In the judgment of the House of Lords, then, factors governing or constraining the use of water revolved around the concept of reasonableness, around distinctions between ordinary and extraordinary uses, and around the comparable interests of other proprietors.[67] These considerations bore little obvious relationship to the underlying environmental values of rivers, now associated with such ideas as "in-stream uses," sustainability, or ecological integrity. The pernicious influence of this omission persists today.

Where natural flow had once held sway, reasonable uses now governed; what might be reasonable then fell to be determined. This complex task entailed a range of factors, some of virtually routine character, others involving an almost infinite variety of intricate considerations, making them a "question of degree" in the minds of judges ultimately called upon to resolve conflicts. "Reasonable" meant reasonable from the perspective of river utilization, not reasonable from the perspective of water systems as natural systems, for usage – that is, human usage – was expected and encouraged. Indeed, it was strongly suggested that when water was unused by man its divine purpose was unfulfilled.

The complexities of ascertaining how such "questions of degree" might be resolved were fully elaborated late in the nineteenth century by a Minnesota court. The factors to be considered in determining whether a particular use was reasonable included, of course, the nature of that use. What was the subject matter of the use? When and in what manner did the use occur? For how long, to what extent, and for what purpose – personal benefit or commercial gain? What claim of necessity might be made in favour of the use in question? Widening their scope beyond these matters, courts concerned themselves with the nature and size of the stream affected and with the extent of injury to other parties. The wider context might also affect the reasonableness of any particular use, with the consequence that courts interested themselves equally in "the state of improvement of the country in regard to mills and machinery, and the use of water as a propelling power; the general and established usages of the country in similar cases." Not to be accused of overlooking some vital element, the Minnesota court expressed its willingness to assess the fitness and propriety of a given claim to the use of water in light of "all the other and ever-varying circumstances of each particular case."[68]

To subsequent observers, the appropriate application of the reasonable-use principle was less than self-evident. Robert Abrams, for example, suggests that riparianism's "reasonable use" principle "approaches the ultimate in amorphousness," offering "debilitating unpredictability of outcome."[69] A modestly less denunciatory assessment of reasonable use refers to it as "a doctrine that lent itself to considerable latitude of interpretation."[70] The scope of that latitude had significant implications for water quality, especially in light of the burden that complex court proceedings might place upon private parties to safeguard waterways against contamination.[71]

Reinforcing private rights in nineteenth-century waterways, as well as the quality of the waterways themselves – at least potentially – were a variety of public interests. Some, such as fishing and navigation, were of long-standing importance; others, such as recreational activity, appeared more novel. Each of these was invoked at times as indirect means to preserve water quality against contamination. These other interests also occasionally underpinned legislative measures directed toward the same end. Yet proxies for water quality had to struggle for recognition against popular preferences and values that often proved hostile or dismissive.

2

Navigating Aquatic Priorities

EARLY-NINETEENTH-CENTURY CONTESTS OVER water quality were frequently indirect, with the interests of some well-recognized water use or purpose such as fishing or navigation serving as a proxy or surrogate for environmental standards. Uses or purposes, some commercial and industrial, others recreational and aesthetic, clashed repeatedly with each other. Increasingly, the subject of those contests was pollution, an activity that, as we shall see, emerged in its own right as an acceptable use of waters. Thus, as community values and preferences evolved, established purposes jostled each other for position on the hierarchy, sometimes with reference to the common law but increasingly, as the nineteenth century progressed, in the context of legislative developments. The process was not unlike a familiar children's game. Would boats beat rivers? Could sawdust beat fish? Who would prevail if swimmers clashed with sailors?

POISONING A FEW FISH

The impacts of early-nineteenth-century urban and industrial development on fisheries were readily apparent. Fine salmon were still being taken from the Thames in central London during the 1820s and at Hammersmith and Barking Creek a decade later, but they were soon entirely displaced by coarse species more able to tolerate rapidly deteriorating conditions.[1] When local fishing interests attributed visible degradation to the discharge of wastes into the waterways, they entered the front ranks of the opposition, where, by and large, they have remained.

In 1832, two years after the establishment of the Equitable Gas Company, which was instrumental in illuminating the dank and narrow streets of late-Georgian London, an assistant to the city's water bailiff surveyed the condition of the Thames. Thousands of dead fish could be seen floating near Equitable's premises amongst "nasty stuff fit to poison a horse." Local fishermen, no longer able to support themselves from the once-rich fishing waters, joined the City of London in instigating a prosecution against various senior officers of Equitable Gas, the principal cause, in their view, of the virtual collapse of the fishery and of the fact that the river's wholesome water had become "corrupted, insalubrious, and unfit for the use of his Majesty's subjects." Another member of the water bailiff's staff provided technical insights into the characteristics of the effluent: the stuff "smelt ready to knock anybody down."[2] This evaluation of local conditions was informally corroborated by a court reporter who declared for the benefit of posterity that a sample produced during the trial fully justified the statement.

Although the abominable consequences of gas, coke, and coal tar residue, as well as other unsavoury liquids in the Thames, were apparent to all, the prosecution had to establish that company officials were responsible for the presence of these contaminants in the river. These gentlemen denied knowledge of the situation but the prosecution asserted that, as the gasworks were carried on by order of the directors, the directors were legally responsible. In this respect they were allegedly no different from the proprietor of a newspaper who would be accountable for the contents of the paper on the general understanding that "a person is liable for what is done under his presumed authority."[3]

The defence barely questioned the responsibility of senior officers for whatever Equitable's workmen had done to divert industrial wastes to the Thames. Instead, in a manner that would later be much more forcefully asserted, the gas company invoked the values, preferences, and expectations of the general population of London to account for an admittedly unfortunate occurrence. Those who enjoyed the benefits of the gasworks were presumed to have consented to any accompanying degradation of the river; in the absence of consent, it was assumed that the advantages outweighed any attendant losses. Diversion of the effluent was necessitated, this argument began, by "some imperfection in the machinery. If it had not been adopted, the company must have ceased for a time to light the district."[4] However regrettable, the contamination of the river must have been authorized or at least permitted by implication: "As the people of England are resolved to have gas-lights and steam-boats, and copper-bottomed vessels, they must be content also to bear the inconveniences which will occasionally result from the use of them." To underscore the point, it was explicitly stated that where a particular facility promoted "the comfort and security of society," it would be "absurd" to impede such advances on account of "the poisoning of a few fish."[5]

If Equitable's pollution of the Thames could not be excused by trivializing the destruction of the fishery, legal counsel for company officials insisted that responsibility for injury to the resource should be widely shared: "If the diminution of fish is to be considered the criterion, then every proprietor of a copper-bottomed vessel, every maker of a sewer, every proprietor of a steam-boat, must be found guilty, as they contribute to such diminution, and it is not a question of degree."[6] These arguments were directed to the technical niceties of two prominent and enduring legal preoccupations: Whose conduct was responsible for any particular eventuality? and, Were those responsible in any sense at fault? More fundamentally, though, they invoked a moral claim that activities undertaken on such a wide scale and so generally perceived as beneficial could hardly be judged wrongful. Degradation evidently served the greater good.

It fell to Chief Justice Thomas Denman to summarize certain essential considerations for the benefit of the jury. On the question of the liability of the officers and directors of the gasworks, Denman expressed the opinion that both common sense and law suggested that those who employed others to carry out industrial operations for their benefit and advantage should be answerable for the conduct of their servants and employees. In finding Equitable's chairman, deputy chairman, superintendent, and engineer legally responsible, the jury evidently agreed. Of equal importance, however, and of more direct relevance to the matter of contamination, are Denman's observations on the condition of English rivers. "It is quite clear," he remarked, "that in great rivers of this sort there must be many inconveniences, arising from a variety of causes."[7] Many of those inconveniences would simply be endured by the river and those dwelling along it, unless, of course, the results of some particular conduct were wrongful, amounting in the circumstances to what the law called a nuisance. The jury's determination confirms its conviction that in this instance the notional threshold separating inconveniences from nuisances had been crossed.

Yet Denman identified a further category of threats to the waterway which would not, it appears, merit legal intervention. "With respect to copper-bottomed vessels, it seems to me that a great number of trifling objects may produce a deleterious effect, though the individual instances may not be the subject of indictment."[8] This left the common law singularly ill-equipped to address what later generations would come to know as cumulative impacts. In isolation, such actions might appear harmless, but they were devastating through repeated individual acts.

The fisheries of North America were also experiencing the effects of nineteenth-century industrialization. Many New England waterways were transformed in the first half of the century, as canal and waterpower control schemes extended upstream from the vicinity of the Atlantic shore to interior headwaters.[9] The process was well advanced in 1839 when Henry David Thoreau and his brother John journeyed downstream from Concord, Massachusetts, to join the Merrimack River by way of the Middlesex Canal before proceeding up the Merrimack past the cotton mills of Nashua, New Hampshire, to Hooksett, whence the brothers continued on foot to the White Mountains. The locks and dams, Henry Thoreau remarked, had proved "more or less destructive to fisheries," with shad, alewives, bass, and particularly salmon suffering the consequences of water-based industrialization.[10] The initial damage resulted largely from interruptions to the flow or obstruction of migratory passageways. As fish struggled upstream, the downstream flow increasingly contained the outpourings of textile manufacturing, paper-making plants, tanneries, and the surrounding industrial communities.[11]

Even in the vast Great Lakes, fisheries were vulnerable to changes in the condition and flow of water. By the 1890s the populations of several major commercial fish species had declined to a critical state. Although widely attributed to overfishing of an unregulated resource, the situation was certainly exacerbated by agricultural and industrial contaminants. Orchards along the shorelines of Lakes Erie, Ontario, and Michigan had begun to use arsenical insecticides in the 1870s, and general agriculture and dairy farming were responsible for growing volumes of pesticides, manure, and fertilizer runoff.[12] Industrial discharges from the iron and steel mills of Milwaukee joined the effluent of breweries, tanneries, and meat-packing operations, burdening rivers flowing into Lake Michigan and threatening aquatic life.[13] With its dark brown current of sewage extending six miles into the lake, the Milwaukee River by 1887 was "perhaps as black, foul and offensive as any stream in the country."[14] Lake Ontario suffered from Rochester's contributions to the Genesee River, as well as from industrial wastes carried by the Oswego and the Black. Fish-stocking experiments proposed for the latter were severely discouraged by reports of its condition: "The water is contaminated by refuse from paper mills situated not far from its mouth, and the acid used is said to kill pike, bass, and other fish, and would prove equally injurious to salmon."[15]

It was one thing to suspect that effluent flows contributed to the decline in fish populations, and another to establish the certainty of that effect in judicial proceedings. Fishermen campaigning against objectionable practices had to demonstrate the relationship between an activity to which they took exception and actual injury to fish stocks. This problem of causation had several aspects, attributable in part to limitations in scientific research and understanding, as well as to the possibility that other factors, including poorly diagnosed environmental changes – and even fishermen's own conduct – contributed to declining harvests. In addition, complaints typically had to proceed through a legal maze whose procedural quirks and doctrinal meanderings often impeded direct resolution of the conflict.[16]

Such difficulties were evident in nineteenth-century British efforts to save salmon. Already in 1830, according to Charles Fowler, the River Aire, near Leeds, was "charged with the contents of about 200 water closets and similar places, a great number of common drains, the drainings from dunghills, the Infirmary (dead leeches, poultices for patients, etc.), slaughter houses, chemical soap, gas, dung, dyehouses and manufactories, spent blue and black die, pig manure, old urine wash, with all sorts of decomposed animal and vegetable substances."[17] Although deterioration along the Aire and the nearby Calder had been under way for some time, observers pointed to the years between 1850 and 1866 as the period in which the greatest increase in pollution occurred. As late as 1850, salmon were seen at Allerton Bywater on the Aire at a time when a few men who had caught trout in the Calder above Dewsbury were still alive. Some trout were still being caught in the Calder after 1839, and between 1844 and 1854 the owner of Esholt Hall in the Aire Valley had enjoyed a regular daily supply of eels, trout, and other fresh fish from the Aire. In sequence, however, the salmon, followed by the trout, and then the coarse fish such as the dace and roach, began to disappear from the Calder. By about 1856 none remained.[18]

Scarcity and rising prices soon directed attention toward diminishing salmon populations throughout England and Wales. However, localized and isolated efforts to stop the deterioration foundered on lack of an agreed technical mechanism to reconcile the interests of industrial water users, including mill owners, with a fish species whose migratory patterns and spawning requirements were still poorly understood. A more comprehensive approach was needed.

The complicated challenge of reconciling the competing interests of the fishing community with the demands of commercial, industrial, and other riparian claimants up and down the salmon streams was referred to a royal commission for investigation in 1861. Recommendations ensued to consolidate some thirty-odd statutes, to control pollution, to suppress weirs and fixed engines, and to regulate close times and seasons on a uniform basis. The royal commission also called for a central inspectorate, a licensing system to finance river management, by-laws to adjust national standards to local conditions, and compulsory fish passes at mill dams. Critics attacked specific aspects of the proposals, or more generally lamented the intrusion of centralized statutory control into another realm of civic life. One parliamentarian, closely allied with industrial interests, frankly protested (as had counsel for the Equitable Gas Company) against the risk of "damaging the great commercial interests of the country for the sake of preserving a few fish."[19]

The first English salmon legislation to emerge from the parliamentary process in 1861 prohibited the future construction of mill dams and weirs without fish passes, yet it failed to require the installation of fishways at pre-existing dam sites. With mechanisms for enforcement representing something of a compromise between central and local authority, the effectiveness of the legislative initiative was of some importance. One of the original salmon inspectors, William Ffennell, assessed the significance of the salmon legislation shortly before his death in 1867. He condemned riparian landholders for their failure to assert their interest in the fisheries, describing their submission to "every species of abuse to their rights" as a "grand mistake."[20] Those entitled to enforce the common law for the purpose of securing whatever assurances about water quality it might have offered had virtually abandoned their claim to do so. The lesson that individual riparians, or even groups, could not necessarily be relied upon to assert their theoretical interest in the well-being of fish and waterways would be learned many times over, even though individual litigants – as we shall see – must often be credited with virtually heroic action to preserve or restore the quality of lakes and watercourses.

By contrast, in the most comprehensive account of the English salmon initiative, Roy MacLeod offers an entirely positive assessment of the early years of statutory and administrative intervention. He argues, somewhat paradoxically, that by intervening in the realm of private rights, the state and the inspectorate actually reinvigorated and defended those rights. Not only had official encroachment upon individual liberty to fish or to interfere with the fishery highlighted ways to improve fish yields and finance fish passageways and other structures, but, in addressing a wide range of contending interests, it had identified a broad legislative base for the betterment of them all. Most important in the long run, argues MacLeod, in a valuable reminder of the role of public opinion, "By force of practical instruction and education, the inspectors had created local reservoirs of further law-making opinion."[21] With modest adjustments, MacLeod's appraisal of the salmon administrators' legacy embodies the essential features of a continuing defence of intervention or regulation.

Building upon the foundations established by their predecessors, the next generation of salmon inspectors raised their objectives: to bolster the scientific basis of the salmon enhancement initiative, to monitor and promote enforcement, and to campaign more explicitly against pollution. In discussing devices to permit the migration of salmon up the course of industrialized rivers, one of these inspectors, the natural historian and amateur scientist Francis Trevelyan Buckland, made a vital observation: "The ultimate verdict of whether a salmon ladder be good or bad [must] be left to the decision of the salmon themselves; all we can do is to find out what they want and accommodate them."[22] The price for disregarding that elemental insight about passageways and corresponding measures regarding water quality is still being paid.

Buckland's colleague Spencer Walpole directed his attention to another dimension of the problem, the question of incentives. Walpole proposed penalties for industrial operations whose discharges of solid or liquid wastes rendered portions of a river uninhabitable.[23] Thus, the question of responsibility entailed the equally problematic matter of appropriate sanctions for the offence of pollution. As established in the original salmon legislation, it was an offence to put solid or liquid matter into any waters containing salmon to the extent that the fish might be poisoned or killed. The penalties prescribed increased from a maximum fine of five pounds for a first conviction to twenty in the event of a third or subsequent conviction.

The potential severity of these provisions was offset by an important limitation designed to accommodate an apparently acceptable level of pollution. Penalties were not applicable to those who had employed "the best practicable means, within a reasonable cost" to render their effluents and deposits harmless.[24] Where the exercise of best practicable means failed to make the discharges harmless, it fell to the salmon to employ best practicable means to look after themselves.

The fisheries protection regime, despite ongoing legislative amendment, failed to advance. As the century drew to a close, the bold initiative declined into "a pathetic history of indifferent half measures."[25] Notwithstanding the labours of 593 bailiffs in fifty-one fishing districts, the central objectives of the 1861 legislation had not been achieved: "the preservation of salmon during a fixed close time, the free ascent of salmon, and the prevention of pollution – were still widely disregarded."[26]

As their report for 1878 suggests, salmon fisheries officials were resigned to the legislature's limited interest in the impact of river pollution on fishing grounds.[27] Inspector Buckland nevertheless carried on a personal crusade, publicly lamenting that "when there are so many mouths to be fed, and so many rivers comparatively salmonless, manufacturers and mine owners, who form a relatively small portion of H.M. subjects, should be allowed to inflict, directly and indirectly, such a vast evil on the public in general."[28] If the fisheries, a long-established commercial undertaking with a quasi-constitutional pedigree dating back to *Magna Carta*, were so hard pressed to defend their interests against the adverse consequences of industrialization, one wonders how navigational or recreational interests might fare in resisting threats to the condition of the waterways.

Sawdust on the Floor of "The Log-Drivers' Waltz"

As its operations extended along such river valleys as the St. John in New Brunswick, Quebec's Saguenay, and the vast Ottawa system stretching northward from the St. Lawrence to the Hudson Bay watershed, the nineteenth-century Canadian timber trade connected interior forest lands to markets, both domestic and international. The situation was much the same in the United States as the industry moved westward across New England and into the mid-continental forests of Michigan, Minnesota, and Wisconsin. Along the waterways, as well as within the North American forest itself, this expansion entailed significant environmental change. By the late nineteenth century, pollution resulting from lumber industry operations had stimulated growing concern and inspired legislative action.[29] Yet the public interest in waterways remained closely associated with their direct contribution to commerce and prosperity.

As Boyd, Caldwell and Company's annual complement of eighteen thousand logs drifted downstream in 1880, Peter McLaren sought and obtained a court order to prohibit their passage through his works at High Falls at the head of Ontario's Dalhousie Lake. McLaren, a riparian owner and rival lumberman with as much as $250,000 invested in stream improvements,[30] asserted his "absolute and exclusive right to the use of the [improved stream] for the purpose of floating or driving saw logs and timber down the same."[31] He wanted to control the use of the river where he had made improvements, and not merely be paid by other users such as

Caldwell, whose competing operations were likely to interfere with his own use. This Precambrian controversy was destined for London, where the Judicial Committee of the Privy Council wryly observed, "[It did not seem that] the private right which the owner of this spot claims to monopolise all passage there is one which the legislature were likely to regard with favour."[32]

To the extent that the McLaren-Caldwell controversy embodied a significant conflict between private rights and public interests in the use of provincial waterways, it is worth pausing to note how limited in conception those public interests still seemed. In part because it contributed to provincial revenues, the industry's interest was considered to be virtually congruent with the public interest. With waterways viewed essentially as highways in an economic system, little, if any, consideration was given to the environmental or ecological implications of forest industry practices. Yet these were significant. The removal of the forest cover altered runoff patterns and rates of soil erosion, and where shoreline cutting eliminated shade, stream temperatures rose. River improvements, designed to facilitate timber drives, increased the scouring of riverbeds and shorelines. The decomposition of immense quantities of bark and sunken logs placed heavy demands on oxygen supplies in the waterways and thereby undermined aquatic habitats.[33] With these matters still poorly understood by members of the scientific community, navigation took its turn as a surrogate champion for water quality.

To its tradition of exploiting waterways for power and transportation, the lumber industry customarily added the practice of employing its own mill shorelines to remove wastes in the form of bark, sawdust, slabs, and edgings. In New Brunswick, aptly described as a timber colony, the same waterways that floated logs and powered lumber mills in the early 1800s were expected to carry away industry debris. Offcuts might find their way to the Bay of Fundy or the Gulf of St. Lawrence. More often, refuse became lodged in narrow gorges, along river banks, or on low-lying ground. Sawdust, also dumped unceremoniously into the rivers, became sodden, sinking to the stream beds where its accumulated mass disturbed river ecology and obstructed navigation.[34]

Complaints about the adverse impacts of mill refuse on navigation, notably along the St. Croix, led to a legislative attempt to deter the practice of dumping lumber waste in the river. In the absence of a corresponding initiative in neighbouring Maine, however, no attempts seem to have been made on the New Brunswick side to enforce the prohibition and to collect the five-pound fine. Further measures, all equally ineffective, were introduced in other parts of the province, although by mid-century some mills had begun to burn their solid refuse. The situation in St. John harbour was such that one commentator alleged that it was impossible for wood waste to float long enough to be carried away by the tide, with the result that refuse simultaneously undermined water quality and threatened to impede navigation as it accumulated on the bed.[35]

Attempts by riparian owners to invoke common law principles against lumbermen met with mixed success at best. Occasionally such efforts even substantiated claims by sawmill operators to the right to use waterways for waste disposal. On New York's Hudson River, neighbouring sawmill owners clashed early in the century when a newly constructed dam interfered with downstream operations by diverting floating sawlogs away from a lower mill while local currents cluttered its works with refuse from the upstream facility.[36] The plaintiff brought forward traditional arguments emphasizing his earlier occupation and the length of his tenure, both designed to establish priority, if not exclusivity. The New York court, in a decision that foreshadowed subsequent developments in riparian law, rejected these assertions in favour of a river doctrine rooted in the correlative rights of the riparian community. Carol Rose has succinctly enumerated the emerging principles of this "commonly owned resource" approach to the waterway: "First, riparian owners had limited but more or less equal rights to use the stream; second, their various uses could cause some inconvenience to other owners; and, third those inconveniences were not actionable if they were merely minor."[37]

Increasing acceptance of reasonable use suggested that some level of wood waste, bark, and sawdust accumulation might have become an acceptable – and legal – consequence of sharing waterways. When a Vermont court was asked in mid-century to determine the lawfulness of a tannery's habit of dumping spent bark into the stream, it expressed a willingness to be guided by industry norms: "The uniform practice, the convenience, and in some instances the indispensable necessity, would seem sufficient to decide such cases." The court considered it to be beyond dispute that "one must be allowed to use a stream in such a manner as to make it useful to himself, even if it do produce slight inconvenience to those below." If irrigation, cattle-watering, and power production had all become lawful stream uses through common law or obvious necessity, the same might occur for industrial waste from tanneries, as had already occurred, the judge insisted, in other industries. It was both lawful and reasonable for soap, dyes, and other manufacturing materials, including sawdust from sawmills, to be discharged to some extent because, or so it appeared, it was nearly indispensable to do so.[38]

Roughly a decade later, a court in neighbouring New Hampshire had to consider whether mill owners who threw sawdust waste from manufacturing into a stream were entitled to do so when the refuse ended up in the fields of a farmer below. Judge Bellows, noting that riparian entitlement to reasonable use was flexible in view of "the growing and changing wants of communities," indicated that waste removal was akin to the abstraction, detention, and diversion of water. It was incidental to other legitimate uses – manufacturing, for example. Wherever you had sawmills, fulling mills, and cotton and woollen factories, "there must be thrown into the stream more or less of the waste." Pollution had its purposes.

Judge Bellows firmly denied the two central contentions of the aggrieved New Hampshire farmer. He dismissed the suggestion that the discharge of the waste could be distinguished from the operation of the mill. It was certainly not, as the farmer's counsel had contended, like depositing the waste directly upon the farmland by means of teams of machinery provided for that purpose. Instead, Bellows insisted, the discharge of the waste into the stream, "so far as it is reasonable, must be regarded as an incident of the right to use the stream for the manufacture which produces such waste." The question of the reasonableness of this incidental use in any given situation then fell to be determined in light of local circumstances, including "the extent of the benefit to the mill owner, and of the inconvenience or injury to others." There would be, in other words, some willingness to balance the interests of the contending parties, possibly even inviting consideration of the broader social or community consequences. Here again was a specific rejection of the farmer's claim that the mill owner "had no right to discharge his sawdust and shavings into the river, unless such discharge was necessary to the running of his mills."[39] Acceptance of mill waste as an incidental by-product of lumber manufacturing, and a willingness to factor into the judicial calculus any alleged advantages of water-borne waste removal enjoyed by the discharging industry, may have afforded some relief to a generation of sawmill operators. However, the courts' willingness to allow reasonable volumes of pollution did not confer *carte blanche* on the sawmilling industry.[40]

When slabs, bark, sawdust, and what were referred to as grindings (parts of slabs which have been crushed into pieces before being thrown in) obstructed waterways and interfered with stream flow, Ontario mill owners protested in much the same way that they would have objected to diversions or flooding.[41] Continuing deterioration of conditions along the Ottawa River after Canadian Confederation in 1867 inspired Richard Cartwright, MP for Lennox County, to introduce legislation to prohibit lumber mills from disposing of sawdust, edgings, and other sawmill refuse in navigable streams or rivers. The situation cried out for action. As Cartwright pointed out to his fellow legislators, almost every navigable stream flowing into Lake Ontario was threatened with this type of obstruction. Conditions were at their worst along the Ottawa near the nation's capital. Here, the annual manufacture of lumber reached something like a hundred million of board measure, or roughly ten million cubic feet. This level of production resulted in two million cubic feet – the equivalent of twenty thousand cords – of rubbish each and every year, by far the greatest part of which was dispatched directly to the river. Disposed of in this way, the lumber mill waste was "sufficient to block up the river for four miles, to a width of 200 feet, and a depth of one foot." Surely, Cartwright pleaded, rejecting the assumption that the leftovers were really useless waste, some means must be found "of turning this enormous mass of fuel to some use in a cold country like ours."[42] The immediate reaction to Cartwright's bill was a forceful petition from Ottawa lumbermen reminding legislators of the size and significance of their industry, protesting the devastating impact of the proposed legislation, and arguing strongly that annual spring freshets removed sawdust, thereby eliminating any injury or adverse effects on navigation.

When the Honourable Hamilton H. Killaly and R.W. Shepherd and John Mather, Esquires, accepted an official invitation to inquire into the impact of sawdust and other lumber industry waste on navigable streams and rivers late in 1871, they may not have grasped the urgency of the matter as viewed by the minister of public works. The minister, at one of the rare moments in a long, sorry tale when senior officials were actually anxious to proceed with dispatch, wished to table his report at the opening of the April 1872 session. Astonishing as it may seem, he may not have appreciated the simple practicalities, for in response to the commissioners' request for details about the complaints that had awakened legislative interest in the waterways and their polite observation that the pace of investigation "must inevitably be governed by the nature of the weather," official Ottawa telegraphed brusquely in January to ascertain whether the inquiry was under way and when the report might be expected. With winter thoroughly installed, and the rivers in question largely frozen over, the commissioners were initially content – despite political Ottawa's enthusiasm for speed – to pursue their inquiry by correspondence.[43]

After breakup and at suitable intervals thereafter, they embarked upon a series of on-site inspections. Following an initial visit to the St. Maurice and other rivers in the immediate vicinity, Killaly, Shepherd, and Mather directed their attention to the Ottawa, for it was here that controversy about the impact of lumber industry waste on navigation had gained public attention. Dr. E. Van Cortland, the city of Ottawa's health officer, cited the threefold impact of lumber-mill refuse – on spawning grounds, on navigation, and on public health – as reasons for remedial action. This was a landmark early attempt to consolidate a range of community concerns over industrial interference with water quality. But mill owners resisted vigorously, threatening loss of employment if they were forced to abandon waterpower technology for the steam alternative. The fact that a court had some years earlier endorsed the burning of slabs and mill waste as a readily available alternative to dumping and flushing seemed of no particular relevance to the Ottawa situation.[44] In the interests of compromise, officials suspended prosecutions against Ottawa Valley lumbermen while the latter, in turn, agreed to grind mill waste to sawdust before dumping it in the river.[45]

In support of the industry, lumberman H.F. Bronson presented engineering studies and numerous affidavits debunking the threat of lumber industry waste to navigation on the Ottawa. The expert opinions of Professor Greene and the engineer William J. McAlpine were particularly notable for their elaborate technical explanations upholding the reassuring proposition that the velocity of the current and the specific gravity of pine sawdust gave "no reason to anticipate the formation of permanent or troublesome bars, or accumulation of saw-dust" in channels of the Ottawa.[46]

The commissioners acknowledged that Greene's analysis was indisputable, "so far as his theoretic calculations and experiments extend."[47] But contrary opinions from equally distinguished sources asserted the injurious impact of sawdust on navigation and indicated a need for prohibitions. In a communication to the commission, Brigadier General Thom of the United States Artillery, for example, reported that serious accumulations of waste, slabs, edgings, and sawdust had greatly impaired navigation on several American rivers. He described the process of accumulation and welcomed statutory prohibitions against discharges of lumber waste, not yet, he regretted, including sawdust. The Honourable William Muirhead of Miramichi, New Brunswick, gave further evidence based on his personal observations of the river's deterioration from the perspective of fishing and navigation.[48] Muirhead's proposed response, a form of punishment that won devotees in many succeeding generations, involved imprisonment of mill owners rather than fines, which, by comparison, he considered ineffective.[49]

After inspecting several representative waterways, Killaly's team added personal observations. By steamer they had travelled from Lachine to the Carillon Rapids and up the Ottawa past the Grenville Rapids and Hawkesbury to the capital, where they reconnoitred numerous bays and the river mouths. Captain McNaughton piloted the inquiry to several locations where he thought obstructions associated with mill waste might be found. Just opposite the home of the Gilmour lumbering clan below Pine Tree Island, McNaughton approached a shoal which, upon inspection, extended some 250 yards. "On the south side," he wrote, "this deposit of slabs, edgings, &c. in some parts united by sawdust, *extends wholly across the river,* until it reaches near the shore at the foot of the hill."On attempting to approach the lock beneath Colonel By's Rideau Canal, Killaly and his colleagues found the bay thoroughly blocked up with logs, square timber, and other debris.[50] Nearby soundings regularly revealed significant accumulations of sand mixed with sawdust, often thought to have been deposited in a comparatively short period of time. As politely as seems possible, the report cast severe doubt upon the reassuring predictions of Messrs. Greene and McAlpine. Sawdust had certainly built up across bays and along shorelines, obstructing navigation in a number of rivers, including the Ottawa – if one understood navigation to involve anything more than the passage of vessels down the centre of main channels.

Perhaps by way of compromise between two prominent industries, the Killaly investigation endorsed an immediate prohibition against the deposit of sawmill waste other than sawdust. Should Parliament eventually determine upon further evidence that the injurious effects of pure sawdust on navigation supported a more comprehensive prohibition, the commissioners assumed such action could then be taken.

Shortly after the Killaly report was delivered in February 1873, legislation embodying the principles of the earlier Cartwright proposal was approved. This measure prohibited the discharge of mill waste, including sawdust, into navigable waterways. Although the possibility of obtaining exemptions was preserved, the onus lay on applicants to demonstrate that the "public interest would not be unjustly affected."[51]

J.R. Booth, already a prominent Ottawa lumberman whose successful commercial ventures would eventually earn him recognition as a "lumber king," declined to alter his Chaudière mill to ensure compliance. He preferred instead to pay a twenty-dollar penalty when, in 1875, he became the first subject of successful prosecution under the new act.[52] Facing the prospect of further prosecutions, and perhaps ultimately injunction proceedings,

Booth and other mill operators seemed to acknowledge that the era of dumping slabs and other heavy waste materials into the river was at an end. Under the supervision of former commissioner John Mather, almost all of the Ottawa Valley mill owners completed approved renovations – not including the removal of sawdust and grindings – by the end of the decade.[53]

Although exempt from prosecution under the federal statute, the lumbermen were by no means immunized from private claims that might be initiated against them. So, when Antoine Ratté, the owner of a floating wharf and boathouse half a mile downstream from the Chaudière mills, finally lost patience with an eighteen-year accumulation of mill waste around his premises, nothing seemed to stand in the way of his claim against Booth and his fellow lumbermen. Chancellor John A. Boyd, in *Ratté v. Booth* found evidence that "this refuse accumulate[d] in great floating masses, substantial enough occasionally for a man to walk upon."[54]

If refuse from the lumber mills, including sawdust, was a potential impediment to the navigational interests of others such as Mr. Ratté, it was also a contaminant with adverse impacts on water quality, or so it seemed to some observers. Indeed, Chancellor Boyd clearly condemned lumbermen as "wrongdoers who from their mills allow sawdust, blocks, chips, bark and other refuse to fall into the River Ottawa." Water pollution and the obstruction of navigation were inevitable consequences of sawdust accumulations. Not only was water fouled and rendered offensive both to taste and smell but, he pointed out, the persistent depositions of sawdust "produce from the gas generated underneath the surface frequent explosions which are disagreeable and sometimes dangerous."[55] Sawdust dumping on the Ottawa River took its first recorded victim in 1897 when John Kemp, a Montebello-area farmer, was thrown from his boat by a methane explosion. He succeeded in catching hold of the boat, but was dislodged by a second "sawdust explosion" and drowned.[56]

Threatened by injunctions, as in the Ratté claim, owners of the Ottawa River lumber mills sought legislative protection for their industry. In 1885 Oliver Mowat's provincial government took steps to immunize the owners or operators of sawmills who threw sawdust and refuse into the Ottawa. In actions arising from the dumping of sawdust or other mill waste, Ontario's courts were instructed to "take into consideration the importance of the lumber trade to the locality wherein such injury, damage or interference takes place, and the benefit and advantage, direct and consequential, which such trade confers on the locality and on the inhabitants thereof, and . . . weigh the same against the private injury, damage or interference complained of." This legislation, whose principles were later extended to other regions of the province, no doubt served as an effective deterrent to complaints against the flushing of sawmill wastes.[57]

Despite provincial legislation to discourage injunctions that would prevent lumber mills from dumping mill waste in the waterways, however, the Canadian sawdust controversy persisted at the national level, where, for constitutional reasons, navigation and fisheries interests were more effectively represented.[58] The occasional explosion resulting from accumulations of methane made lumber waste difficult to ignore and stimulated public criticism. Moreover, the harmful effects of sawdust on marine and aquatic life were by now more fully appreciated than at the time of the Killaly inquiry. In 1889 a federal fisheries official explained that the effects of the sawdust scourge on water were more ruinous than the impact of sawdust upon land: its floatability allowed the "blasting influences" of the sawdust scourge to spread downstream to vulnerable estuaries, inlets, and bays, either coastal or inland. There, it "kills the sources which give life and food for the smaller races of insects and other marine animals [and] forms a compact mass of pollution all along the bottoms and margins of the rivers and inlets, filling up the crevices on the gravel beds . . . where aquatic life is invariably produced and fed." Sawdust eventually became "a fixed, imperishable foreign matter," adhering to the beds of lakes and streams and forming "a long, continuous mantle of death."[59] These consequences, in all but name, were ecological. Yet it would take several decades before concerns of this nature would be directly addressed through legislative reform.

Ottawa's constitutional responsibility for navigation in the late nineteenth century seemed somewhat more robust than its ostensible authority over fisheries, and so, despite continued agitation from Booth and his lumber industry colleagues, the sawmill provision was rewritten in 1886 with reference to the firmer constitutional foundations: "No owner or tenant of any sawmill, or any workman therein or other person, shall throw or cause to be thrown, or suffer or permit to be thrown, any sawdust, edgings, slabs, bark or rubbish of any description whatsoever, into any river, stream or other water any part of which is navigable, or which flows into any navigable water."[60] With a flourish, that final phrase – "which flows into any navigable water" – asserted the sweeping extent of federal concern, for as one critic of the initiative declaimed, "I defy anyone to find a river in this country which does not flow into navigable water."[61]

There was little evidence of support for vigorous enforcement against the lumber mills from the higher reaches of the federal government, whether in the interests of the fishery or navigation. Prime Minister John A.

Macdonald may simply have voiced contemporary understanding of both forestry and the fishery when he remarked that each should take its turn – when all the trees were cut along the river, it could then be restocked with fish.[62] The minister of marine and fisheries, Sir Charles Hibbert Tupper, was a Maritimer whose priorities were decidedly influenced by his experience of salt water. In debate on fisheries legislation from which the Ottawa River was specifically exempted, he rebuked criticism with the quip, "There are not many lobsters there, at any rate."[63]

Canadian federal resolve may finally have been strengthened by growing provincial interest in the sawdust question. As Ontario explored the responsibilities that may have come its way as a result of constitutional decisions, provincial officials became increasingly vocal in their condemnation of sawdust as a threat to fish stocks. Empowered by an important legal ruling in 1898, fish commissioners and then provincial fisheries bureaucrats repeatedly singled out the sawdust menace.[64] In 1899 prosecutions were launched to encourage compliance, but over a three-year period J.R. Booth nevertheless managed to obtain further indulgences, ostensibly to complete conversion of his lumber mill and to develop a pulp plant where sawdust would be used as fuel. On 11 September 1901 prosecutors finally secured a conviction when Booth pleaded guilty to charges under the *Fisheries Act*. This enforcement effort produced another twenty-dollar fine to accompany the 1875 penalty whose insignificant deterrent effect had been glaringly apparent for a quarter of a century.

Industry allies encouraged the aging lumber baron to battle on. Hiram Robinson, for one, was still smarting from the effects of federal regulation on his own operations at Hawkesbury. When confronted with compulsory – and costly – modifications to his mill, Robinson thundered in response, "This is a most serious matter to us, and it cannot be possible, or reasonable, that Parliament can legislate away our property without full compensation being made."[65] Robinson's understanding of his property evidently included an entitlement to continued use of the river for waste disposal, a practice, he observed, that the firm had carried on for nearly a century.

Robinson urged Booth to invoke the "technicalities of the law" to fend off continuing "persecution" by authorities. "I fancy that a good lawyer would put up a good fight," counselled Robinson, who encouraged Booth to maintain the harmlessness of sawdust, certainly in comparison with sulphuric acid and sewer filth entering the river at Ottawa. As a legal argument, the observation that there were other and possibly worse polluters had little to commend it, but as a claim for public sympathy and a defence of the morality of industrial flushing, it had its uses, as the Equitable Gas experience had shown in London long before. Robinson also advocated opposition to Ottawa's constitutional authority over fish and fisheries, for he made it no secret that he would "like to see these Dominion officials knocked out."[66] Ultimately, Wilfrid Laurier became personally involved in the Booth file. The prime minister was instrumental in securing Booth's firm commitment to cease dumping sawdust at the end of the 1902 season in exchange for relief from further prosecutions and the threat of injunction proceedings to close down the mill.[67]

Conflicts over sawmill rubbish were certainly not eliminated by the federal legislative prohibition; however, the ripple effects could be seen when litigation did arise. By 1913 Ontario courts readily – and at the appeal level "unhesitatingly" – dismissed a sawmill operator's claims that it enjoyed a right to dump sawdust and other refuse into Constant Creek, based either upon the implied consent of those who suffered the adverse consequences or upon the long-standing practice of dumping. The accumulation of wood waste in a millpond, the nuisance originally giving rise to the complaint, was considered by one court to be a violation of the *Navigable Waters Protection Act* prohibition against throwing sawmill waste into any stream that flowed into a navigable water. To the Court of Appeal, the nuisance might even have been criminal under the common law, "for millwork travels far and is an enemy of navigation."[68]

As Brigadier General Thom had informed his Canadian neighbours during the Killaly inquiry, lumber refuse equally threatened navigational interests in the United States. If anything, the sawdust menace in the United States appeared more severe than its Canadian equivalent. Not only were American lumbering operations widespread, they were, in certain districts, extraordinarily intense. Mills established at close quarters to each other to take advantage of the rapid fall of New England rivers burdened their waterways heavily with refuse. Competing users, and therefore potential critics of the situation, were also more numerous. The sportfishing community viewed sawmill practices as the source of all manner of evils. Fish, it was said – at least those which, as eggs, had not been smothered in spawning beds – could suffocate from sawdust clogging their gills. This alleged peril had been brought to public attention as early as 1864 when the pioneering Vermont environmentalist George Perkins Marsh, then serving his country as ambassador to Italy, noted it in his *Man and Nature*.[69] Sawdust was also suspected of depleting the food supply of game fish. Even allegations of direct threats to human health were not unknown in an era when the decay of organic matter was a presumed cause of disease. Citing the work of the New York Fish Commission, the sportsman's journal *Forest and Stream* painted a grim picture in 1884: "Waters polluted by decaying sawdust spread malaria, and make miserable the lives of those who dwell on the banks of plague-bearing streams." The lower course of the Raquette River, "cursed with chills and fever," was offered as an example of one such waterway, making Potsdam, New York, as a consequence "a very undesirable place of residence."[70] Elsewhere, even typhoid was attributed to the stench of decaying sawdust. To such allegations we shall shortly return.

Despite the range and intensity of these alarming suggestions, early efforts to curtail the objectionable conduct of lumber mills were largely fruitless. Progress in cleaning up waterways (to the extent that any was made) was accomplished through US federal measures to safeguard navigational interests from post–Civil War obstructions, including not only dumping, but dredging, bridge building, and the construction of new wharves and landing facilities. This responsibility, involving potential intrusions upon state interests, derived constitutional authority from the US Commerce Clause. Measures were carried out through the work of the US Army Corps of Engineers, an institution whose historic practices and original priorities did not endear it to the environmental community.[71] Indeed, since the early 1800s, when the corps ventured away from fort construction toward navigational improvement, it has been more commonly associated with efforts to remodel nature and with fostering the view that rivers are essentially technological systems subject to more or less unlimited human intervention.[72]

Given its American constitutional origins in the power to regulate commerce, authority over navigation was ordinarily associated with the commercial necessity of shipping or defence, and perhaps more loosely with general assertions of national sovereignty. Any link to water quality was indirect at best. Nevertheless, campaigns promoting federal legislation to deal with refuse were occasionally launched after mid-century, inspired perhaps by the devastation of California's rivers resulting from gold seekers' penchant for hydraulic mining,[73] and by the steadily deteriorating condition of New York Harbor. New York State fish commissioner E.G. Blackford, for example, produced disturbing reports from his investigations of East Coast garbage scows and urban sewers: scavengers were replacing more desirable fish species, shellfish stocks were ruined, and coastal waters in the vicinity of New York City were covered with a "thick glutinous substance of the most vile odor."[74]

Efforts having foundered for years, a breakthrough finally occurred in 1890 in the form of a package of anti-obstruction measures attached to general legislation on rivers and harbours. This initiative, directed against dredging, refuse dumping, and other operations that threatened to interfere with navigation, conferred authority on the secretary of war to issue permits for operations that were otherwise prohibited. And it provided a foundation for legal action against offending industries and municipalities, including sawmill owners with questionable waste-management practices. But this set of measures was of limited utility. No officials were specifically charged with enforcement, and the necessity of linking offences to navigational obstructions or impediments often proved to be problematic, especially in the context of large rivers. In the absence of evidence to suggest actual malice, both judges and juries showed considerable reluctance to attach criminal responsibility to accused parties.

By the end of the century, revised legislation had replaced the pioneering measures. The principal new provision, section 10, broadly asserted US federal power over waterways by prohibiting any change in their "course, location, condition, or capacity" resulting from obstructions that were not "affirmatively approved" by Congress. This requirement soon figured prominently in the contest between the federal government and Illinois over the latter's diversion of water from Lake Michigan to flush Chicago's sewage downstream. Section 13, known as the *Refuse Act* of 1899, contained elements that would eventually support a pollution-management regime. Subject to one exception, this provision made it unlawful for ships, mills, or manufacturing facilities to discharge or deposit "any refuse matter of any kind or description whatever" into navigable waters or their tributaries in the United States if doing so would result in the impedance or obstruction of navigation.[75] Despite this "primitive absolute"[76] – sometimes attributed to a distracted Congress's preoccupation with the Panama Canal – it was nevertheless possible for the secretary of war to permit deposits of otherwise prohibited materials in circumstances where the chief of engineers concluded that no injury to anchorage or navigation would result. In issuing such authorizations, the secretary of war had the power to specify appropriate limits and conditions.[77]

Exempted from the *Refuse Act*'s controls was refuse matter "flowing from streets and sewers and passing therefrom in a liquid state, into any navigable water."[78] The category was a broad one – its rationale not easily discerned. Perhaps legislators sought to accommodate a distinction between dissolved (in solution) and suspended matter, the latter typically organic and generally presumed to be subject to natural decomposition. Or possibly Congress intended to acknowledge and defer to municipal decisions endorsing the flow of wastes into waters across the country.[79]

Congressional records and debates on the question are silent.[80] To understand the sewage exemption in the *Refuse Act* of 1899 and to appreciate its scope and significance requires an understanding of nineteenth-century sewerage, a subject in which readers willing to turn the next few pages will shortly be immersed. It is sufficient here to observe that navigation, like fisheries, seemed unlikely to vanquish growing threats to water quality from either municipal or industrial sources. Perhaps, though, recreational users of waters and waterways might overcome the limitations experienced by fisheries and navigational interests in the struggle for water quality.

Sea Bathers Plunge into Legal Controversy

In season, and subject to their own particular sensibilities, early-nineteenth-century patrons of a small English hotel along the Mersey shore in Lancashire might have engaged the services of one Catterall to transport them to the pleasures of the sea in the establishment's horse-drawn bathing machines. As they had done from at least the middle of the eighteenth century, such contraptions afforded shelter and privacy to persons who, for reasons of health and general sanitation, took up sea bathing and even ocean swimming in growing numbers in the years just after the Napoleonic wars. The trend to the seaside was sufficiently powerful that Bournemouth and Torquay began to displace such famous inland spas as Bath. In Jane Austen's unfinished novel *Sanditon,* Mr. Parker pursued a venture similar to Catterall's Merseyside hotel. "The success of Sanditon as a small, fashionable Bathing Place," Austen writes, "was the object for which he seemed to live, [transforming] a quiet village [into a] profitable speculation."[81] Perhaps we should not be surprised, therefore, that in 1821 Lord Blundell of Great Crosby Manor, where the premises of the real hotel were situated, took exception to Catterall's commercial expeditions across his foreshore lands. At the very least, Blundell took exception to his own lack of participation in the profitable venture.

Lord Blundell looked to the law to redress his grievance against the concessionaire Catterall.[82] Blundell contended that the recreational traffic amounted to trespass, and that the bathing machines were damaging the soil, sand, and gravel shoreline of his riparian lands. He had never authorized such access, and was doubtless astonished by the audacity of Catterall's arguments: legal counsel for the bathing machine operator vigorously asserted the public's entitlement to bathe in the sea, accompanied by whatever right of way would be incidental to its exercise. Indeed, it was inherent in Catterall's response to the trespass claim against his bathing machine traffic that the common law actually afforded members of the public a right of access to any public waters for any lawful purpose they might wish to pursue.

Before the matter was put to rest, a number of experienced judges set out elaborate reflections on the general rightness and wrongness of various uses of water and the manner in which conflicts concerning them ought properly to be resolved. These curial deliberations provide further illumination of a most challenging and durable controversy, the problem of public and private rights in water, and of the ranking, clustering, and jostling of aquatic interests. Each of the judges declared, in effect, that the broader community could expect significant social advantages to result from the particular legal outcome he himself espoused. They offered, in other words, quite clear indications about their own values and priorities rather than anodyne disclaimers rooted in the alleged inevitability of the law. For these insights into the relationship between water, legal culture, and the destiny of civilization, subsequent observers should be grateful.

Chief Justice Abbott, a respected and hard-working jurist who had authored *A Treatise of the Law Relative to Merchant Ships and Seamen* (1802), expressed a profound general reservation about Catterall's expansive claim of access. He remarked, "Every public right to be exercised over the land of an individual is . . . a diminution of his private rights and enjoyments, both present and future." The protection of such private rights, the chief justice insisted, was "one of the distinguishing characteristics of the law of England."[83] For Abbott, the general principle of safeguarding property (in this case Lord Blundell's) would facilitate – probably even encourage – human industry, and thereby promote the welfare of the entire community. Such a vision of beneficial endeavour, the consequence of industrious activity stimulated by property ownership, was by no means uncommon in the era. Nor is it today.

Yet one of Abbott's colleagues, Justice Best, dissented strongly from the chief justice's opinion. Best did not lack sympathy for the rights of property, though he suggested by way of limitation that the principle of exclusive appropriation should be confined to things that human industry was capable of improving upon. "If it be extended so far as to touch the right of walking over these barren sands," Best warned, "it will take from the people what is essential to their welfare, whilst it will give to individuals only the hateful privilege of vexing their neighbours."[84] In opposition to the theorem anticipating advances in general welfare through the productive employment of private property, Best insisted that public welfare would be served directly by permitting people to splash in the sea. He explicitly grounded his opinion on a claim of "public advantage." Best even identified a "right of bathing in the sea," which he deemed to be "essential to the health of so many persons," and "as beneficial to the public as that of fishing." Pursuing the comparison with fishing, Best reasoned that bathing "must have been as well secured to the subjects of this country by the common law."[85] He went too far.

Best's attempt to defend Catterall's clientele by channelling the novel claims of bathers and swimmers toward the shelter of comparatively well-recognized rights of fishing provoked an immediate rebuttal. Justice Bayley, a jurist renowned for his comprehensive legal learning, briskly refuted the attempted assimilation of bathing rights with fishing rights: "It would by no means follow because all the King's subjects have a right to pick up fish on the shore, that they have, therefore, a right to pass over the seashore for the purpose of bathing."[86] Bathing was not among the public rights recognized in *Magna Carta,* and Bayley explained why. Firmly denouncing his colleague's suggestion that the virtues of bathing put this activity more or less on a par with fishing, he insisted on the need to distinguish between the different descriptions of rights, according to his understanding of the benefits each offered to society. "The public may have a right of navigation, which is for the general benefit of all the kingdom; and a right of fishing, which tends to the sustenance and beneficial employment of individuals," Bayley acknowledged, "but it does not thence follow that they have also the right of bathing."[87] What, he demanded, was the good of bathing anyway? It was not, as a claim concerning fishing rights most assuredly was, "a claim for something serving to the sustenance of man." Bathing, he added dismissively, was "a matter of recreation only." Chief Justice Abbott had been equally skeptical about the merits of sea bathing, an activity which he knew to be "until a time comparatively modern, a matter of no frequent occurrence."[88] Neither necessity nor "general usage of the realm," remarked yet another judge, supported the claims of bathers.[89]

Still, Best pursued his defence of sea bathing. This activity, he protested, "if done with decency, is not only lawful, but proper, and often necessary for many of the inhabitants of this country."[90] Invoking the therapeutic benefits of a plunge in bracing waters, Best echoed a good many English physicians and other similar advocates in proclaiming that bathing promoted health.[91]

A still more desperate assertion emerges from Best's judgment, an ingenious attempt to narrow the gap between well-recognized public interests in navigation and current, if still marginal, fashion: "By bathing, those who live near the sea are taught their first duty, namely, to assist mariners in distress. They acquire, by bathing, confidence amidst the waves, and learn how to seize the proper moment for giving their assistance."[92] Swimmers save ships, or at least swimmers are more inclined than non-swimmers to attempt to do so. In a country whose prosperity and independence were so closely tied to navigation, the still-hopeful dissenter rhetorically inquired how something as vital to navigation as "a road for all lawful purposes to the sea" could not have been preserved for the public.[93] It was an attempt at least to assign some measure of social utility to ocean bathing.

Acknowledging the prospect of conflict between a public interest he sought to uphold and private interests he knew merited respect, Best offered the common law as a means of reconciliation: "The law in these, as in all other cases, limits and balances opposing rights, that they may be so enjoyed as that the exercise of one is not injurious to the other."[94] Whether and how the common law might satisfactorily manage this balancing has been as controversial in relation to water as in many other areas of human activity.

Quite progressively, Justice Bayley conceded that sea bathing might contribute to health. The practice, he felt, must nonetheless be confined within reasonable limits to avoid inherent difficulties. A right to bathe, Bayley remarked in a passage that foreshadows much subsequent debate over means of reconciling competing uses for waterways, certainly didn't need to be coextensive with the entire seashore.[95] Bayley accordingly called for regulations to resolve some of the more awkward complications of public bathing. Such regulations might, for example, restrict designated bathing machines to specific locations; bathers who did not use machines might be allocated more secluded spots at some distance from shoreline areas frequented by walkers. It was, he thought, equally important to ensure "a separation of those which are for men from those which are for females." The effectiveness of desirable regulations along these lines, Bayley concluded, would be greatly hampered if a general common law right of access were recognized.[96]

Just as the judges failed to agree upon the law and differed as to which form of intervention – the common law or regulation – might best alleviate any residual controversies or concerns, they also disputed the impact of the law upon the community and the behaviour of its members. The judges equally disagreed on matters of enforcement. Who was most likely to ensure that the law, whatever it might be, would most effectively – and fairly – be invoked to implement whatever water use or access regime had been agreed upon? Such matters remained controversial as the legislative and regulatory infrastructure of modern society evolved alongside continuing efforts by private parties to assert their common law interests in water.

Justice Best, determined to secure for all his countrymen the vital privilege of access to the sea, and convinced that only the issue of public decency blocked their way, was deeply suspicious of private litigation initiated, as he pointedly noted, by lords of manors. The risk of caprice, of self-interested zeal, and of unaccountable disregard for the position of others was too great. Best was willing to accept that ownership provided a powerful incentive to protect property, but concluded that both property and the motivation to protect it were private. Is it wise, he seemed to ask, to rely on private parties to safeguard public interests? Indecent bathing, Best believed, could more appropriately be punished on the initiative of public officials such as disinterested and responsible magistrates. He distrusted the whim of "an interested and irresponsible lord of the manor" who might chose to remove bathers from his own personal view without taking into account the interests or concerns of his neighbours.[97] As concern about pollution and water quality later intensified, this particular tendency to disregard the neighbours after looking out for oneself became all too common.

The chief justice, addressing only one aspect of Best's concerns, offered the reassurance that frivolous and vexatious suits were subject to suitable legal restraints and in any event most unlikely. Experience, he suggested, demonstrated "that the owners of the shore do not trouble themselves or others for such matters." Indeed, he continued – in the common law tradition of not crossing too many bridges in advance – no indication of excessive zeal had been present in this case. "But where one man endeavours to make his own special profit by conveying persons over the soil of another, and claims a public right to do so, as in the present case, it does not seem . . . that he has any just reason to complain, if the owner of the soil shall insist upon participating in the profit, and endeavour to maintain his own private right, and preserve the evidence thereof."[98]

Those who make, interpret, and apply law – whether in relation to fisheries, navigation, or coastal recreation – do so not in isolation but in the context of societies. The values of those societies are inevitably intermingled with the values of the law. Even law apparently preoccupied with the technicalities of property, procedure, and precedent reflects evolving social preferences and understandings. The differences within the court that decided Lord Blundell's claim against Catterall were certainly extensive. They involved a good deal more than divergent readings of earlier chroniclers of the common law whose views on the law and propriety of bathing machines (not to speak of their audacious occupants) would never be known. The nineteenth-century judges who considered this particular variation on the controversy of public and private rights to water did so with reference to values of their own. They demonstrated less than congruent perceptions of the relationship between public and private interests, derived, as an insightful commentator has observed, from various competing philosophies of the age in which they lived. Visions of a free market clashed with advocacy of social welfare regulation, while a tolerant humanism contended with puritanical strictness. "With city life thickening, recreation like sea bathing had begun to take on new importance, but the need had not yet clearly arisen to reallocate diffusely appropriate water resource interests."[99] Providential invocations and purposeful, usage-oriented perceptions of water contributed importantly to the general intellectual context within which decisions about water quality were made, either through private disputes or in the context of a body of legislative controls.

Legal institutions are also confined by the knowledge of the day, with the views of scientists, engineers, public health officials, civic administrators, and other professionals of particular note where the condition of water is concerned. These new perspectives became increasingly relevant as nineteenth-century municipalities introduced modern water supply systems, found new uses for the flow, and eventually began to address the implications of living better aquatically. For its part, the nineteenth-century revolution in water use soon presented the legal order with new challenges.

How, one wonders, would law makers and other legal practitioners, no more likely to be aware of the physical properties of water than their contemporaries, absorb the findings – some promising and others certainly interim or misconceived – of chemists in the Lavoisier tradition and scientists from other disciplines who subsequently pursued inquiry into the natural qualities of water alongside the legal and economic properties that intrigued William Blackstone and Adam Smith? With understanding of the very nature of water only beginning to be divested of mythological and religious overlay, legal controls on its use were coming under extreme pressure to accommodate new and increasingly intense needs. This was true especially in relation to municipal water supply, a development with sudden and profound consequences for water use and waste removal.

3
A Source of Civic Pride

NINETEENTH-CENTURY URBAN AND INDUSTRIAL growth dramatically increased the number of water users who lacked direct access to natural water supplies and had to depend on other ways of meeting their needs. Initial supply options, in addition to urban wells, included the communal pump, commercial carters and water bearers, and rooftop rain-collection systems feeding cisterns and other receptacles. In short order, to meet rising demands, elaborate networks of pumps, pipes, reservoirs, and aqueducts were engineered to provide urban consumers with water – ordinarily drawn from local sources, but not infrequently from more distant locations. As a consequence of water's increasing availability, domestic household arrangements, patterns of usage, and civic expectations were permanently transformed.

The transition to piped water delivery, frequently though by no means universally under municipal control, may be accounted for in several ways. Urbanization itself placed vastly increased demands on traditional sources of supply such as wells and springs. Moreover, high population concentrations compromised the quality of local sources, especially where private wells and privies coexisted in close proximity; where contamination was suspected of contributing to disease – however crude the epidemiological theory – further pressure arose to replace traditional water sources. Community needs for street cleaning and firefighting also highlighted the importance of public access to larger volumes than could be obtained from traditional sources. With urban growth came industrial expansion, similarly necessitating public water supplies to a far greater degree than widely dispersed small-scale trades and manufacturing operations had ever called for. And, as statistics soon clearly demonstrated, the convenient availability of water, combined with rising levels of affluence, dramatically increased per capita consumption.[1]

Watering Cities

By 1800 England's heavily populated capital city had more than two centuries of experience with water supply systems, overwhelmingly in private hands. Although town authorities elsewhere had often exercised a measure of responsibility in this area, London stood out as "a bastion of private enterprise in the water industry."[2] A badly deteriorating network of cisterns, conduits, and reservoirs delivered water from the local districts of Holywell, Muswell Hill, and springs at Tyburn, Hampstead, and Highgate.[3] However, as the nineteenth century began, the principal operations included those of such antiquated private suppliers as the London Bridge Water Works, which dated from 1581 and, as of 1809, distributed four million gallons of Thames water each day through wooden pipes to ten thousand consumers.[4]

London's New River Company initially drew its supply from Chadwell Spring to the north, later supplementing this source with water from the River Lee. When bargemen on the Lee protested this interference with their ability to navigate the waterway, the conflict was resolved in favour of the water company. Christopher Wren and a committee of investigation determined that, by taking only "about one part in thirty parts," New River's pipes "were very little prejudicial to navigation."[5] Construction of the channel and aqueduct from the source to London, some twenty miles as the crow flies but a good deal longer overland, had been initiated in 1609 under the direction of an enterprising goldsmith, Hugh Myddleton. Myddleton had the good fortune to recruit as a co-venturer King James I,[6] whose involvement proved instrumental in securing the consent of a number of landowners who had previously objected quite vigorously to the prospect of the New River pipes crossing their domains.[7]

From 384 subscribers in 1615, the New River Company had by 1809 extended its customer base to 59,000 homes. This remarkable expansion occurred despite the hostile cries of independent water carriers who, disregarding centuries of abuse of the Thames, continued to champion "fresh and fair river water" over "your pipe sludge."[8] The New River Company was then London's largest water company by far, even if the volume actually delivered to consumers fell substantially short of the eleven million gallons entering the distribution system each day.⁹

Other ventures, the East London Water Works Company, for example, also entered the market. A sermon by the Reverend Edward Robson highlighted ceremonies to mark the opening of the enterprise's facilities. Drawing inspiration from the biblical text, "Behold, I will stand before thee there upon the rock in Horeb; and thou shalt smite the rock, and there shall come water out of it, that the people may drink," Robson gave thanks to "the Supreme Being for the power thereby vested in the Company of dispensing to the numerous inhabitants of the Eastern District of the Metropolis the Blessings of Health, Security and Domestic Comfort." Robson's invocation of the split rock with its spring-like vision of water emerging from some eternal reservoir may have induced a favourable reaction amongst the dignitaries and potential consumers who were on hand for the proceedings or others able to read newspaper coverage of the events. But there was no split rock. In the early 1820s, the Thames – its condition rapidly deteriorating due to a formidable stream of wastes whose devastation of the fishery we have already noted – remained the source of roughly half of the twenty-eight million gallons by then distributed daily to business and residential premises throughout London.¹⁰

In the short term, at least, matters other than quality dominated a growing litany of complaints and protests. Water companies, though willing to increase the quantity of water provided to the warehouse districts, resisted the costs of laying down larger mains to improve delivery to areas where prospects of remuneration were uncertain. Even where service was available, delivery was notoriously unreliable, encouraging charges that water companies deliberately maintained the complexities and inconvenience of the London cistern system. The failure of water suppliers to provide constant access was not infrequently attributed to "a scheme of tricking" allegedly devised to restrict availability. In the absence of vigorous intervention by municipal and health authorities, Sunday supplies were non-existent, despite the profound needs of working-class consumers for water for cooking and other uses on their closest approximation to a day of rest. Speculation also suggested that the companies harboured a devious tendency to open the mains supply at times when the general population was otherwise engaged and thus not readily able to attend to the opening and closing of their own domestic taps.¹¹

Water pressure within the London systems was inadequate for any number of reasons. Wooden pipes, upon which the established suppliers continued to rely even in the early nineteenth century, were in constant need of repair. Missing and ineffective stopcocks and valves further undermined the operation of distribution arrangements that had been intended to support ground-level delivery only.[12] Other technical deficiencies accentuated the pressure problem, a most worrisome limitation in view of the persistent threat of fire. When residential cisterns ran down, for example, they had to be refilled before pressure could be restored at the hydrants, then referred to as fire plugs. Even at full pressure, London mains could rarely deliver a jet farther than 120 feet. The mains could seldom be used until a fire had already been checked, at which point firemen could only cool the ruins with jets from the mains – disparagingly described as "dummies."[13]

However unreliable, the introduction of piped distribution systems contributed to immediate and substantial increases in per capita consumption. Those who had depended upon water delivered manually or by cart from surface sources, wells, and ponds, or who relied upon cisterns to retain rainfall, had an incentive to conserve or to moderate usage that was inherent in the means of supply. Under these conditions, early consumers either required or made do with roughly three to five gallons per day per person.[14] Yet, basing his calculation on a total daily distribution of seventeen million gallons to ninety-two thousand London premises in 1809, and assuming eight residents per household, one analyst estimates average "consumption" (not personal use) in the capital at twenty-three gallons per person per day.[15]

Despite some evidence of competitive touting such as the West Middlesex Company's claim to offer "water of the purest quality,"[16] the quality of water was only intermittently a focus of concern. Yet the issue eventually came to the fore for water consumers as it had for hundreds of displaced Thames fishermen. A precipitating event was journalist John Wright's denunciation of one water company's arrangements. The Grand Junction Water Works Company obtained its supply from a facility known as the Dolphin, which operated alongside the Thames near a sewer outfall. In a celebrated pamphlet, Wright condemned the liquid delivered to fellow Londoners from the Thames as offensive and destructive to their health, and called for the water companies to abandon this discreditable source of supply. More than 130 public common sewers, he wrote in 1827, contributed "the drainings from the dung-hills and lay-stalls, the refuse of hospitals, slaughter-houses, colour, lead gas and soap works, drug-mills, and manufactures, and . . . all sorts of decomposed animal and vegetable substances" to the Thames in the short distance between Chelsea Hospital and London Bridge.[17] Why it might be considered objectionable to draw such a potion from the sewer-charged Thames – or why it might not – will shortly be considered.

Municipalities in Britain became increasingly involved in water supply as the century advanced. Glasgow, Manchester, and Liverpool, for example, were among the first to replace local water companies with municipal systems; in the second half of the nineteenth century, roughly fifty British cities and towns introduced municipal supplies each decade.[18] Efforts to remedy the London situation encountered resistance from the water companies and were the subject of intense debate among interested professionals. However, in 1842 Edwin Chadwick's classic *Report on the Sanitary Condition of the Labouring Population of Great Britain* brought a powerful impetus for change. Chadwick, a pioneering sanitary and social reformer, recommended that government ownership and public administration take the place of privately owned water companies, and that springs and gathering grounds north and south of London replace the Thames as a source of supply.[19]

In the years following Chadwick's recommendations, several private suppliers did make improvements to the London water system. Some relocated their intakes upstream from the Thames' most heavily polluted stretches.[20] A few even introduced filtration, although there was yet little theoretical appreciation of such factors as depth, rate of flow, and media on the effectiveness of the treatment.[21] Yet the companies resisted external reform efforts and defended their interests even against reinvigorated challenges incited by a cholera epidemic in 1848-49 – during which, it was widely noted, recipients of distant New River supplies fared rather well in comparison with consumers of Thames river water.[22] Further public inquiry in the form of the Health of Towns Commission created favourable conditions for the *Public Health Act* in 1848 and its formation of a General Board of Health. As commissioner under the new legislation, Chadwick pursued his quest for reform.

The private enterprises continued to fend off a series of competitive ventures that proposed tapping remote sources. In 1851, by undertaking to provide water on a continuous basis, they were equally successful in countering legislation calling for their amalgamation under government control. An alternative measure designed to place the water companies under the control of local parish vestries fared no better.[23] When water supply legislation for metropolitan London was enacted after a change of government in 1852, it still assumed the continued presence of private suppliers in delivering water described from the perspective of its quality or condition as "pure and wholesome."[24]

Amphibious Humanity

Several North American cities were similarly engaging questions of water supply and its private or public provision. Philadelphia, thanks in part to Benjamin Franklin's foresightful gift, became a prominent early exception to the general pattern of privately financed waterworks. At the time of his death in 1790, the well-travelled Franklin predicted on the basis of his international observations that the wells and springs then serving the citizens of Philadelphia would ultimately deteriorate. The implications were clear enough to the author of *Poor Richard's Almanac*, who was on record to the effect that "when the well is dry, we learn the worth of water." Franklin bequeathed £1,000 to Philadelphia, recommending that the accumulated capital eventually be used to render the nearby Schuylkill River navigable and to lay pipes for water delivery from Wissahickon Creek.[25]

Sooner than Franklin might have imagined, a civic committee on water supply sought the guidance of Benjamin Henry Latrobe, a young engineer and architect of Huguenot descent with professional experience in his native England, who had recently emigrated to North America. Latrobe outlined a strategy to ensure timely and reliable delivery of water, presenting his recommendations to the community in 1799 in his *View of the Practicability and Means of Supplying the City of Philadelphia with Wholesome Water*. His projection that a steam-powered system drawing water of "uncommon purity" from the Schuylkill River could become operational within a year proved overly optimistic. However, despite the challenge of financing at the municipal level, the antagonism of the promoters of an alternative supply source involving a canal, and practical difficulties of weather and limited supplies of certain materials, the project was completed by 1801. Philadelphians were then able, if they so chose, to avoid local pumps and wells already suspected of harbouring, communicating, or otherwise contributing to disease.[26] It was, in fact, an epidemic of yellow fever that had encouraged Philadelphia's civic leaders to address the water supply question – although precisely how disease might be transmitted through water was not yet understood.

Limitations in Latrobe's plan became apparent as early as 1811, prompting inquiries to identify alternatives. Attention turned to a pumping station with a new reservoir at Fairmount, on the banks of the Schuylkill. These arrangements, operational as of 1815, in turn exposed inadequacies in the distribution network, for although the new engine could pump three million gallons per day to a reservoir with roughly that capacity, the hollow logs that served as conduits throughout the city could distribute no more than one million gallons daily.[27] Philadelphia then turned to iron pipes, although there are indications of wooden logs being laid as late as 1832, and it was 1849 before the city completed the conversion.[28] This was but one of several cumbersome transitions in the evolution of municipal infrastructure initiating (if we may borrow the phrase) the real underground economy and altering the human connection to the hydrological cycle.

The saga of the New York water supply was even more prolonged than Philadelphia's experience. Also threatened with yellow fever in the late eighteenth century, New York City residents set out to secure a greatly expanded and publicly controlled supply of water superior in quality to their severely degraded local sources. Instead, they obtained a modestly improved water system in the hands of the Manhattan Company.[29] The lengthy act of incorporation contained a curious provision permitting the Manhattan Company to use its surplus capital in lawful "monied transactions," including banking.[30] Thus, the Chase Manhattan Bank had its start. This arrangement was largely attributable to the machinations of assemblyman Aaron Burr, who shepherded legislation through the state legislature at Albany to supply the city with "pure and wholesome water."

Even while prospering in the realm of banking, the Manhattan Company remained vulnerable to a provision in its charter that made the privilege of banking contingent upon supplying New Yorkers with wholesome water. Interruptions in supply – to the point that customers in the southern portion of the service area were unable to fill their pitchers for nearly a week in August 1811 – and increasingly common complaints about the quality of water actually delivered, stimulated company officials to contemplate alternative arrangements, perhaps even sale of the waterworks to the city. Amendments to the charter granted by the state legislature gave the company temporary breathing room, but the Manhattan Company's continuing deficiencies as a water supplier encouraged competing initiatives.[31]

Threatening competition emerged during the 1820s in the form of a proposal by the New York and Sharon Canal Company to link the Hudson River to the city by means of a waterway to Sharon, Connecticut, connecting to a canal tied in to the Housatonic River.[32] The New York Water Works Company then entered the fray, vigorously defending competition as a way of promoting the public interest.[33] Neither venture came to fruition, leaving New Yorkers dissatisfied and vulnerable to disease and fire, such as the 1828 conflagration that destroyed $600,000 worth of property.[34] Meanwhile, as calculated in 1829 by enterprising researchers associated with the Lyceum of Natural History, city residents were depositing more than a hundred tons of excrement each day into the soil above heavily used groundwater supplies. Recent New York historians have perhaps understatedly described the fluid extracted from downtown wells as a "tainted brew."[35]

In the early 1830s, as the city's population approached a quarter of a million people whose activities consumed roughly twelve million gallons of water each day,[36] water commissioners recommended municipally owned works. They did so in part on the basis of the London experience, or at least on their often critical understanding of the London model. The Croton River, a tributary of the Hudson, emerged as the favoured source. Once state legislative approval had been granted in 1834, supervising engineer John B. Jervis (having risen through the ranks from canal labourer) and a panel of government-appointed commissioners oversaw the 1837-42 construction of a dam and aqueduct system about forty miles long. The receiving reservoir (Yorkville) was located on Manhattan Island in what would become Central Park, with a distributing reservoir at Murray Hill – then "a short drive from the city"[37] and later the site of the New York Public Library.

The aqueduct leading into the city was engineered to facilitate the flow of ninety-five million gallons of water per day along its 12.6-inch-per-mile slope.[38] The arrival of the *Croton Maid* at Yorkville in June 1842 – despite some portaging around tunnels that had facilitated her journey – was heralded as "indubitable evidence that a navigable river was flowing into the City for the use of its inhabitants."[39] A year after the arrival of Croton water, New York socialite George Templeton Strong gleefully embarked upon "an amphibious life" with the installation of a bathroom, where he soon found himself "paddling in the bathing tub every night."[40]

In contrast with that of New York, where the scale of population demanded a comprehensive response early in the century, Chicago's water supply developed incrementally. A core system, established during the 1850s under the direction of consulting engineer and surveyor William J. McAlpine, serviced residents within the existing city limits between 1851 and 1861. Direct connection to the tap remained a luxury, so public hydrants served as points of access until the end of the Civil War. Outside the core, suburban and rural areas, whose development was frequently encouraged by real-estate promoters or industrialists such as railway magnate George Pullman, struggled to establish autonomous systems. These drew water from Lake Michigan across neighbouring communities, or exploited underground supplies. More than a dozen independent systems in outlying counties were eventually consolidated (chiefly through annexation) and incorporated within the confines of the city of Chicago. As the century drew to a close, more than 320 million gallons flowed daily through 300,000 taps serving the residents of Chicago, who were then connected to 1,847 miles of water mains.[41]

The story, generally with more similarities than variations, was repeated across the United States, and not only in major centres.[42] Smaller communities, often seeking to stimulate rather than to respond to the forces of expansion, and perhaps more anxious to foster nurturing environments than to clean up festering sources of filth and the associated social perils that typified large cities, also introduced waterworks in the late nineteenth century.[43]

Canadian cities also anxiously explored the options available to them, often soliciting guidance from the same small cadre of engineers whose designs so heavily influenced American developments. Montreal faced challenges similar to those of other mid-nineteenth-century urban centres. After several private ventures had failed to implement viable arrangements for water supply, and following vigorously contested municipal elections, Montreal acquired the entire existing operations. The facilities, for which the city paid £50,000, included a set of steam engines and about fourteen miles of pipe.[44] Severe fire damage in 1852 dramatically demonstrated the need for improvements,[45] although other signs of the unsatisfactory state of the city's water supply had been evident for some time.

To address the water supply requirements of their ambitious northern metropolis, civic leaders, like their counterparts elsewhere, sought professional advice. They turned to a young practitioner whose reputation even at the early stages of his career portended his subsequent status as the most celebrated hydraulic engineer in Canada. Thomas C. Keefer considered gravitational and artesian options before recommending that Montreal locate its intake at the head of the Lachine Rapids on the St. Lawrence River. He firmly dismissed the Lachine Canal as "the common cess-pool for the offal of boats of every description passing through it."[46] Although the substantial financial implications and the proposal to abandon steam engines for hydraulic pumping initially caused hesitation, Montreal's Water Committee accepted the young engineer's proposal. Having been authorized to obtain the services of consulting engineers, Keefer called upon American veterans John B. Jervis and William J. McAlpine. Jervis's engineering career combined railway and canal construction with extensive waterworks experience in connection with New York's Croton Aqueduct and the development of a water supply for Boston. The Montreal project came comparatively early in what would become a very extensive waterworks practice for McAlpine. An apprentice of Jervis, he had made his engineering reputation in the construction of the US Dry Dock at Brooklyn, and by the mid-1850s had worked on water supplies for Albany, Chicago, and Brooklyn.[47] Both consultants, albeit with minor modifications suggested by Jervis, endorsed the Keefer plan.

Although Montreal's new arrangements offered enormous benefits to the city, they were not without drawbacks. When ice formation restricted access to the water supply in winter, puncheons returned to service intermittently. Once consumption had reached five million gallons per day – only a decade after construction – a steam engine had to be added in 1868 to supplement capacity.[48] This temporary measure proved to be more enduring; by the end of the century, steam pumps provided well over half of Montreal's daily consumption of forty-three million gallons.[49]

With its own population expanding, Toronto, too, set out to tackle a mid-nineteenth-century water-supply challenge. Perhaps "resumed the challenge" would be more accurate, for the water question had surfaced from time to time in previous decades, and since 1841 at least portions of the population were supplied by the Toronto Gas, Light & Water Company. Despite the implications of its name, this enterprise was a private franchise under the exclusive control of Albert Furniss, a Montreal businessman who was also the proprietor of that city's gasworks.[50]

The decision to place Toronto's waterworks in private hands had not been taken lightly. After much deliberation, civic leaders concluded that Furniss's capital and expertise offered significant attractions, most notably the modest level of municipal financing required. Complaints soon surfaced, however: the Furniss distribution system provided insufficient volume and pressure to deal with fire, and consumers were dissatisfied with the quality of the supply. Furniss denied responsibility for the latter difficulty,[51] noting curtly that his waterworks intake had already been in operation at the foot of Peter Street when the city chose to locate the outlet for its main west-end sewer there in 1845.[52] Despite more or less continuous protest and any number of unfortunate incidents, even in June 1854 with cholera in Atlantic cities and at Grosse Isle, Toronto was taking water directly from a section of the bay that was only five feet deep.[53]

Toronto's Standing Committee on Fire, Water and Gas sponsored a public competition in 1854 to solicit proposals on the city water supply. The distinguished panel, composed of engineers Walter Shanley and Casimir Stanislaus Gzowski and architect Fred Cumberland, was charged with evaluating the entries. The plan favoured by the evaluators, identified as the work of "A Red Cross" but penned by George Kent Radford, proposed taking water from Lake Ontario. To avoid interference by steamers and other lake vessels, Radford suggested locating the intake within the line of projecting wharves in a depth of at least ten feet. He acknowledged alternative locations, but did not think any better supply could be procured "with a due regard to economy."[54]

Basing his consumption projection on the advice of English authorities, Radford provided for twenty-five gallons per head to be distributed from the central main, running from a reservoir at the top of Spadina Avenue along a concession road to the head of Yonge Street. Radford's innovation was to deliver water to the exterior of every house through wrought-iron pipes instead of lead, which was then commonly used. Since joints – apart from a simple screw joint – would be eliminated, a considerable saving was anticipated, both in material and labour. Radford's estimates, therefore, included the cost of furnishing every house listed in the Toronto rate book with a supply at the exterior wall, a location that would become an increasingly important junction between the realms of public water distribution and private water consumption. Further installations would be the responsibility of tenants or landlords, although Radford recommended that the city might also undertake this work, charging a rental fee to relieve the poor from the burden of an immediate capital outlay. To respond to other civic needs, Radford noted, the cocks and pipes would "afford the means of watering and cleansing the streets, by means of spreading jets . . . thereby saving a great expense in carts and horses, and affording streets . . . suffering from dust the benefit of water."[55] Radford's sensible plan was, however, destined to relative oblivion as Toronto officials balked at the overall expenditures involved and became entangled in protracted negotiations to purchase the competing Furniss system.

By the mid-nineteenth century, Toronto, by means of a provincially approved charter, had joined a growing list of cities, including London, New York, and Philadelphia, in which the new option of public ownership contended with a continuation of the private approach to municipal water supply. However, as of 1858 only 900 of 9,500 houses in the city had accepted private or public service.[56] Whatever momentum municipal ownership might have been developing in Toronto was interrupted by the economic crash of 1857, and the city continued to rely for a time on the Furniss company, with its modestly growing customer base. In the mid-1870s, however, dissatisfaction with the private supplier reached new heights. "We have neither the quantity nor the quality necessary to secure the health and comfort of the citizens," protested the *Toronto Globe*, declaring the city to be "equally destitute of what is indispensable for the safety of our houses from fire, the flushing of our common sewer, and the watering and cleaning of our streets."[57] These sentiments were shared by the legislature, which only a few months earlier had introduced legislation authorizing municipal waterworks with specific reference to "grave and frequent complaints" about existing supplies.[58] Furniss's death in 1872, and the willingness of Toronto residents to assume substantial long-term financial obligations, paved the way for successful negotiations between the city and his estate. Between 1874 and the end of 1877, the year in which Toronto's system finally came under the direct control of city council, the number of houses enjoying the public water services rose from 1,375 to 4,518.[59]

Viewed more comprehensively, statistics on the nineteenth-century expansion of municipal water systems demonstrate the scope of the transformation in Britain, the United States, and Canada. Although barely half of Britain's fifty largest communities relied upon publicly authorized waterworks systems, by 1914 only twenty-nine of more than eleven hundred urban districts outside London lacked piped water supply. Municipal suppliers serviced about 80 percent of these communities.[60] From 1800, when sixteen installations supplied water to 2.8 percent of the US population, waterworks systems began to proliferate across North America. At midcentury, eighty-three operations provided water to 10 percent of the population. A decade later the number had grown to 136 waterworks systems, and it reached 598 by 1880. That year witnessed the formation of the American Waterworks Association, signalling a still more dramatic increase in municipal water services. By 1897 more than 3,100 waterworks supplied 41.6 percent of Americans.[61] A quarter of a century later, a mere 3 percent of urban US residents continued to rely on private wells.[62] At this point, slightly more than two-thirds of urban residents still relied on surface-water sources (open rivers, impounded streams, and lakes), roughly a quarter were supplied by mixed surface and groundwater, and fewer than 10 percent depended exclusively on groundwater.[63] Accelerating development of waterworks systems in Canada got under way somewhat later in the nineteenth century than in Britain and the United States. From a mere thirty systems in 1880, however, by the eve of the First World War Canada had more than five hundred waterworks suppliers. Roughly 80 percent were municipally owned.[64]

This reconfiguration was of some consequence. As wells or cisterns, necessarily fixed in one location, were replaced by networks of underground pipes delivering water to widespread and growing populations throughout Europe and North America, a broadened perspective was called for. "Such a project forced citizens to look beyond their individual spheres of interest to the less familiar terrain of a larger community-wide interest."[65] In time, the collective dimensions of waste removal would call for a similar transformation in outlook or perspective, a challenge that has been less successfully met.

The quest for public water supplies had involved legal, financial, and political – as well as technological – hurdles. In London, Toronto, and elsewhere, entrenched franchise holders resisted alternative arrangements. Particularly contentious disputes arose in the United States, where property enjoyed a heightened degree of constitutional security.[66] Financing, too, presented obstacles, given restrictions on the ability of local governments to borrow or to incur debts above limits prescribed by legislation.[67] As historians Chris Armstrong and Viv Nelles state about Canadian cities, "Water supply passed from private ownership to private investment secured by public credit."[68] Important issues arose even at the household level, as arrangements were established that provided for municipal interference with any number of presumably private matters in order to pursue the community interest in water supply. On this score, legal historian Earl Murphy insists that the legislative foundations of public sanitation, though humble in appearance, were instrumental in transforming the general legal landscape so as to "make out of the events, needs, desires of the nineteenth century the legal order of the twentieth."[69] Underlying the adoption of municipal water supply systems and sanitary initiatives was an emerging vision of social and community advancement, which paralleled many of the themes associated with rising industrial demand for water.[70]

A Philosophy of Waterworks

No clear point of departure marks the quest for municipal waterworks. After acknowledging the inadequacy of prevailing arrangements, communities confronted an array of difficult decisions, many unique to their individual circumstances. Each step called for some attempt to rationalize the path taken in preference to other available options. For communities with options, the selection of a source was fundamental. Surface supplies and groundwater each had their advocates, and more distant sources often contended with proximate supplies. The quality of water – its purity, as the matter was customarily put – had some importance, but the tendency of closely crowded wooden buildings to incinerate their residents and commercial contents suggested that, in the interests of fire protection, water volume would often be the decisive requirement. Despite growing interest and frequent controversy, major developments in water quality knowledge still lay ahead.[71]

The technologies of collection and distribution equally demanded attention, for, as nineteenth-century inventors and innovators patented the fruits of their imaginations, competing systems offered varying advantages. Some appeared better suited to future expansion, and others promised very appealing interim economies; certain technologies were capable of delivering a constant supply, but others could provide only intermittent service. (Although it may seem curious to later generations of observers, each of these options had some advantage over the other.) Further unknowns also challenged engineers and officials: how much water would consumers use, and what would they pay for the privilege – if it was not yet a right – of having water piped to residential, commercial, and industrial premises?[72]

Such questions were so interrelated that the proper sequence for their resolution was hardly clear. And, to those who found these water supply decisions insufficiently problematic in their own terms, the rapidly changing views of interested professionals contributed exponentially to their complexity. Engineers, chemists, medical practitioners, and eventually a new generation of public health officials, voiced firm – and sometimes dishearteningly contradictory – opinions on many of the matters to be resolved. Among the common themes emerging from municipal deliberations, in addition to the important one of adequate supply, civic autonomy and a cluster of concerns relating to quality are noteworthy.

Despite, or perhaps because of, his Montreal waterworks involvement, Thomas C. Keefer had declined to prepare a formal entry for Toronto's 1854 waterworks competition, arguing that the modest prize offered as an inducement grossly underestimated the time, trouble, and expense of responding thoroughly. Keefer was already known as a gifted publicist whose widely circulated essay *Philosophy of Railroads* (1849) did much to promote "the *idea* of railroads."[73] As a biographer of Keefer explains, the consequences of railways – their social, economic, and moral implications – would ultimately be more far-reaching than the modest appearance of a simple line of track might have suggested.[74] Keefer, McLuhanesque in his time, crystallized the impact of the railroad on human progress: "Steam has exerted an influence over matter which can only be compared with that which the discovery of Printing has exercised upon the mind."[75] Keefer's celebration of the railway, in particular his vision of its promise for Montreal, had greatly pleased his audience. Could he do the same for water?

A philosophy of waterworks seems somewhat lacking in the qualities that allowed railway visionaries to stir the imaginations of investors, civic boosters, and travellers. Yet Keefer's conception of municipal waterworks as a chain of consequences was ultimately no less expansive than his earlier vision of the railway. His contribution to a philosophy of waterworks combined assumptions about the abundance and utility of water with calculations respecting its power – and at least a measure of purity (an enduring if widely variable notion) in the service of humanity. Associated with the challenge of securing generous supplies was a widespread civic preoccupation with autonomy.

"There is no class of works which require to be more thoroughly and efficiently constructed than the water works of an important city," he pontificated presciently, catering to the ambitions of his select but influential readership. "It is better to be without them than to have them on an inefficient scale, because where they exist the population become in a short time so wholly dependent upon them that unless they be adequate to every demand the consequences will be disastrous in the extreme."[76] In a preliminary engineering survey for Montreal, Keefer had considered two options – a gravitational system and an artesian alternative – and handily dismissed both, opting instead for a pumped supply from the head of the Lachine Rapids. Pumping was the only practical solution, and a choice to be welcomed, he argued. "One of the largest and purest rivers in the world flows at the very feet of your city," the young engineer proclaimed, "affording not only an illimitable supply for consumption, but the cheapest power for elevating this supply into the highest parts of the city."[77] The St. Lawrence River presented, and Keefer embraced, an extensive surface supply whose own flow promised to facilitate its delivery.

Keefer reiterated the significance of Montreal's proximity to a major waterway. Being "inexhaustible" as a fountainhead for water supply and power, the river would allow successive generations of Montrealers to enlarge their works as necessary "so long as the St. Lawrence flows toward the sea."[78] More than a few readers would have recognized the teaching of Ecclesiastes 1:7: "All the rivers run into the sea; yet the sea is not full; unto the place from whence the rivers come, thither they return again." Professional opinion and spiritual guidance combined to affirm that nature's bounty would be – in terms of water, at least – costless, unlimited, and eternal.

Keefer's philosophy of waterpower as a God-given gift also contributed to the growing nineteenth-century association of a bountiful water supply with civic pride and autonomy. As Boston mayor Josiah Quincy had expressed it in the 1820s, "If there be any privilege, which a city ought to reserve, exclusively, in its own hands and under its own control, it is that of supplying itself with water."[79] In a similar vein, the 1842 report of the New York water commissioners had anticipated remarkable results from the Croton project. New York's water supply system was not designed to safeguard citizens against an external foe, but rather to make the "whole population happier, more temperate and more healthful." Countless millions who would enjoy such benefits "will have clear heads, correct eyes, strong arms, and instead of walls, present breasts so strong and hearts so brave, that in a just cause our city may defy all foreign foes."[80] Keefer's report, in more modest terms, concurred: "In no matter should a city be more thoroughly independent than in its water supply."[81]

Using, Consuming, and Wasting the Flow

The unlimited supply of the St. Lawrence, which had figured centrally in Keefer's Montreal analysis, influenced his recommendation that existing arrangements for steam pumping be replaced by waterpower: "If the quantity required were fixed, and not to be increased – and if the cost of procuring it were to be a permanent burden upon the present generation," he allowed, "there would be good reasons why a known annual cost in steam power should be preferred to the investment of a large sum, and payment of interest thereon, in order to secure ultimate economy of management." However, he dismissed this approach on the grounds that since "such an enterprise as the water supply of a city, will be a self sustaining one . . . the burden of any investment, will be but temporary." This was true, Keefer argued, even in the context of increasing consumption: "And as the consumption will increase – not merely by the natural growth of the City, but *with the facilities for extending the supply* – it is evident that nothing but absolute poverty can justify the continued use of steam, where another power is within reach – which promises not only ultimate economy – but (what is far more important) an extension of the consumption, and all the advantages which flow from it."[82]

Water supply, whether seen as a costless resource or one essentially capable of paying for itself, offered many attractions. The more water a growing city might consume, the greater it would appear were the benefits. There was little indication in such an approach to urban water supply that the constraints applicable to ordinary riparian water users or consumers of well water would be applicable to an ambitious urban community. But how much water did a city actually require?

Average mid-nineteenth-century levels of per capita consumption for several major North American cities with piped supplies range from Cleveland's eight gallons per day (1857) to fifty-five gallons in Detroit (1856).

Chicago's average of thirty-three gallons (1856) and New York's estimate of forty gallons of daily per capita use for public, industrial, and domestic purposes in 1850 fall in between. Toronto's mid-century requirements were estimated in the same range, that is, about thirty gallons for each of the city's forty thousand inhabitants.[83] By the time of Keefer's 1854 report for the city of Montreal, the scale of works in contemplation was for a daily supply of five million imperial gallons delivered to a reservoir (on Mount Royal) at an elevation of two hundred feet above Montreal Harbour. Montreal's position, Keefer speculated, might differ somewhat from circumstances faced by other major centres: "With you, therefore, it may prove the wiser policy to provide for that *waste* of water so much complained of in New York and Boston – rather than to rely upon your future ability to check or prevent it."[84] Keefer's original estimate of thirty gallons per head as a reasonable requirement for Montreal was well below average consumption levels in New York and Boston.[85]

As the nineteenth century progressed, water consumption levels vastly exceeded mid-century norms. Per capita daily water usage in Chicago and Detroit had reached more than 140 gallons by 1882. By 1872 Cleveland residents consumed fifty-five gallons per day, and in New York the per capita daily water consumption figure had risen to seventy-eight gallons by 1880, with sixty gallons of that attributed to domestic use.[86] A decade or so later, a Boston study conducted by the State Board of Massachusetts suggested that provision should be made to allow 100 American (or 83.3 imperial) gallons. Moreover, actual consumption in the sense of volume pumped per capita in Toronto, based on 1894 data, was already 107.7 imperial gallons.[87]

By the early 1900s, several American cities, including Chicago, Pittsburgh, and Philadelphia, were distributing more than 200 gallons per resident per day, with Buffalo providing an extraordinary 334 gallons. New York, Washington, Detroit, Cleveland, and St. Louis delivered between 100 and 200 gallons. Only in a few smaller communities such as Lowell, Worcester, St. Paul, Minneapolis, and Hartford did consumption levels between 40 and 80 gallons per day still prevail.[88]

These volumes greatly exceeded levels common in Britain and major European cities. After northern British cities assumed municipal control over water supply in the early 1850s, average consumption per head increased substantially, from 4.8 gallons per day in 1841 to 32.6 in 1875 (Manchester), from 8.1 in 1851 to 20.5 in 1871 (Leeds), and from 30 in 1845 to 49.7 (Glasgow) in 1871.[89] The average per capita supply of eighty British towns and cities in 1882 was a little over twenty-seven gallons.[90] In France, water requirements fluctuated and expanded rapidly from an estimated 10 to 20 litres daily per person in Paris in the early 1800s, to the figure of 1,000 litres (265 gallons) recommended by learned specialists convening at international conferences on hygiene in the final decades of the century.[91] However, as calculated for the 2,700,000 Parisians of 1900, actual per capita daily consumption was about 250 litres (66 US gallons), roughly 20 percent of which was attributed to waste.[92] The Paris supply, reported at 65 US gallons per capita daily in 1901, was well above norms in other European centres such as Amsterdam (37), Berlin (22), Copenhagen (29), and Liverpool (44).[93]

To British dam builder James Mansergh, advising Toronto on supply options, it was "a matter of common knowledge that in the United States, Canada, and other of the Colonies, the consumption per head is very much greater than it is in England." A royal commission on which he had served established that for 1891 London's daily per capita rate of supply was 38.68 gallons, an amount adequate for all purposes – domestic, sanitary, trade, and ornamental.[94] Mansergh had a fair idea as to why Toronto was pumping three gallons for every one supplied in London: "I have a shrewd suspicion that by far the greater part of the excess is due to misuse or waste in one way or another."[95] Ultimately – and recognizing that some discipline and effort would be required to hold to his target level – Mansergh in 1896 proposed fifty gallons a day per person as the supply level to be met in Toronto. For Canada as a whole, daily per capita consumption in 1915 was 111 imperial gallons.[96]

Keefer's Montreal report offered an unsolicited analysis of the benefits of liberal water usage. The uses of water, he reminded his readers, were by no means limited to domestic purposes, for "its abundance and cheapness give rise to manufactures, and thus become sources of immediate profit to the community." Moreover, given (as Keefer assumed) "an overflowing amount and a sufficient head pressure on every street," fires would most certainly be expeditiously extinguished.[97] This increased security of water supply would undoubtedly bring about a decline in insurance costs, and, as the consultant's upside forecast projected, these lower costs should attract immigration and capital.

Water, Keefer further urged, "should be used most unsparingly" as a source of comfort for citizens of the community and to preserve property, given the intense heat of the summer and the inordinate quantities of dust generated in urban settings. He advocated a generous supply of fountains – in parks, in public gardens, and in pleasure grounds.[98] A partial rationale, focusing on practicalities, had previously been advanced by Benjamin Henry Latrobe: "[Fountains] are the only means of cooling the air. The air produced by the agitation of water is the purest kind, and the sudden evaporation of water, scattered through the air, absorbs astonishing quantities of heat, or to use the common phrase creates a great deal of cool."[99] Keefer went still further, assuring civic leaders that "abundance of water" in such locations "must be considered not merely as a luxury, and a purifying the air" because there was also a positive commercial value associated with every "object of interest and beauty" such as fountains surely were. Tourists, "always the wealthier classes of society," would be drawn to such landmarks, spending handsomely for travel and accommodation.[100]

The most significant civic incentive for enhanced waterworks, an interest reinforced by profound practical and financial implications, was fire.[101] After the Great Fire of London in 1666, when losses from an outbreak in Pudding Lane were estimated at between £7 and £11 million, the market for fire insurance was opened up.[102] Such companies as the Hand-in-Hand, the Sun Fire Office, Union Insurance, the Phoenix, and the Westminster assumed responsibility not only for providing financial compensation in the event of actual loss, but also for protecting – with their own fire brigades – the insured premises (identified by plaques or medallions affixed to the exterior). However, the companies' notorious rivalry severely compromised the effectiveness of such arrangements.

In increasingly crowded urban centres, lack of cooperation between insurers was extremely hazardous. The desire for municipal control of fire services became overwhelming. Factoring in the impact of readily available water supplies on insurance rates made civic initiatives to combine municipal fire brigades with new public waterworks systems highly attractive to consumers. When ineffective coordination among Edinburgh's eleven private insurers severely undermined firefighting efforts, the city moved to implement a municipal fire service in 1824. A few years after the Edinburgh initiative, London insurers merged their firefighting efforts into the London Fire Engine Establishment; but here too a publicly supported Metropolitan Fire Brigade finally replaced the insurers in 1866.

For reasons explained by the superintendent of the London fire brigade, other communities were increasingly interested in similar arrangements. Liverpool had seen its mercantile insurance rates climb drastically in the 1840s in the aftermath of warehouse fires between 1838 and 1843. In response, legislation was introduced to regulate construction materials and distances between buildings.[103] An additional statutory measure designed to bring water into the city for fire and street watering helped restore premiums to more affordable levels.[104]

A more general message, emerging from studies of fire response times, was that water tapped quickly and directly from the mains rather than from fire engines would greatly reduce fire losses. Edwin Chadwick pointedly noted that whereas "a pail of water might suffice within the first minute after ignition, in the second minute more than a ton weight would be required." But in the absence of infrastructure, even the two-minute response would be exceptional. Demonstrating a boundless enthusiasm for measurement, Chadwick induced the London police to calculate the time elapsed between the first sounding of the fire alarm and the arrival of the engines. For a fire located a mile or so from the station, some twelve minutes might pass as the inbound message of alarm crossed the city at five miles per hour. Five minutes later, horses, engines, and men were in motion, but at ten miles per hour, allowing for obstacles, another six minutes were lost before arrival at the scene – of what was by then most likely a conflagration – of an *empty* fire engine, yet to be filled in a process that could easily consume five minutes more. This fatally slow proceeding of twenty-eight minutes would reduce to only about twenty minutes if the distance to be covered was halved to half a mile.[105] Mid-century analyses such as this, combined with the experiences of Liverpool, Hamburg, and other communities, supported Chadwick's conclusion that for the purpose of suppressing fires, modern municipalities needed a constant supply of water at high pressure to be applied through hoses directly from mains.

Toronto's interest in upgrading the municipal water supply was also directly linked to its experience with fire. A major fire in April 1849, which caused damage estimated at £100 thousand over twelve acres, was followed by numerous smaller outbreaks, continuing reminders of danger. In 1851, twenty-seven separate fires consumed forty-two buildings, leading to losses totalling £7,712, well above their insurance coverage of £2,485; the cost of water to extinguish the fires was a mere £157. The next year seventy-eight buildings valued above £10,000 were destroyed in nearly three dozen fires, with insurance covering roughly a third of the losses and the cost of water put at £196 10s. In 1853, twenty-one fires caused losses of £12,711 including, this time, sixteen buildings. Insurance proceeds amounted to £8,634 and water costs to £67 10s. were associated with the annual fire experience.[106] Little wonder, then, that Toronto municipal officials longingly envisaged the contribution to "the comfort and repose of the inhabitants of a great city" that increased security from such a high level of risk would furnish. Not only could they anticipate having "a salutary check upon the incendiary," but, as rates of insurance fell correspondingly, they could readily imagine the impetus to new construction, whether of buildings or of ornamental structures.[107]

When the United States National Board of Fire Underwriters published its first nationwide rate sheet in 1872, a significant rate differential between communities with water systems and those without could be noted. Toward the end of the following decade, premium reductions ranging between 20 and 50 percent were common for towns where waterworks had been established.[108] Especially after steam fire engines were introduced in the 1850s, the fire and insurance perspective on water supply put a premium on volume or quantity rather than quality (even to the extent that quality was then understood).

In the interests of economic advancement, fire protection, civic pride, and general concerns with sanitation, nineteenth-century cities responded to the demands of growing urban populations for municipal systems of water supply. As Keefer and others had predicted, the convenience of public supplies greatly stimulated consumption. A variety of new uses for water soon emerged or gained popularity, including the use of water to remove human and domestic waste, at the household level and on a more comprehensive scale. The stream of consequences continued to flow thereafter, profoundly altering relationships between individuals, between communities, and with the natural environment.

4

The Water Closet Revolution

THE DOMESTIC FLUSHING REVOLUTION – symbolized by the water closet – rested on the increasing availability of running water and the dogged ingenuity of a handful of inventors. The devices that proliferated in the nineteenth century owed their conceptual origins to Sir John Harington, godson of Queen Elizabeth I and author of *The Metamorphosis of Ajax* (1596), in which the construction of an early valve closet – a "necessary" – is described.[1] An interval of two centuries, however, preceded a flurry of patents by the likes of Alexander Cummings (1775), Samuel Prosser (1777), and the very inventive cabinetmaker Joseph Bramah (1778).[2] The celebrated Thomas Crapper, who served as the royal sanitary engineer for many members of England's royalty, eventually took his place in this procession. Although there was some US use of the water closet in the early 1800s, there were no patents there until 1833.[3]

Adoption of the new technologies was influenced by, and in turn shaped, the broader social context, including its legal dimensions.

The Domestic Flushing Frontier

No statistical series comprehensively records the introduction of the water closet to general use, a phenomenon that initially proceeded with "hesitant irregularity."[4] English cabinetmaker and inventor Joseph Bramah reported six thousand sales of his patented WC in his first two decades of production to 1797, a total equivalent to six installations per week.[5]

Reports dating from 1844 for the Old Church district of Lambeth (one of the few areas then serviced with sewers) indicated that water closets, although in more general use than previously, were largely confined to the residences of the wealthy. A few years later, it was evident that even prosperous citizens remained loyal to cesspools and privies: "Water closets are very rare, even in the better class of houses."[6] The persistence in London of intermittent water supply undoubtedly prolonged hesitation to adopt the new technology. Nor did royal example encourage modernization: Windsor Castle had fifty-three overflowing cesspits in 1844. Although the prince consort had encouraged the replacement of Hanoverian commodes with water closets and drains, the process of modernization was more or less abandoned at the time of his death in 1861.[7]

By the mid-1850s, with some thirty thousand flush toilets in London, an accelerated transformation was under way in Britain.[8] From the late 1850s, though more slowly in Wales, the north of England, and working-class districts generally, the adoption of the water closet accompanied the introduction of water systems. One sanitary inquiry conducted in the wake of local epidemics revealed that "in one part of Manchester in 1843-44 the wants of upwards of 700 inhabitants were supplied by 33 necessaries only – that is, one toilet to every 212 people."[9] By 1871 Manchester boasted 10,000 of the new installations. In Liverpool, where 2,639 privies had been converted to water closets by about 1865, there were 15,000 flushing toilets in 1871; Edinburgh had 28,000 in 1873, and Birmingham 8,000 two years later. More than half the residents of Glasgow flushed by 1870. Dublin, with 743 water closets in 1880, was late to embark on the conversion process, yet in only two years the city's 15,000 WCs outnumbered its remaining privies.[10]

The forerunners of the organic porcelain toilets with which Europeans and North Americans are now universally familiar have been described as "complicated, often eccentric, masterpieces of mechanical complexity."[11] Mid-nineteenth-century water closets, for example, took a variety of forms, divided generally into hopper- and pan-style appliances, but all involving an array of flaps, valves, levers, and hinges. The hoppers, ordinarily simpler in operation, consisted of a funnel tapering to a connection with a soil pipe and trap. The entire apparatus might be rinsed by a sudden or continuous application of water, although the effectiveness of the typical "meagre spiral squirt" left much to be desired.[12] Pan models were identified by the presence of a hinged pan located beneath the bowl and above a receiver. The pan, usually copper, retained a shallow pool of water. By means of a handle that tipped the pan into the receiver en route to the soil pipe, users simultaneously engaged a valve permitting fresh water to enter the device – some for the purpose of encouraging wastes along their way and some to replenish the pan upon its return to the horizontal state. The more elaborate pan closets were truly cantankerous and prone to malfunction.[13] Introduced in the 1870s and 1880s, the wash-down model, sealed by water rather than a mechanical valve, and refinements such as flushing rims, represented important advances toward the units now in common use.

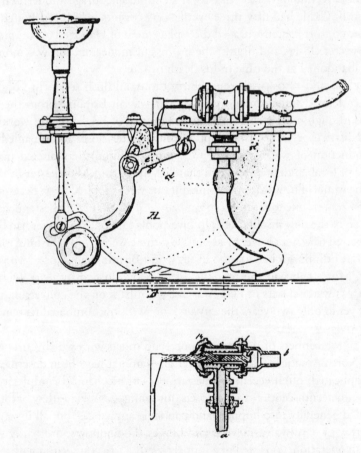

Patent application drawing of Carr's 1859 pan water closet. These closets contained many movable—and fragile—parts. The pan itself is indicated by the small set of double dotted lines in the upper drawing. The smaller drawing at bottom is a detailed side view of the valve mechanism attached to hose "b" in the upper drawing. U.S. Patent 25,092, William S. Carr, "Water-Closet Valve," 16 Aug. 1859.

In the early stages of technological development, the practicalities of the pan water closet were challenging, as is illustrated in this 1859 patent application.

That water closets such as this "Unitas," with their lovely fittings, were quickly and widely adopted during the late nineteenth century was due in part to their aesthetic and ornamental potential, which may have distracted consumers from reflecting on the appliance's essential function of removing bodily wastes for discharge into local waters.

West of the Atlantic, domestic appliances including the water closet also developed an avid following. In 1823 the Watering Committee of Philadelphia recorded that that city's residents were still in possession of 401 portable bathtubs; the first installed or fixed bath with plumbing appeared in the city three years later.[14] In the next quarter century, bathtub installations proceeded at an explosive rate: by 1849 there were 3,521 in the city, although with the population well above 300,000 this number was hardly generous by later standards.[15] In Boston, with a mid-century population approaching 150,000, the registrar for the Cochituate Water Board reported that customers of the water supplier owned more than 27,000 plumbing fixtures, of which about 2,500 were water closets. Thereafter, Bostonians' adoption of the new convenience was exponential: usage climbed to 6,500 by 1857[16] and more than doubled to 14,000 out of approximately 87,000 water fixtures by 1863. By 1870 the total number of fixtures had reached 124,000.[17] As of the 1850s, New Yorkers put Croton water to use in some 14,000 baths and more than 10,000 water closets. Chicago, a young community seemingly well supplied with plumbers at mid-century, soon boasted plumbing fixtures in homes all over the city.[18] In 1874, with a population of about 118,000, Buffalo supplied water to 5,191 dwellings, 3,310 with water closets. The best general estimate, admittedly imprecise, suggests that by 1880 water closets could be found in roughly a quarter of all urban households in the United States.[19]

Domestic appliances appealed for a variety of reasons. The first of these, convenience, is readily understood: the devices facilitated domestic labour in some way, or otherwise promised to improve home life.[20] But more remarkably, in America at least, the quest for household convenience was mingled with and reinforced by a sense of civic and national responsibility. As Maureen Ogle argues in *All the Modern Conveniences,* her study of the evolution of household plumbing in nineteenth-century America, convenience "stemmed from a broader desire to create excellent domestic environments and thereby effect national progress."[21] According to her analysis, citizens of good character were most likely to have been nurtured in family settings; whatever improved the American family supported the foundations of American civilization.[22] A more comforting rationale to support domestic consumerism would be hard to find. Other observers, in a more general account of cleanliness, have remarked that through a complex of associated values, cleanliness came to represent "a measure of a society's rank on the scale of civilization."[23] However cautiously one might wish to approach the ranking of civilizations, the mutual influence of culture and technology must be acknowledged. Law, too, shaped and experienced the consequences of flushing.

Despite the combined attractions of convenience and national advancement, plumbing was not universally welcomed without further encouragement. A period of gradual and voluntary adoption of appliances in the first half of the nineteenth century was superseded by a disciplined and systematic campaign to promote near-universal use of certain domestic fixtures within urban settings. Increasingly stringent legal and administrative mechanisms were brought to bear to promote a more uniform, regulated, and scientific approach to domestic arrangements, including water-using appliances, and the concomitant matter of drainage.[24] Public institutions were marshalling to exert a significant influence over personal and domestic behaviour, initially as promoters and subsequently as regulators of the culture of flushing.

In this respect, as in so many others in this history, legislation in Europe and Britain somewhat preceded North American developments. In Britain, extraordinary bodies – the commissioners of sewers – had for centuries exercised significant powers over the English landscape and its inhabitants. As expressed in legislation dating from the reign of Henry VIII, it was the responsibility of sewer commissions "to survey the walles, streams, ditches, bankes, gutters, sewers, gotes, calcies [causeways], bridges ... and the same cause to be made, corrected, repaired, amended, put down and reformed."[25] Most notable for the Great Level, an extensive reclamation of the fenlands around Cambridge and Ely in the early 1600s, the impact of the sewer commissioners' administration was most keenly felt by those required to contribute financially to designated works, and by those who otherwise believed that their interests were poorly served by the intervention of this eccentric institution of English government.

In an opinionated account of London's sanitary evolution, the physician Henry Jephson, an exemplar of the rising professional specialist, portrayed the commissioners of sewers, among other ancient institutions, as perpetrators of incredible stupidities. As late as 1845, Dr. Jephson records, no drainage survey existed of the metropolis. There was, accordingly, a different drainage level in each local district, and no attempt to coordinate the works of the several districts to an overall plan. Large sewers, Jephson lamented, discharged into smaller; some of the conduits were higher in elevation than cesspools they were supposed to drain, implying – it would seem – an assumption that sewage might be inclined to flow uphill.[26] His widely shared assessment was that some central authority was essential to overcome the obstacles to an adequate and efficient general sewerage system created by the various bodies of commissioners still operating within the London area.

Crowded, literally covered in generations of cesspools, intermittently supplied with water by private companies whose apparent indifference to considerations of quality was widely noted, and governed by a patchwork of parishes and local agencies, London was ready for reform. As mid-century approached, London's drainage was in the hands of eight independent commissioners of sewers with separate responsibilities.[27] The established local commissions, which had been interested chiefly in flood control and navigation rather than domestic sewerage, were replaced by a Metropolitan Commission of Sewers, whose role was to supervise the construction and maintenance of new sewers and levy rates.[28] The commission's mandate involved more than abolishing local authorities and promoting the integration of a more comprehensive metropolitan management area: it was also to ensure that a survey of London sewerage would at last be undertaken and that every new house in London would be equipped with either a water closet or a privy. Legislation to this effect empowered a new breed of professional public administrators, often with contemporary qualifications in public health or engineering, to promote and encourage reform.[29]

The new administrative structure had its champions and popularizers, one of whom, J. Toulmin Smith, was particularly outspoken in favour of allowing administrative authorities the requisite leeway and flexibility to fulfill the wide range of responsibilities conferred upon them. Smith insisted that effective application of legislation depended upon a proper understanding of its origins and objectives. Perhaps somewhat injudiciously, in an era when emerging principles of statutory interpretation would often be applied to reconcile assertions about the absolute sovereignty of parliaments with equally amorphous claims about the inviolable rights of subjects, Smith contributed the opinion that "the mere words of an Act of parliament" provided little practical guidance. Only "the intelligence of those who have to carry it out" and their awareness of the task to be accomplished could infuse legislation with life and purpose. On account of its "peculiar nature and the very important interests involved," Smith felt that the newly passed *Metropolis Local Management Act* of 1855 represented a classic example of this principle. Originating in remarkable circumstances, the act had to "depend for its good result upon the comprehension by its administrators of those circumstances, and their continual regard to them."[30]

One of the act's intended "good results" was an upgrading in the character of domestic conveniences. In designated parishes, any new housing and certain renovated or reconstructed premises were to be equipped with "a sufficient water-closet or privy and ashpit." Proper doors and coverings were required; in the case of water closets, a suitable water supply, its apparatus, and a trapped-soil pan were to be furnished so as to ensure "efficient operation."[31] Such requirements clearly indicated the strength of official endorsement of the new domestic plumbing arrangements that flowed from improvements in water supply.

The promotion of flushing at the household level represented a significant change in policy, for the practice had formerly been discouraged. In the 1770s, for instance, at least one resident of Bath had been threatened with disconnection of his water supply unless he abandoned his water closet.[32] At a more general level, especially at the time of the *Waterworks Clauses Acts* in the late 1840s, legislation somewhat deterred adoption of water closets by placing an extra charge on the owners of such appliances.[33] The Grand Junction Water Works Company had, around 1820, employed a waiter in a Piccadilly hotel to provide information on water use in the establishment, and had thereafter treated the number of times the water closets were flushed as a surrogate for metering. The hotel's annual rate rose sharply.[34]

The extension of more effective sewerage under the auspices of local authorities encouraged replacement of privies with water closets, although initial progress was sometimes hesitant. The forces of inertia played a part in this, but the campaign to abolish cesspools, encourage adoption of the water closet, and supply adequate sewers encountered other formidable obstacles.[35] Pockets of recalcitrance even emerged, and objections to the new conveniences, especially the water closet, soon made their appearance on the judicial agenda.

The Campaign for Flushing

The District of Wandsworth, the product of the consolidation of six separate parishes south of the Thames,[36] faced an exceptional challenge from its rapid population growth. This eighteen-square-mile jurisdiction, with nine to ten thousand dwellings and a population of fifty thousand at the time of its amalgamation, rapidly increased by twenty thousand, with a further fifty-five thousand residents to be added within a decade. As might be expected with new legislation for a changing environment, local authorities became deeply embroiled in conflict on the water closet frontier. Judicial intervention, just as J. Toulmin Smith had anticipated (though wishing to forestall), was instrumental in the outcome.

In a determined effort to eradicate the more disagreeable residual aspects of the privy and cesspool era, and by corollary to advance the cause of flushing, the Board of Works for the Wandsworth District adopted a general practice of requiring cesspools to be filled in and privies to be converted to water closets wherever sufficient sewerage could be provided. An order to this effect, applicable to thirty-nine rented cottages at Ford's Buildings, Battersea, reached their felicitously named proprietor, Mr. Tinkler. Although he completed certain improvements as requested, Tinkler resisted the installation of water closets, insisting that the existing privies satisfied his tenants and were neither unwholesome nor a nuisance. He added – somewhat pointedly – that there was no sewer into which the drains required by the order might discharge their contents.

These objections failed to impress the board, whose chairman, the Honourable Mr. Curzon, reiterated that the abolition of privies was consistent with the "spirit" of the legislation. He also offered assurances that a new sewer would indeed be completed. When Tinkler still declined to avail himself or his tenants of the new conveniences, the Wandsworth board dispatched a crew of its own to excavate and install pipes in the cottage gardens. When the landlord sought to halt further ameliorative efforts by the Board of Works, the ensuing legal clash checked the pace of sanitary change.

Lord Justice Knight Bruce, adapting himself to the nineteenth-century administrative revolution at a leisurely pace, found some of the provisions of the *Metropolis Local Management Act* remarkable, but nonetheless inadequate justification for Wandsworth's initiative. "The question," he carefully explained, "is not whether they have power to cause or order privies within their district to be put in a proper and decent state, if not in that state." Rather, he emphasized in language conveying a certain suspicion of unfamiliar appliances, the issue to be resolved was whether officials had authority to compel Tinkler to install "the mechanical contrivance of water-closets, with their requisite apparatus, for which he is to find water supply as best he may, instead of the privies (sufficient as privies if kept in a condition proper for such conveniences are) which are upon his land for the purposes of his cottages there."[37] Lord Justice Bruce thought not.

Lord Justice Turner was also unable to find legislative authorization for Wandsworth's resolute campaign against privies. Even if local authorities enjoyed the power to order the replacement of privies with water closets in cases where a requirement to do so could be established, nothing in the act provided any foundation for a general rule to this effect. The board, having acted on the basis of its own policy rather than by statutory prescription, had exceeded its legislative powers. Its actions in the Tinkler matter were therefore judged unlawful. Justice Turner further admonished the Wandsworth administrators: it was "their bounden duty to look well that they keep strictly within their powers, and not . . . be guided by any fancied views of their own as to the 'spirit' of the act by which they are to be governed."[38] This sharp rebuke was by no means the last sign of resistance to Wandsworth's campaign to advance the cause of domestic plumbing.

In the ensuing years of continued expansion, Wandsworth officials repeatedly encountered buildings commenced, or even completed, without previous notice being given to them, as the legislation required.[39] Such circumstances, the Board of Works noted in its annual report of 1861, made it difficult if not impossible to verify drainage arrangements, and threatened to leave some doubt as to whether the foundations would actually permit drainage to occur. Lack of notice also largely precluded any timely rectification order the board might wish to make in response to unsatisfactory installations. Simply disregarding the omission was one option; the statutory alternative in such circumstances was the decidedly crude measure of demolition, a power the board itself recognized as "most arbitrary." Only once, in an extreme case, had it resorted to this remedy, but despite several years during which the legislation had been in force, certain builders continued to embark upon construction without troubling to notify official Wandsworth. Perhaps – regrettably – the demolition authority would need to be exercised once more.[40]

On three separate occasions in connection with housing erected on Lower Atkins Road, Henry Cooper Sr., seemingly having dispensed with the particular nicety of advance notice, received correspondence regarding his omission. On each occasion, Henry Jr., also active in local building, represented his father in negotiations. The third round ended with the board communicating its intent to demolish forthwith in the event of any further violation, in order, one presumes, to preserve something of its authority and to facilitate the inspection process necessitated by the ongoing transition to water closets and central drainage throughout London.[41] In view of this experience, the Cooper family, with father and son allied in local building, were hardly unfamiliar with their obligation to provide notice to the Board of Works in advance of construction.

Yet, in the summer of 1862, Mr. Johnson, a Wandsworth district surveyor, learned – and not by way of the statutory notice – that the Cooper clan had a fourth house under way. Johnson inspected the premises and reported to the board. Taking the view that Cooper was trifling with its authority, this earnest body instructed Johnson to effect the demolition.[42] Johnson immediately engaged a contractor who, accompanied by some thirty men, devoted the evening of 30 July to the task. Between 7:00 p.m. and midnight, Johnson's recruits reduced three weeks' worth of Cooper's work to a pile of bricks and rubble. To add insult to injury, Wandsworth pursued Cooper at the local police court for the cost of the demolition – six pounds, fifteen shillings, and eight pence.

Official triumph over the demolition was short-lived. With the able counsel of A.S. Edmunds, Henry Cooper Jr. asserted that the board's entry upon his building site had been unlawful, a trespass in effect.[43] He testified that he had sent a notice, even if it had never been received. He insisted, in fact, that proper notices had also been sent on the three previous occasions, even while maintaining that he had nothing to do with those houses. Make what you will of Cooper Jr.'s reliability as a witness, and, assuming you are prepared to believe him (a generous concession, I think), set aside any questions his evidence might raise about his failure – given the demolition warning issued after the third of the previous contretemps – to ensure that the fourth notice didn't also go astray![44] These matters were not at issue when *Cooper v. Board of Works for the Wandsworth District* reached the Court of Common Pleas in April 1863, for the judicial search for truth can be remarkably selective.

The trespass proceedings ultimately focused on the legality of surveyor Johnson's entry. The demolition crew entered the Lower Atkins Road property, counsel for the Wandsworth board explained, because the *Metropolis Local Management Act* authorized the demolition. The legislation seemed clear enough in providing that, where a builder had failed to give notice of construction, the board or local vestry authority was authorized "to cause such house or building to be demolished or altered, and to cause such drain, or branches thereto, and other connected works and apparatus, and water supply, to be re-laid, amended or re-made, or, in the event of omission, added, as the case may require."[45] Further provision was made for the officials responsible to recover expenses they might have incurred in carrying out their work. Wandsworth readily acknowledged the arbitrary nature of the power, while explaining its necessity for purposes of ensuring the public good. The public interest required provisions of this nature, the board insisted, to facilitate inspection and thus to compel proper drainage of buildings. The statute provided no monetary penalty, and because of that the builder "set the Board at defiance" and proceeded to build. In the board's view, it had become absolutely necessary to make an example for the public good. It further contended that the legislature had given this power to the board and that its members alone were the "judges of the fitness of this exercise."[46]

Though the statute was clear to read, to the judges it was incomplete. The "mere words" of the legislation, as Smith had described them, had to be understood in relation to a deeply rooted legal principle requiring that anyone who might suffer a loss of property should first have an opportunity to be heard in the matter, rather than from the perspective of officials charged with implementing the act. This was especially so where a judicial body was concerned. Here, to some degree, the Board of Works experienced the same treatment previously accorded the commissioners of sewers, that ancient body whose ostensibly judicial functions exposed it to supervision by the common law courts. In the words of one of the justices, "Although there are no positive words in a statute requiring that the party shall be heard, yet the justice of the common law shall supply the omission of the legislature."[47] Thus, on behalf of a good many eager sanitarian bureaucracies, Wandsworth learned a costly lesson at the hands of the Cooper family. The final tally of £325 in damages and nearly a further £300 in legal costs was a formidable sum; its bitterness only increased with the thought that a comparatively modest procedural interlude might have exposed Cooper's brash disregard for the sanitary agenda and avoided the harsh judicial consequence of the demolition.

Cooper, for his part, seems to have taken the paperwork more seriously after the demolition, using the good offices of his solicitor to provide the requisite statutory notice. By October of 1862, when Cooper was rebuilding, he wrote to advise that the drains would be open for inspection whenever required. Foolishly perhaps, Wandsworth instructed its clerk to insist that "the Board cannot approve of any drains constructed before notice given to them."[48] If there was any consolation to be had, it lay in a subsequent amendment to the *Metropolis Local Management Act,* which provided a monetary penalty in the case of buildings constructed without notice. This more moderate remedy at least reduced the likelihood that confrontations over the interface between private property and public interests at the nexus where private residential drainage met public community sewers would escalate so rapidly,[49] but it did not eliminate sources of friction.[50]

THE PUBLIC FACE OF FLUSHING

The proliferation of public lavatories and public sanitary conveniences, often in especially prominent locations, reinforced flushing's grip on the popular imagination.[51] By mid-century, sewer commissioners were in search of suitable locations around London for a projected 154 urinals.[52]

When George Jennings installed public flush toilets and decorative urinals at the Crystal Palace for the Great Exhibition in 1851, roughly 14 percent of the six million visitors to the exhibition demonstrated a willingness to "spend a penny" for such amenities. Jennings went on to promote "conveniences suited to this advanced stage of civilization." As he explained in his catalogues, "false delicacy" ought not to prevent immediate attention being given to the health and comfort of the populace. Putting the matter still more boldly, Jennings asserted – as his American counterparts were equally inclined to do – that "the Civilization of a People can be measured by their Domestic and Sanitary appliances." He anticipated a day when well-serviced "Halting Stations" would be distributed generously along public thoroughfares. Jennings's own catalogue eventually boasted such installations in thirty-six English towns and in the facilities of thirty railway companies. His public urinals also graced the streets of Paris, Berlin, Florence, Madrid, Frankfurt, Hong Kong, and Sydney, New South Wales, and inspired similarly elegant facilities in other cities.[53]

Sanitary conveniences were by no means as uncontroversial as Jennings might have anticipated. Any number of communities proposing to install public lavatories faced challenges from nearby homeowners and others who anticipated or experienced detrimental impacts on their properties. The Vestry of St. George encountered such an objection in 1863 when it approved construction of a urinal in Grosvenor Place, adjacent to Buckingham Palace. Vice Chancellor Stuart accepted the argument of a neighbourhood resident that the proposed facility would constitute a nuisance, and issued an injunction against it. On appeal, local officials challenged the nuisance finding and succeeded in having the court's order set aside.[54] Nevertheless, the judicial boundary between private nuisance and public convenience was frequently tested in similar circumstances for decades thereafter.[55]

Notwithstanding setbacks such as *Tinkler, Cooper,* and urinal controversies – where courts insulated property owners against the incursions of any public health authority lacking a clear legislative mandate[56] – the continued advance of the water closet frontier was assured. Not being up to the task in the *Tinkler* controversy, the particular section was repealed. New provisions explicitly precluded the use of the privy as an alternative to the water closet, except where a sufficient water supply was "not reasonably available" and in other limited circumstances. And there were positive developments even from the courts. Where an existing privy was inadequate for the task at hand, the sanitary authority could indeed prescribe a water closet,[57] even though authority to specify a particular type of fixture was denied.[58] Given Lord Justice Knight Bruce's endorsement in *Tinkler* of privies as "sufficient . . . if kept in a condition proper for such conveniences," it was of considerable assistance to the water closet movement that magistrates were denied jurisdiction to assess the sufficiency of existing accommodations. The sanitary authority's decision in that regard was binding. For their part, magistrates may have been grateful to be relieved of this sensitive responsibility.[59]

In the United States as well, devotees of the privy system who challenged official and administrative measures aimed at depriving them of facilities to which they had become accustomed were ultimately overwhelmed by the advance of the water closet. Dr. Henry Bixby Hemenway, a prominent American public health analyst and physician, was therefore pleased to report in his 1914 *Legal Principles of Public Health Administration* that legislation providing for summary destruction of privy vaults, even pending appeal, had survived late nineteenth-century constitutional scrutiny in the United States.[60]

The privy pit or vault nevertheless remained a familiar and formidable presence for decades after vigorous assaults upon it in Victorian Britain and the United States. In inventorying the sanitary landscape in the 1880s and 1890s, for example, the Illinois State Board of Health found that inside water closets were making little headway in towns such as Rockford, Galesburg, and Decatur. Most of the population of Champaign, living in close proximity to a university where sanitary issues were under intense scientific investigation, still resorted to outside toilets in 1895. The same was true of Urbana.[61] Indeed, on the wider American scene, a 1918 survey indicated that roughly 40 percent of houses and 20 percent of apartments retained outdoor privies.[62]

The gradual pace with which water fixtures, closets in particular, were adopted, and the presumption that they were essentially private in nature and without implications for the wider community, at least partially accounts for the general absence of plumbing standards until fairly late in the century.[63] Yet by about 1890, after a decade in which the number of municipal waterworks had trebled in the United States, and with running water increasingly regarded as a household necessity,[64] plumbing had become an integral component of interconnected systems of urban sanitation. Their safety and successful operation was a matter of public importance. Concerns included the prospect of organic matter from cesspools oozing out to permeate wells; toilet soil settling around construction, causing drain pipes connecting the soil pipe to the cesspool to rupture; and interior pipes, damaged by settling or shrinkage or inexpertly joined, allowing "sewer gases" to escape. Under the leadership of sanitarian reformers, the campaign for scientific plumbing became something of a crusade to secure expert supervision of domestic flushing technology and design.[65]

In this atmosphere, the legal framework governing domestic appliances quickly extended well beyond the realm of common law principles and general legislation. An array of by-laws, regulations, and ordinances often prescribed in detail the technical requirements for water closets. Other regulations addressed the manner in which particular installations might be undertaken and by whom. Having provided public lavatories or sanitary conveniences, local authorities also sought power to regulate the management of these facilities. They set out to enact by-laws "as to the decent conduct of persons using the same," to lease them out, or to charge fees "for the use of any lavatories or water-closets provided by them as they think proper."[66] This concession to pay-as-you-go elimination and flushing recognized a principle that had been pioneered at least as early as the Crystal Palace Exhibition.[67]

Having raised the alarm, sanitarian reformers set out to identify solutions. Traps and ventilation systems became an important focus of attention, as these techniques, when properly implemented, offered dramatic increases in domestic protection against the widely feared perils of sewer gases. An earlier generation of water closets plagued with mechanical idiosyncrasies – and often dangerously inefficient – also came under close scrutiny. Ameliorative reforms in design and construction involved simplification and the elimination of superfluous moving parts, the replacement of metal components with streamlined china and earthenware, and a severe dose of flushing involving "a very copious use of water."[68]

In this context, plumbing practices could no longer be regarded merely as matters of private preference. There were social, structural, and organizational implications, at least at the civic level. Sanitarians therefore pressed municipal leaders to oversee plumbing, their goal being "a scientifically designed water carriage sewer system," in which "wastes flowed directly and smoothly from point of creation to point of disposal." The preconditions of success were a generous supply of water carried through an integrated system combining waterworks and sewer mains joined by carefully made domestic plumbing connections.[69] The authors of a medical commission report for Boston in the 1870s articulated the principles of efficient flushing in typical fashion. It was considered desirable that "all the various arrangements which have in view the removal of sewage should, from the beginning to the end of their course, constitute, as far as possible, parts of one common whole, systematically devised for harmonious co-operation, and placed under one common jurisdiction."[70]

Efforts to promote common standards, consistency, and reliability gained strength. These involved inspections and ultimately compliance requirements for domestic plumbing arrangements. In Britain, for example, the London County Council undertook the detailed regulation of water closets with considerable zeal. Each water closet (or urinal) was to be furnished with a flushing cistern separate from any cistern allocated to drinking water; the service pipe had to connect to the flushing cistern rather than to any other part of the closet; for the pipe connecting the cistern with the pan, a minimum diameter of one and a quarter inches was prescribed; the pan or basin itself was to be fabricated of non-absorbent material that was shaped, sized, and constructed in order to contain sufficient water and to allow "all filth to fall free of the sides directly into the water." Further specifications sought the "prompt and effectual" removal of filth, requiring for this purpose "an efficient siphon trap constructed to maintain a sufficient water seal between the pan and the drain or soil pipe," forbidding other arrangements, and insisting upon suitable ventilation by means of an anti-siphonage pipe "not less than 2 in. in diameter, and connected to a point not less than three and not more than 12 in. from the highest part of the trap."[71]

Regulation frequently reflected distinctive local influences. The proliferation of localized and often idiosyncratic regimes eventually inspired efforts to promote greater consistency. Thus, the British Institute of Water Engineers constituted a committee in 1900 to standardize and codify plumbing regulations. The Worshipful Company of Plumbers joined the effort to promulgate common practices in 1903, and within half a decade published model by-laws. Britain's Local Government Board followed suit in 1911 with recommended by-laws of its own. These, detailing the thickness of copper to be used in the ball float and specifying the material of the lever, among other particularities, provoked sufficiently widespread condemnation that the British Waterworks Association, the Ministry of Health, and then the British Standards Institute all took their turns at formulating common standards.[72]

Local regimes prevailed equally in the United States.[73] In 1891 the New York City Board of Health announced the successful eradication of cesspools and outdoor privies.[74] New York's model *Tenement House* legislation of 1901 required a water supply to be made available on each floor of new buildings and subsequently for each apartment. Other major cities – San Francisco, Baltimore, Cleveland, Pittsburgh, Chicago, Boston, and Philadelphia – soon emulated the initiative by requiring a water closet for each family or for every three rooms.[75] An elaborate framework regarding health and building in New York City incorporated provisions from the Sanitary Code, 1903, and the New York Charter of 1897. These operated in conjunction with legislation concerning plumbing in New York City, a much-amended statute falling within the purview of the buildings department.

Although generally similar in substance to requirements then current in London, American by-laws on the construction of water closets tended to prescribe more rigid controls on the way the necessary work was executed, with New York offering the most stringent example. To obtain approval from the Department of Buildings – a precondition of repairs and alterations as well as original construction – complete drawings and particulars of the materials to be used had to be submitted. The work itself could be carried out only under the supervision of a registered plumber – registered, that is, by means of an annual certificate of competency from the city Examination Board. To emphasize the mandatory nature of these drainage and plumbing regulations, a fine of $250 and the prospect of three months' imprisonment might be imposed on anyone convicted of violating the requirements. A certified master plumber would, in addition, forfeit his certification.[76] In due course, plumbers elsewhere were brought within the scope of licensing regimes, and were expected to apply for approvals in connection with any openings they had contracted to install.

There was much educational work yet to be done, with implications ranging from public health outcomes through the architectural profession and construction trades. William Spinks reminded any readers of his 1895 treatise on drainage (a somewhat select audience, one imagines) that plumbers, like other tradesmen, worked with specifications supplied by others, and "so long as architects are content to specify in general and vague terms, so long will there be defective internal sanitary arrangements." Nor were plumbers responsible for drainage, a service plagued, in Spinks's view, by entirely too much "jerrying." Spinks longed for the day "when our sanitary authorities will have codes of regulations for drainers quite as stringent as those in vogue for gas and water purposes." Hopefully, he anticipated a time when "we shall hear less of putrid sore throats, typhoid, and other drain-begotten diseases." Spinks insisted that plumbers were "not an unimportant wing of the army which is waging war against zymotic diseases," but were unlikely to enjoy the justice and respect due to them "until the public can be induced to see the rottenness of the tender system in [this] branch of the building trade."[77]

Municipal sanitary enforcement was by no means wholly reliant on the regulation of plumbers or on the voluntary compliance of residents.

Door-to-door inspections permitted a high degree of follow-up monitoring. Ohio legislation, fairly typical of the genre, mandated sterner alternatives when orders from health officials were neglected or disregarded, either in whole or in part. In such circumstances, the health board had authority to arrest and prosecute offenders, or even to employ its own personnel to undertake repairs. Before proceeding to escalate the level of intervention in this way, officials were required to follow procedural niceties. A citation concerning the cause of the complaint and inviting attendance at a hearing on the matter was to be served personally on responsible parties residing in the jurisdiction of the board, or delivered by registered mail to non-resident owners.[78] Evidently, the lesson the Coopers taught Wandsworth had been learned.

New York around the turn of the century was among the cities that most systematically enforced the sanitary impulse. Upon his election as mayor in 1894, William L. Strong recruited the charismatic Colonel George Edwin Waring Jr. to spearhead reforms. Waring, who had earlier gained prestige for innovative sanitary sewer systems in Lennox, Massachusetts, and in Memphis, Tennessee, put his staff in uniform and placed householders and other waste producers on notice as he industriously attacked an array of urban problems.[79] In 1896, for example, forty-seven patrolmen, two roundsmen, and one sergeant from the regular police force temporarily supplemented the ranks of the sanitary force for their house-to-house visits. Following the inspection of 42,909 houses, 38,858 nuisances were reported abated.[80] Enforcement and implementation, however, remained decidedly uneven, both between cities and within them, for custom and cost presented formidable obstacles. Investigators concerned about the disproportionately high incidence of typhoid amongst the immigrant population in the vicinity of Chicago's settlement service, Hull House, for instance, determined in 1902 – well after the formal abolition of privy pits and undrained vaults – that 48 percent of dwellings in the area lacked modern plumbing.[81]

In Toronto the remedial agenda included an array of "nuisances," defined initially with reference to common law principles as "anything which is injurious to health, or which is materially offensive to the senses, or which interferes with the enjoyment of life and property."[82] The statistical record suggests that local boards of health relied heavily upon inspections, educational measures, and warning notices to prompt remedial activity or abatement measures without going to court. In 1889, 6,000 privy complaints prompted 18,000 visits by Toronto's eight sanitary police officials, leading to 1,400 notices and 5,500 post-notice inspections. Of the

234 prosecutions initiated, 170 were withdrawn on the grounds that the necessary work had been done. Two thousand privies were cleaned, 113 repaired, and 143 closed.[83] In 1893, there were 3,473 privy pit complaints in Toronto. Of these, 2,369 were abated on notice; 473 were abated on prosecution; and in 222 cases, officials concluded that the complaints were unfounded.[84] Perhaps indicating the success of such measures, only ninety proceedings were instituted in police court the following year on the basis of 1,474 complaints about privy pits.[85]

Toronto officials found the situation so difficult to resolve on a case-by-case basis that, like their counterparts elsewhere, they resorted to prohibitive by-laws as a more comprehensive response to privy pit complaints.[86] The necessity for abolition came to the attention of Dr. Norman Allen, the medical officer of health, when he realized that approximately two-thirds of the complaints that were received concerned privy pits. The warning notice, the visit, and the ensuing abatement, however, produced only short-term amelioration, causing Dr. Allen to observe, "Sixty to sixty-five per cent of the work of this department is of no permanent benefit."[87] He therefore argued that "privies should not exist in cities." Dr. Allen could find no objection other than expense to his crusade, and handily discounted that obstacle when set against the saving of hundreds of lives. In his plea for a by-law prohibiting privy pits, he was supported by several associates – "eminent sanitarians" – who petitioned city council to attend to these "disgusting relics of barbarism," or, in the words of a colleague, to adopt the principle "'tout à l'égout' everything to the sewer."[88] In a remarkably short time, the privy pit, piously endorsed by judicial pronouncement at mid-century at the centre of the Victorian empire, had become a "disgusting relic of barbarism" slated for wholesale removal from the confines of what passed for urban civilization in colonial Toronto. These, despite familiar pockets of resistance, were ultimately to be banished in the interests of public health.

The water-borne removal process that replaced the privy vault and the cesspool systems of waste disposal differed significantly from its predecessors in important respects. Whether efficient or not, the cesspool model was locally based, labour intensive, and in general dependent on private initiative to ensure any degree of maintenance.[89] By comparison, the water-based removal system operated automatically, thereby eliminating dependence on the householder's decision to have the wastes removed. Public services replaced personal responsibility and independence.

With the advent of municipal water supply – either public or private – and especially once water-borne domestic waste removal had been adopted to replace privy pits and cesspools, such autonomy as had been inherent in the individualized water supply and household waste arrangements of the early nineteenth century underwent massive erosion. The associated sensitivities, accompanied by sometimes formidable expenses, underlay skirmishes such as *Tinkler* and *Cooper*.

Of equal if not greater importance, the replacement of household waste management by the water-borne removal of domestic sewage had implications at the community level. The new arrangements, because of their expense as well as their significance for community health, were public undertakings, usually at the municipal level.[90] The shift had significant consequences, certainly apparent in Dr. Hemenway's celebration of the demise of the privy, but even more graphically expressed by Henry Jephson, in whose mind the *Metropolis Local Management Act* had ushered in a period of war. The "very just and necessary" conflict had as its goal "to enforce upon land-owners, and house-owners and house-middlemen, obedience to the principle that 'property has its duties as well as its rights,' and that those individual rights should not be exercised – as they had hitherto so cruelly been – to the mortal injury of vast numbers of the community."[91]

These were important principles involving the priority of individual and collective interests, and engaging the obligations as well as the rights of property holders. The transition implied activity at virtually every level where the interests of citizens were directly involved. But it was much less influential as a factor in protecting waterways. An ironic consequence of the municipal water supply systems, often touted as enhancements to civic autonomy, was a dramatic increase in municipal interdependence as the water closet displaced the privy. Nineteenth-century communities, having adopted water supply systems at an accelerating pace, were then challenged to address the implications of removing wastewater.

5

Municipal Evacuation

MAJOR EUROPEAN CITIES OFTEN SEEMED caught off guard by the problem of removing wastewater from their confines, or were apt to disregard the challenge. Similarly in the United States, despite the proliferation of water-using appliances in hundreds of communities during the first three-quarters of the nineteenth century, "no city simultaneously constructed a sewer system to remove the water."[1] Things were much the same in Canada: wastewater arrangements seldom accompanied the introduction of water supply. When Dr. Henry Bixby Hemenway, a leading American analyst of public health, casually remarked in his 1914 *Legal Principles of Public Health Administration* that "the subjects of water supply and of sewage and garbage disposal are intimately associated," his matter-of-fact understanding of the relationship between water supply and sewage reflected a significant transition.[2]

Although urban infrastructure was altered profoundly by the engineering of extensive sewerage channels, the transformations brought about by flushing were more than merely physical. The decision to flush municipal waste into available waterways – with all its consequences – enjoyed support in most quarters. But there were strong voices of dissent among those wishing to develop or preserve an alternative – the conservancy model of waste recycling. Some observers and economists especially regretted the loss of valuable manurial resources. Others objected to seeing waterways, including coastal waters, burdened with unremitting waste flows. New legal and technological measures to ensure the control and treatment of sewage reflected these perceived losses – both the foregone benefits of human waste and the adverse impacts of waste streams on the interests of downstream riparians.

Filth, Miasma, and Urban Reform

Why were the interconnections between water supply and sewerage not more apparent at the outset? Or, if they were recognized, why did they fail to command greater attention? The question is worth a moment of reflection.[3] Perhaps it was assumed that a continuous municipal water supply would not actually produce a wastewater problem. The limited usage previously associated with cisterns, wells, and manual deliveries had offered few hints about the enormous levels of water consumption that lay ahead. In these circumstances, conventional disposal arrangements may have appeared satisfactory, for long-established drainage practices, though frequently problematic, were not universally inadequate. Unpaved nineteenth-century surfaces had a significant capacity to absorb liquids. And, insofar as household refuse and domestic waste were concerned, teams of scavengers and professional cleaners had proven reasonably reliable for some time. Privy pits, cesspools, and refuse heaps drained gradually into the soil below and were accessible for occasional servicing and waste removal – a household chore, or a matter for the nightman.

These arrangements, sometimes referred to as the conservancy system, were consistent with an outlook that was still widely entrenched as new water supply systems and domestic appliances came on the scene. In an agricultural context, animal wastes, linked through their nutrient value to the natural cycles of growth, decay, and renewal, constituted a valuable resource. The bodily wastes of growing urban masses, it appeared to a significant constituency, could also be connected to the natural processes of regeneration and renewal, and, not incidentally, to economic opportunity. The nineteenth-century abandonment of the conservancy model in favour of water-borne waste removal substantially altered the natural, social, and legal environments.

The advantages of water-borne waste removal seemed compelling from the perspective of contemporary knowledge. Most notable were arguments derived from an influential stream of medical or public health understanding which associated filth and the miasma that arose from it with the origin and spread of disease. A description of the effects of an irrigation project near Edinburgh illustrates how miasma had taken hold of the popular imagination. It reads like a fairy tale, albeit of the darker variety:

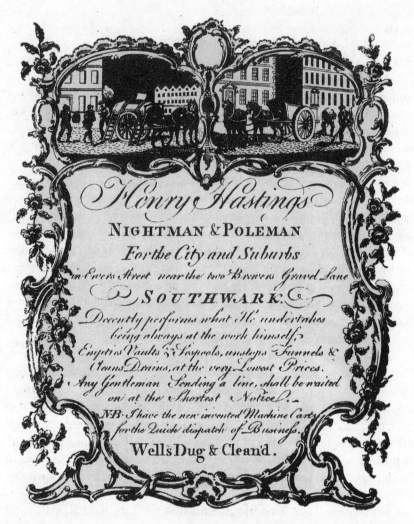

This advertisement for the services of an individual who undertook to clean privy vaults and cesspools illustrates the degree of personal responsibility involved in such labour prior to the introduction of municipally financed water-borne waste removal systems. It reads: "Henry Hastings, Nightman & Poleman For the City and Suburbs in Evers Street near the two Brewers Gravel Lane Southwark Decently performs what he undertakes being always at the work himself, Empties Vaults and Cesspools, unstops Funnels and Cleans drains at the very lowest prices. Any gentleman sending a line shall be waited on at the Shortest Notice. N.B. I have the new invented Machine Carts for the Quick dispatch of Business. Wells Dug & Clean'd."

"These rank and fetid exhalations . . . poison the air for miles around. They are insufferable to passengers, and to those living in the neighbourhood. They are carried by the winds into the City – into the Palace – and into the Barracks; and after being condensed in the atmosphere by the evening's cold, they fall down in the form of damps, bringing with them sickness and disease."[4] Martin V. Melosi deftly summarizes the sometimes confusing relationships among several associated descriptions of disease theory in this era: "The so-called filth, or miasmatic, theory dominated the thinking of sanitarians until late into the century. Because disease was understood to arise from putrefying organic wastes, bad smells (miasmas), and sewer gases – and could not be transmitted from person to person – the filth theory is described as anticontagionist."[5] Because decaying organic matter was perceived to be harmful – either poisonous in itself by causing disease directly or weakening resistance – the prospect of removing such materials or "filth" from centres of population became a most desirable goal for sanitarian health reformers.[6]

The strength of the convictions underlying the diversion of waste into waterways was rarely challenged by contemporary understanding of water quality. A notable early forum on water policy was provided, however, by a public inquiry into the water supply of London established in 1828 in the aftermath of John Wright's condemnation of the Dolphin facility. This royal commission represented "the first significant discussion in Britain of what standards of quality ought to be expected of a public water supply."[7] By helping to frame fundamental considerations of policy, the process served, in historian Christopher Hamlin's words, as a dress rehearsal for subsequent debate: What level of quality, explicitly if not yet conclusively linked to concerns for life and health, could urban residents expect? To what were they entitled? And how would a standard of quality, once agreed or ascertained, be met? Well-recognized advocates were already in evidence: water companies, civic leaders, scientific specialists, and a few medical men, soon to be joined by municipal reformers, engineers, economists, and the proponents of various forms of public finance. Their debates about water reflected the divergent streams of contemporary thought.[8]

Dr. William Lambe was a prominent advocate of the theory that decaying organic matter (with which the Thames was plentifully supplied) was responsible – directly or indirectly – for extensive ill health throughout London. The water quality studies of the day were generally preoccupied with the significance of dissolved salts in mineral waters, but Lambe urged a shift in focus to the problems of organic contaminants in public water supplies, notably the Thames. Convinced that organic matter – or the processes of its decomposition – caused or in some way contributed to disease, he further argued that the techniques of conventional analytical chemistry were inadequate to the task of detecting the presence of organic matter. He was adamant in taking the latter point to its penetrating conclusion: "It is vain . . . to say, that where nothing is discovered there is nothing wrong."[9]

As much as Lambe's medical perspective on organic contamination – his insistence on the importance of water quality for public health – appears to mark something of a breakthrough, as Hamlin observes, its actual impact was modest. Lambe was by no means the first to speculate about such linkages, and his forerunners had failed to transform conventional practices. Moreover, even to the extent that warnings about putrefaction were accepted, the remedial response was limited indeed.

Why fouled water supplies were not more vigorously protested begs some explanation beyond reference to limitations in scientific understanding.[10] Preference for – or easy recourse to – other beverages must at least partially account for lack of concern. And people were perhaps sufficiently familiar with different waters, with all their various impurities, to appreciate that, even if unfit for drinking, grossly impure water might be suitable for industrial uses such as dyeing, bleaching, and tanning. Few people had any real expectation of "pure" water, for purity was a matter of degree. It may also have been assumed to be within the competence of ordinary folk to assess water quality simply by using their senses of sight, smell, and taste. *Rees's Cyclopedia* reinforced this supposition with the anodyne assurance that the classification of waters "according to their sensible properties, coincides likewise, as well perhaps as the present state of the subject will admit, with their chemical and medicinal properties." Moreover, it was assumed that concerns with quality could be addressed on the premises at the domestic, household, or manufactory level through filtration and chemical purification schemes.[11]

What may have been intended by "pure and wholesome" as the phrase took its place in waterworks legislation is equally of some interest. The metaphorical standard, "pure and crystalline in appearance as that of a mountain spring," was certainly in use, but often already well beyond realistic expectation.[12] Had controversy put the phrase before the courts, several alternative approaches might have been advocated. To the common law, evidence that cows were willing to drink it was indication enough of a source's wholesomeness. Alternatively, the ordinary and plain meaning of the words might have been urged on the bench; common sense – or more to the point, the common senses of sight, smell, and taste – would answer the question. Or, in litigation, some party might have pressed the court to accept a technical and specialized meaning, suggesting that the advice of chemists of the day to water companies should govern the matter. Courts might even have been persuaded to defer to legislative intention as to the meaning of "pure and wholesome." Should the latter approach have prevailed, despite its rather abstract character and the enduring conceptual challenge of determining how a legislative body composed of numerous individuals intends anything, we would then face the task of infusing legislative intent with more tangible content. Let us credit our mid-nineteenth-century parliamentarians with a defensible, perhaps even commendable, desire to adopt the best professional advice available. What would their advice on the vital question of water quality have been?

At mid-century, even in the aftermath of the latest outbreak of cholera, attention remained focused on the hardness of water.[13] Many advocates of water supply reform invoked this criterion as the basis for finding alternative sources to the Thames.[14] On the other hand, to a new breed of specialist, the devotees of microscopy exemplified by physician and chemist Arthur Hill Hassall, the perilous nature of Thames water could most readily be appreciated from an inventory of the organisms inhabiting it. The theory held that such organisms were harmful in themselves or that they signalled the presence of other harmful matter. At the very least, the presence of microscopic organisms in the water supply was disgusting. A most effective publicist for his cause, Hassall magnified his findings in graphic illustrations, rendering the latter proposition almost irrefutable.[15] Yet another expert perspective on the quality of Thames water was advanced by a panel of eminent chemists specifically commissioned to advise government; their report, delivered in June 1851, found no indication that current sources of supply in the Thames and Lee represented any danger.[16] Confronted with this contradictory array of professional opinion[17] several years before pioneering epidemiologist Dr. John Snow was to demonstrate the devastating impact of water from London's Broad Street pump in an 1854 cholera outbreak, the government of the day finessed the quality question.

A North American distillation of the water quality knowledge of the age appeared in the submission of "Ke-see-nah Zi-bing" to the Toronto waterworks competition. The authors turned out to be none other than Henry Youle Hind and Sandford Fleming, not yet the prominent figures they would become in the wake of Hind's explorations of the Canadian North-West and Fleming's contribution to transcontinental railway building and standard time. These gentlemen launched their proposal on a high note, underlining the importance of their subject with rhetorical poignancy:

"No want of society has been so urgently advocated, so thoroughly investigated, or so wretchedly neglected," they began, lamenting those situations where "the indolence or the cupidity of man refuses the small exercise of his energy which would suffice to bring it to the threshold of the poor, as well as the rich; of the hospital as well as the palace."[18]

In addressing quality, Hind and Fleming invoked the opinion of Henry Croft, the former's colleague in the Department of Chemistry at the University of Toronto. Torontonians, he said, though lacking plentiful mineral springs, could congratulate themselves on possessing "lake and mineral water of a greater degree of purity than almost any other part of the world." Croft described Lake Ontario water as "of most extraordinary purity,"

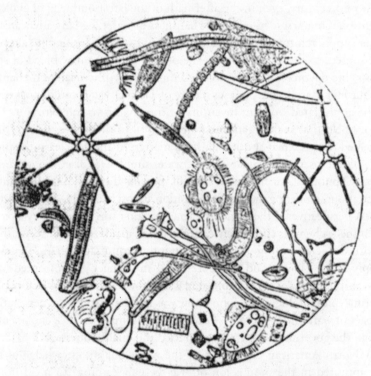

This engraving represents the chief animal and vegetable productions contained in the water as supplied by the *Grand Junction Company.*

The work of microscopists such as Arthur Hill Hassall, who magnified his findings in graphic illustrations, as shown here, gave rise to considerable apprehension and alarm about the condition of mid-nineteenth-century water supplies, yet scientific understanding remained limited, and the discovery of bacteria still lay ahead.

DEATH'S DISPENSARY.
Open to the Poor, Gratis, by Permission of the Parish.

"Death's Dispensary," 1860. Even if the actual process of disease transmission remained unclear, urban populations were apprehensive about their vulnerability to contaminated water supplies.

adding that the waters of the upper Great Lakes would be "found to be still more free from extraneous matters."[19] Hind and Fleming added the caution that "It is impossible to judge by the eye alone of the purity of water," signalling the unreliability of personal judgment. The authority of the expert should have precedence over the untutored preferences of ordinary consumers, for, "A perfectly clear and apparently pure water, may, and frequently does, contain, in solution, pestilential organic matter which breeds dysentery and other dangerous diseases."[20] Hind and Fleming proceeded to draw their readers' attention to the more revolting manifestations of the "vital vigour" to be found in Toronto's open reservoir. Myriad frogs and an abundance of aquatic beetles in all stages of development disported about with "countless millions of microscopic animalculae." The presence of these intruders was explained with the suggestion that they "were pumped into their present congenial abode from the bay, a nauseous comment upon the purity of the water which has been doled out to the Queen City of the West for years past." Unable to resist a more graphic portrayal of the unsatisfactory condition of Toronto's water supply, the authors pointedly referred to the occasional "passage of a leech from the tap."[21] Such disconcerting experiences were not unknown to other communities.[22]

To the economic and salutary dimensions of their philosophy of waterworks, Hind and Fleming introduced a moral element. The influence of water supply upon morality, they argued, was as powerful as its influence upon health. "Among the masses of mankind, uncleanliness is always associated with vice; and where unavoidable uncleanliness exists, demoralization is sure to prevail. There is no more positive indication of human progress – in the simple and rational acceptation of that commonly misapplied phrase – than a due attention to those inestimable blessings which accompany a copious and unrestricted supply of pure water."[23]

It was not uncommon during this period to associate moral deterioration with surroundings, although the direction of the causal compass was subject to fluctuation. To the Reverend Charles Kingsley, author of *The Water Babies*, unsanitary practices might well provoke the intervention of God: "Filthy and unwholesome habits of living are in the sight of almighty God so terrible and offensive, that He sometimes finds it necessary to visit them with a severity with which he visits hardly any sin; namely, by inflicting capital punishment on thousands of His beloved creatures."[24] Yet, as American urban historian Stanley K. Schultz compellingly explains, a reformist variant reversed the polarity: immoral conduct might itself be transformed through improvements in the conditions of life. Moral environmentalism offered solutions, or at least an optimistic course of action, in response to problems that appeared intractable to those who found the roots of social misconduct in human nature. "If the physical environment strongly influenced moral behaviour, genuine social change was possible. Human action could alter the face of the city and reshape the moral health of the urban populace."[25] Against the backdrop of evolving yet inconclusive scientific appreciation of water quality and its relationship to human health, and alongside various strains of agitation for civic and moral improvement, water-borne waste removal secured deep urban foundations.

The Great Stink

In his influential report *On the Sanitary Condition of the Labouring Population of Great Britain* (1842), the industrious Edwin Chadwick, responding to apprehension about the miasmatic implications of accumulating waste and refuse, championed a solution. He envisaged a vast publicly supported network of arterial drainage through which a continuous flow of water would carry off sewage and organic waste before the perilous process of decay, decomposition, or putrefaction – with its attendant atmospheric dangers – got under way.[26] Although conceived on a vast metropolitan scale, Chadwick's infrastructure plan built on earlier foundations. As early as 1791, Jeremy Bentham, the utilitarian social critic and philosopher, had advocated a network of glazed tile pipes to promote waste carriage. Three decades or so later, around 1820, John Roe began implementing his "arterial-venous" plan for the Holborn Sewers Commission. Round and tubular pipes with smooth tile lining and a small circumference ensured that curves replaced sharp corners and could be flushed regularly.[27] Chadwick had been personally tutored by Roe, who introduced him to some more notable features of London's deteriorating underground drainage, and to the mechanics of providing a constant flow of water to remove waste.[28]

The scale of the technological challenge necessitated an elaborate and centralized administrative structure. In the words of the leading legal historians of modern England, Chadwick's plan was "the most uncompromising proposal of the mid-century years for collective governmental action and it attracted opposition of a fervour correspondingly unbridled."[29] Chadwick's determination to replace local officials with a central authority was a source of continuing friction until he was dislodged from the General Board of Health in 1854. While there, however, he succeeded with gratifying results in introducing underground hydraulic sewers on an experimental basis to replace conventional surface drainage.[30]

London's common sewers, having been designed to serve as storm drains, were reserved for surface water, and city dwellers had been prohibited from using them to discharge any waste other than kitchen slops. As of 1815, that prohibition was lifted, and London became the first metropolitan area to allow its inhabitants to connect their cesspools, where privy wastes accumulated, to the common sewers.[31] When combined with official measures to promote the adoption of water closets in the rapidly increasing residential stock of the capital, the London initiative contributed greatly to the continuing deterioration of the Thames. The 1840s alone saw the construction of forty-three thousand new houses.[32] At the time of the 1851 census, roughly two and three-quarter million Londoners lived along the banks of the Thames or within the scope of its natural drainage. With water closets discharging from downtown London almost directly into the river through historic common sewers originally intended just for surface waters, it was estimated at mid-century that the Thames served as depository for 250 tons of fecal matter each day.[33] Given the tidal character of the river at this point, actual dispersal was limited: much sewage simply accumulated on tidal flats in the vicinity of the outfalls.[34]

Having authorized sewerage of domestic wastes, London officials and sewer commissioners soon proved incapable of alleviating the chaos. Henry Mayhew, a dedicated observer of London social life who became one of the founders of *Punch* magazine, described underground conditions in 1848. The accumulated deposit consisted of all manner of waste from London's breweries, gasworks, and chemical and mineral manufactories. There was no shortage of dead dogs, cats, kittens, and rats, together with offal from slaughterhouses, "sometimes even including the entrails of the animals." Vegetable refuse abounded alongside stable-dung, refuse from pigsties, night-soil, and ashes, all mixed with dirt and horse-droppings washed from the streets. Approaching the more solid end of the spectrum, the inventory included tin kettles and pans, broken stoneware, building bricks, slabs of wood, rotten mortar, and other assorted rubbish. Mayhew reported further on one wretched sewer leading to a thirty-foot chamber. From its roof, he wrote, "hangings of putrid matter like stalactites descend three feet in length." At the end of the chamber, the sewer passed under public privies whose ceilings were visible from the author's vantage point below. "Beyond this it is not possible to go."[35] Nor, for atmospheric purposes, need further description be added here of the city's offering to the Thames.

The Metropolitan Board of Works, successor in 1855-56 to the ancient sewer commissions, soon solicited proposals to restore the Thames to something more closely resembling water. The competition garnered 116 submissions, none of which proceeded to implementation.[36] Parliamentary attention was not firmly re-engaged until 1858, the summer of the Great Stink, when the stench from the river rendered legislative activity decidedly unpalatable. With the windows of the Houses of Parliament draped in sheets drenched in chloride of lime (generous portions of which were also dumped directly into the river) and with some talk of relocating the Law Courts to Oxford,[37] decision makers empowered the Board of Works to address London's plight and to do so expeditiously. Queen Victoria welcomed the statutory measure passed "for the purification of that noble river, the present state of which [was] little creditable to a great country, and seriously prejudicial to the health and comfort of the inhabitants of the metropolis." Her Majesty, ensconced for much of the Great Stink on the Isle of Wight, emphasized that "the sanitary condition of the metropolis must always be a subject of deep interest."[38]

The same year, Benjamin Disraeli, as chancellor of the exchequer, introduced an amendment to the *Metropolis Local Management Act* that made way for a scheme to re-engineer the Thames shoreline through central London. The resulting embankment of the Thames created a corridor suitable for drainage facilities, and, by confining the river to a narrow channel, had the further effect of improving its flow.

The actual job of supervising the massive sewerage undertaking fell to the energetic and visionary Joseph Bazalgette, first chief engineer of the Board of Works.[39] In this capacity he directed the board's effort to construct arrangements to eliminate the discharge of sewage into the Thames within the confines of the metropolis.[40] While *Punch* observed at the expense of the board that progress was "slow but sewer" and that from the perspective of finance a "Stinking Fund" might be of assistance,[41] Bazalgette and six thousand labourers pursued the undertaking at an ultimate cost of £4.6 million. The project resulted in a network of main sewers north and south of the Thames, designed to intercept discharges that had previously flowed directly into the river within the city and to convey the wastes to downstream reservoirs at Barking and Crossness that were emptied according to a schedule calculated to catch the outgoing tide.[42] In 1860, before the outfall sewer was completed, the Metropolitan Board of Works had created stormwater outlets that entered the River Lee directly. Due to delays in constructing the outfall sewer, these stormwater outlets were injudiciously used to discharge sewage into the Lee rather than the Thames. The resulting accumulation of waste in the Lee provoked the trustees of the River Lee to take legal action based on two concerns: the threat of sewage sludge to navigation on the Lee and the growing incidence of disease in the neighbourhood of the discharge.

Vice Chancellor Sir Page Wood declared an unfavourable disposition toward London's sewerage enterprise. Even on the assumption that members of the board were acting to benefit the public and sought no personal pecuniary advantage, the judge found their manner of operations entirely unacceptable. He was certain that these gentlemen had been advised that it was doubtful that they could, without authorization from the Lee conservators, divert the entire flow of sewage into that waterway. They had acted, he believed, in a clandestine manner and without communicating with the conservators. Attributing their conduct to the all-too-common character of public agencies, the vice chancellor remarked scathingly upon their "disposition to act in an arbitrary manner, and to treat it as an impertinence on the part of others to interfere with their proceedings."[43] It was therefore not uncommon for such an institution to disregard "that reasonable course of conciliation, and free and frank communication of their plans which this Court always looks for and generally finds on the part of private individuals."[44] However extensive it may have been, he judged, the statutory authority of the Metropolitan Board of Works was far from adequate to excuse its performance. Bazalgette nevertheless completed his assignment, earning respect from his peers – if not the gratitude of future generations.[45]

Five hundred guests gathered for a salmon dinner to mark the formal opening of London's main drainage system. Among the dignitaries was the Prince of Wales, who presided over the inauguration of steam pumps at Crossness, the southern outfall. The archbishops of Canterbury and York joined those celebrating a feat of engineering that, although benefiting the metropolis, signalled the further desecration of the Thames below London.[46] By the end of the century, the completed infrastructure was flushing 150 million gallons of London sewage daily into the Thames at the Barking and Crossness outfalls. This volume of sewage, constituting roughly a sixth of the river's flow, rendered the restoration of the Thames fishery highly improbable.[47]

While Bazalgette and his associates were pursuing the sewering of mid-nineteenth-century London, counterparts elsewhere were undertaking projects of their own. By the late 1840s, national legislation on municipal governance and public health provided a broad impetus to the sewerage of English towns.[48] European cities, sometimes with the direct encouragement and collaboration of British engineers, also began to prepare sewerage schemes. In Hamburg, for instance, after a fire in 1842 had created the

Sir Joseph Bazalgette supervised the engineering and construction of the original London metropolitan sewerage system. The project, while greatly improving sanitary conditions in central London, culminated in massive downstream outfalls, where raw wastes poured into the Thames.

need and the opportunity for municipal reconstruction, London-born William Lindley used endorsements from Chadwick to promote a costly municipal sewerage scheme over strong opposition from local taxpayers. A decade later, the city sewerage scheme became operational, also flushing municipal waste directly into the harbour.[49] Other European communities converting to the water-carriage system included Paris, Berlin, Brussels, Frankfurt-am-Main, and Danzig.[50]

In North America, as water supply systems extended their reach into municipalities, the common initial tendency, as elsewhere, was for households newly supplied with water to make connections with existing surface-water drainage facilities. However, widespread statutory prohibitions were soon enacted – and reinforced by substantial fines – to deter municipal residents from connecting water closet or privy wastes to municipal drains.[51] New York City prohibitions against depositing any type of garbage, for example, dated from as far back as the introduction of drainage in Manhattan in 1676 and were explicitly reformulated in 1819 to prohibit fecal matter.[52] There may well have been an understanding in the United States that drainage – that is, the removal of runoff from rain and snowfall and the drying out of low-lying, swampy ground – was a communal obligation, in contrast with the disposal of domestic and household material, which remained the personal responsibility of those who produced it.

American understanding of the effects of sanitation drew heavily upon thought and practice from across the Atlantic. Inspired by Chadwick's example, the New York physician and city inspector John H. Griscom scrutinized sanitary conditions in his community. Drawing the attention of his fellow citizens to the thirty thousand cesspools still present on Manhattan Island in 1849, well after the influx of Croton water, he renewed the call for underground sewerage and proposed a corps of "health police" to uphold "a law of domiciliary cleanliness." Clearly a subscriber to the miasmatic theory of disease, Griscom, in his analysis *The Uses and Abuses of Air,* discussed atmospheric contributions to the production of disease, a process in which those cesspools were implicated for their role in "filling the atmosphere with nauseous gases."[53]

The North American sanitarian movement, a loose conglomerate of physicians, municipal officials and managers, and church leaders, among others, found considerable intellectual support for its enterprise in the anti-contagionist theory of disease, which identified reformable environmental conditions as a major source of disease. Although the American sanitarian movement may have been dominated by "specifically American concerns and values: anxiety about the present and future, optimism about the potential for change, and an emphasis on the individual as the root of the problem and the source of the solution,"[54] American sanitary engineering was also heavily influenced at mid-century by practice overseas.

In search of greater understanding of the sewering movement, and desperate to "improve and preserve" the city's health after devastating experiences with cholera and dysentery, Chicago hired Ellis S. Chesbrough. The new city engineer, whose early career had included water work in both New York and Boston, was soon dispatched to Europe to report on arrangements for handling organic wastes in centres such as London, Liverpool, Paris, Amsterdam, Hamburg, and Edinburgh. Chesbrough's observations resulted in an important synthesis of European practice as of 1858, later described in America as "the first really exhaustive study which the subject has received on this side of the water." Chesbrough's advice involving the elevation of city streets – and entire adjacent buildings – to facilitate drainage toward the Chicago River and Lake Michigan set in motion a massive and transformative undertaking. Although Chicago's adoption of systematic sewerage flowing toward the city's water supply in Lake Michigan almost immediately presented difficulties, the initiative was influential and widely imitated.[55]

By mid-century, Brooklyn and Jersey City were also sewered (by 1855 and 1859, respectively), and New York was taking steps toward a community-wide approach to waste removal, awkwardly echoing – without the ceremony that had accompanied the *Croton Maid*'s festive voyage – aspects of the earlier program of water delivery from the Croton. Those on sewered streets were obliged to connect their buildings to drains, and efforts were made to restrict new construction to "improved" – that is, sewered – lots. Ad hoc sewerage initiatives by local resident groups gradually gave way to a more coordinated waste-removal program in which block funding supplanted private financing, and the idiosyncratic decisions of the Common Council were transferred – not without struggle – to a more centralized and powerful mayoralty where engineers and city managers exercised greater professional influence.[56]

When construction resumed after the Civil War, a growing number of communities adopted similar administrative arrangements. A medical commission appointed by the Board of Health for the City of Boston in the 1870s illustrates the reasoning typically underlying such initiatives. Reporting in 1875 on a number of "filth diseases" (with typhoid the leading exemplar), the physicians who conducted the Boston study advised that "in the case of all these diseases a partial dependence upon filth-infection [had] been suspected, if not actually demonstrated." They speculated further that other fatal diseases, not then shown to be causally connected to filth, might nevertheless "find easier victims in those sufferers whose general health has been previously undermined by filth-poisoning." Remedial and preventive action – the prescription – demanded "energetic measures designed to prevent all possible contamination of . . . air, water and food, by the putrefying organic matters of all kinds which constitute 'filth.'" Sewage, if not handled properly, remained a primary source of concern on account of "poisonous vapors . . . which convey filth infection in all its forms." The purification of Boston could be accomplished, it was then explained, only by means of "a rapid and continuous translation of . . . sewage from its initial starting points, the water-closet or the privy, and the sink." All this involved an elaborate series of intermediate channels leading to "terminal outlets," ideally to be "situated far beyond the confines of the city."[57] Thus was the end-of-pipe solution envisaged, however little actually ended at "terminal outlets."

Earl Murphy's account of Wisconsin drainage legislation suggests how deeply embedded the culture of flushing had become by the final decades of the nineteenth century, and how profoundly it contributed to the transformation of civic life. Municipal charters that had been silent on the question of sewerage gradually came to embody special authorizations and general legislation extending powers to construct sewers. Giving this evolution a sense of scale, Murphy argues that it was a greater step for legislators in the nineteenth century to authorize local governments to compel residents to comply with public plans than it was for their twentieth-century successors to require individuals and local governments alike to conform to a state plan.[58]

By 1890, the first date for which aggregate figures are available for the United States, cities of over 25,000 residents had installed more than 6,000 miles of all types of sewers. Two decades later, in 1909, roughly 25,000 miles of sewer lines could be found in cities of over 30,000 people. The larger the cities, the better they were equipped. In the same year, 67 percent of residents in cities of 30,000 to 50,000 had service; 71 to 73 percent of those in cities of 50,000 to 300,000 were sewered; and in the largest cities, those with over 300,000 residents, 85 percent of the population had sewage service. From another perspective, it could be said that though a mile of sewer served 1,832 persons in 1890, by 1909 each mile of sewer served 825 people.[59]

By the end of the nineteenth century, the use of water to remove domestic and industrial wastes was the rule rather than the exception in population centres of Europe and North America – more than 70 percent of cities and towns had sewerage.[60] Per capita rates of discharge by 1905 already approximated 100 gallons in several Massachusetts cities and 250 gallons in the south metropolitan district of Boston – the equivalent of "a fair-sized river."[61] Commenting on the pace of sewerage in Britain, sometimes thought to have been inadequate, historian Anthony S. Wohl argues that, in view of the attendant problems, major municipalities were extremely active: "rather than criticize the slow adoption of large-scale sewer systems we should perhaps wonder at the adoption of so many in the mid- and late-Victorian years."[62] In the early 1900s, the average daily English flow of sewage was much more modest, estimated at twenty-five gallons per capita, though reaching thirty-four gallons for each resident of London.[63]

The intentional transfer of domestic waste from cesspools, privies, and closets to waterways was fully consistent with prevailing theories of disease relating to miasmas and zymotic processes of infection or which viewed filth as a contributor to the spread of illness. Not only was it acceptable for wastes to enter the waterways, it was desirable because the perils of putrefaction and miasmas were thereby removed from population centres. Impressive as sewage flows became, however, little thought was given to the possibility of treating them in any way. Why only a very modest proportion of the sewage outflow was treated is of some interest. Costs played their part, as did uncertainty about the precise effect of particular treatments that might have been adopted. A comforting and remarkably resilient belief that treatment might not actually be necessary also discouraged implementation.

Self-Purification and the Political Economy of Flushing

Flushing may have appeared convenient and compelling for two reasons: the proximity of waterways and the rate of expansion of so many nineteenth-century urban centres. William Ripley Nichols and George Derby, in the context of a sewerage and pollution survey of Massachusetts in the 1870s, attributed its popularity to human weakness: "The temptation to cast into the moving water every form of portable refuse and filth, to be borne out of sight, is too great to be resisted."[64] Some observers resigned themselves to the phenomenon, at least on an interim basis: "Until we have better means of disposal of our refuse than at present," the Massachusetts State Board of Health concluded in connection with a study of the Nashua River, "some of our rivers must be used, more or less, to scavenge the country."[65]

If a few people expressed mild regret as flushing became habitual, others championed the practice explicitly and without apology, defending it as essentially unobjectionable and harmless. In the first half of the nineteenth century, New York's Common Council, though apprehensive about disease-causing miasmas that might arise if organic wastes accumulated in sewers designed for stormwater removal, did not hesitate to permit direct disposal of such materials into the river. Tanners, butchers, slaughterhouse operators, distillers, and those responsible for particularly offensive privy pits were encouraged to find direct access to the waterways to avoid offending fellow residents.[66] Within the engineering community, and indeed, more generally within influential professional circles, depositing municipal sewage in nearby waterways was considered safe due to a unique characteristic of waterways – their capacity to purify themselves. Methodology used in the chemical analysis of water quality through to the 1870s supported the theory, which was not without a certain degree of truth.[67]

King Street Sewer, Midland, Ontario, 1901. Following the introduction of water-borne waste removal systems in the major centres of Europe and North America, underground sewerage arrangements were installed in smaller cities, where they facilitated the discharge of domestic and industrial wastes to nearby waterways.

Organic materials can indeed be broken down in surface waters through the action of micro-organisms in the presence of dissolved oxygen. Thus, not only household sewage but a good deal of nineteenth-century industrial waste, such as animal oils, waxes, and resins, was subject to a process of natural decomposition that would ultimately render it harmless. In this respect, some waters might indeed purify themselves. Limitations, though occasionally acknowledged, were unexplained by the science of the nineteenth century.[68]

With the unknowns by definition excluded from the calculus, and given a certain inclination to discount limitations, the efficacy of self-purification was subject to exaggeration. John Nichol and Charles Macritchie, consultants on a sewerage scheme for Quincy, Illinois, for instance, advised their client that the rapid current of the Mississippi River would, "quickly carry away the discharge from the sewers, and from the quantity and nature of the river water in constant and rapid motion . . . quickly dilute and deodorize the sewage, so that all traces of it [would] disappear in a short distance below the point of discharge." Similarly, as late as the 1870s, the consulting engineer for Lawrence, Massachusetts, confidently concluded that, despite serving a population of close to a hundred thousand,[69] the community waterworks required no sewage treatment: "One of the most remarkable qualities of running water is that of self-purification," he rationalized, adding, "When the most noxious matter has been thrown into a running stream, all traces of it have disappeared in the course of a few miles."[70]

Similarly, the temptations of ocean dumping often proved irresistible to seaside communities, or those with ready access to the sea. In Britain, for example, London, Manchester, and Salford by late in the century employed specially designed steamers to ship sludge for discharge into the sea a few miles from the coast. On Canada's Atlantic coast, residents of Nova Scotia congratulated themselves on avoiding large expenditure, pleased that "practically every community discharges its sewage directly into tidal waters, rendering sewage treatment works unnecessary."[71] American practice was often similar, for, to put the matter starkly, "the ocean furnished seaboard cities with the most favorable conditions for disposal in water." Boston took advantage of this understanding in establishing its main drainage arrangements between 1877 and 1884 before subsequently installing pumping stations to remove sewage to deeper water.[72] The practice was so popular in Massachusetts that around 1900 the sewage of nearly half the state's inhabitants was encouraged to find its way to the Atlantic.[73]

Another striking example of the commitment to flushing was the plan for a sewer running from Concord, New Hampshire, to the Atlantic, "to take the refuse matter from the many cities on the Merrimack and carry it to the ocean, where the chances of its ever becoming a menace to the public health would be exceedingly small."[74] Along the coast a commission charged with advising Baltimore on sewage disposal in 1899 highlighted dilution as "the natural and most economical method," one that would give it a leading place among communities. If Baltimore chose to disregard that recommendation in favour of another course of action, this would leave "the economies of the dilution method to be enjoyed by our neighbors."[75] The Pacific Ocean served equally as a repository for sewage, sometimes indirectly, as in the case of Portland, Oregon, which discharged untreated municipal waste into the Columbia and Willamette Rivers through fifty-eight separate outlets. Los Angeles, too, as of 1887 undertook a vast sewerage scheme to transport its wastes twenty miles to the beach at Santa Monica. Initially flushed at the low-water mark, sewage was eventually conveyed nearly a mile out to sea in a seven-foot-diameter pipe.[76] On a national basis in the first years of the twentieth century, the sewage of six and a half million Americans was discharged directly to the sea, while fresh waters received the untreated effluent of a further twenty million people.[77]

The rationale for this means of handling the residues of urban civilization was confidently endorsed by a certain element of scientific opinion at the close of the nineteenth century. As Charles V. Chapin, the health commissioner of Rhode Island, explained, "If sewage is discharged into a large enough body of water in such a manner that it quickly becomes diffused through it, it is so gradually oxidized as not to give rise to any serious offense." Noting a comparative locational advantage in such arrangements, he added, "Many of our largest cities are fortunately so situated that this can readily be done, as New York, Philadelphia, St. Louis and New Orleans."[78] Cleveland, not quite so fortunately situated, barged its waste to Lake Erie.[79] Detroit and Buffalo were similarly accustomed to barging municipal wastes, including dead animals, excreta, and slaughterhouse wastes, into Great Lakes waters.

Occasionally these vessels reached the Canadian side. Here, Canadian authority, in the personage of the redoubtable Captain Edward Dunn, customs official and master of the *Petrel*, ably assisted by a local constable and the local health officer from Amherstburg, brought one dumping episode to its sudden conclusion with a shot across the bow of an American tug towing a garbage scow for the Detroit Sanitation Company. Dunn's triumph, vindicated in the course of an inquiry carried out by the

United States Department of State, led to an invitation to him to direct his attentions toward Buffalo's habit of dumping in the vicinity of Fort Erie, Ontario. In each case, modest fines were secured against the wrongdoers.[80]

To urban populations, now readily able to disperse their domestic wastes without further thought, the virtues of flushing seemed undeniable. Even among those who were attuned in principle to the importance of monitoring the quality of lakes, rivers, and streams and to the possibility of degradation, limitations either in measurement or in understanding of the transmission of disease or the processes of ecological degradation sometimes led to assessments that amounted effectively to a continuing go-ahead for flushing. "The condition of the rivers is not yet very bad" was the finding of a study for the Massachusetts State Board of Health published in 1876. In other words, carry on.[81] Further attractions were to be found in the observation that the overall costs of sewerage, including capital and maintenance requirements, might actually be lower than expenses associated with the privy vault/cesspool system. Moreover, or so it appeared on the basis of a certain view of the causes of disease, sanitary improvements resulting from waste removal would lower morbidity and mortality rates. The further consequence of removing waste from municipalities would be a greater degree of attractiveness to prospective residents and industry.[82] Notwithstanding the general attraction of arguments promoting sewerage en route to water-borne waste removal, there were reservations.

Excremental Values

Although mid-nineteenth-century fear of miasma forcefully underpinned the sewerage movement from a human health perspective, and belief in self-purification alleviated many attendant concerns, municipal flushing was not yet universally endorsed. Pockets of resistance and proposals for alternatives were associated with the thought that valuable materials were being lost and should really be conserved.

To Cuthbert W. Johnson, English barrister-at-law, editor of the *Farmer's Almanac,* and author of a leading treatise on manure, "What is a nuisance in London is a source of revenue in Brussels."[83] By this he meant that human waste was a valuable commodity. Such was the sense of loss that even in the early twentieth century Albert E. Lauder referred in his manual on local government to "a wail of regret [that] is continually going up from sanitarians and economists at the valuable manurial products annually finding their way, by more or less devious routes, to the sea."[84]

Among promoters of conservancy, theoretical authority for the proposition that human waste was valuable and ought not to be dissipated in the waterways derived, at least in part, from the work of Justus von Liebig, a renowned German researcher in the fields of organic and agricultural chemistry. Putrefaction, he wrote in the 1840s, was a prerequisite to agricultural productivity. It was accordingly improvident, to say the least, to discharge sewage, effectively a national treasure, to the rivers and the sea. With disdainful reference to water closets, von Liebig remarked in his much republished volume *Familiar Letters on Chemistry in Its Relations to Physiology, Dietetics, Agriculture, Commerce and Political Economy* that "contrivances resulting from the manners and customs of the English people and peculiar to them, render it difficult, perhaps impossible, to collect the enormous quantity of phosphates which are daily, as solid and liquid excrements, carried into the river."[85]

British commentator J.J. Rowley, vexed, as he put it, by the problem of town sewage, set out to assess the economic consequences of late nineteenth-century sanitary practices. Ships travelled thousands of miles across the Atlantic, he pointed out, to scrape up guano from the Peruvian islands of Chicha in order to support vegetable cultivation "on the barren fields of Albion." Guano prices had fallen off from twenty to thirty pounds per ton to eight to ten pounds per ton, but for an inferior product often combined with sand. The entire effort was wasted, Rowley concluded, given the availability of better-quality guano – in the form of human excrement – at home. Estimating the value of human excrement at "one shilling per head per annum for every individual," Rowley calculated that Sheffield lost some £12,500 worth of valuable agricultural manure each year to the water closet. "If the price is raised from 1 s. per head to 2 s. 6d., which is still merely nominal, the value of waste in Sheffield alone, destroyed by the W.C., will be upwards of £30,000 per annum." All this was readily available without the necessity of shipping, and had to be disposed of in any event.[86]

The protests of Rowley and others rested on foundations articulated by a series of distinguished predecessors. Even Chadwick, a leading champion of urban sewerage, had hoped that human wastes might be recaptured and their value retained through a marketing process that would actually help to finance arterial sewers.[87] Although ultimately unsuccessful in this respect, Chadwick was not alone in imagining the flow of sewage as an income stream. In the late 1840s, young Joseph Bazalgette, then an aspiring engineer still a decade away from his appointment as chief engineer to London's Metropolitan Board of Works, passionately insisted on the importance of public water closets and urinals. He envisaged a franchise scheme to employ operators throughout London and anticipated significant revenues from the recovery of waste. Extrapolating from his personal observations of traffic levels at existing facilities and the advice of experts on the value of urine, Bazalgette forecast a respectable profit.[88]

Although calls for waste recovery went unheeded in the metropolis, the conservancy impulse sparked challenges to the water closet movement. Henry Moule, Cambridge graduate and for much of his career the vicar of Fordington, Dorset, endures biographically as "divine and inventor." Arguably he saw himself as a divine inventor, for after patenting the dry-earth system of sewage handling in 1860, Moule promoted it indefatigably as part of an alternative vision: the substitution of earth closets for waterborne waste removal as a way to replace disease and waste with national health and wealth.[89] Supporting his invention with a series of promotional pamphlets, he argued that water merely removed sewage without either absorbing or deodorizing it. He recommended dry earth as a superior alternative. Essentially a composting toilet, Moule's dry-earth system employed the oxidizing effect exerted by dry earth, a porous substance: sewage mixed into it came into intimate contact with the air contained in its pores. Moule's system garnered a mixed but nonetheless favourable review from the 1911 edition of the *Encyclopaedia Britannica:* Insofar as it left "other constituents of sewage to be dealt with by other means," it was rather limited in application, but excellent "so far as it goes." In the absence of a general system of water-carriage sewerage, the earth closet could deliver "perfect satisfaction" when used "in careful hands."[90]

The earth closet concept had its supporters elsewhere as well. It caught the attention of the innovative Colonel George Waring, who invested in the concept and for a period promoted it in North America.[91] Rowley, another passionate disciple of the dry closet movement, continued to view it as the solution to town sewage as late as the 1880s: "The only way and the most proper way to dispose of the sewage of towns is to make none." Acknowledging the paradox, Rowley offered the dry closet as a solution that provided an automatic separation of recoverable materials.[92] Before the arrival of sewerage in Ottawa in 1874, residents of Canada's capital were much taken with the earth closet as a source of relief from intolerable stenches.[93] And with Toronto on the verge of abolishing its remaining twelve thousand privy pits, the dry-ash closet found favour as an alternative to the water closet, where, for reasons of expense or because of the need to safeguard them from frost, water closets could not readily be installed.[94]

Commitment to the regenerative contribution of organic waste was resilient for some time. New York conducted profitable sales of manure to surrounding counties in the early 1800s, and more than a hundred American cities were still using privy vault waste as fertilizer as late as 1880.[95] A number of European centres collected excrement in airtight vaults, to be pumped into tank carts as the need arose.[96] Given the bulk-commodity characteristics of the subject matter, nineteenth-century entrepreneurs came to appreciate the impact of transportation costs and storage capacity on the agricultural market for human waste. In certain circumstances, farmers might even be induced to use their own tank wagons to deliver city sewage to rural destinations: "The transfer . . . can be most easily made by running the city carts upon a raised platform below which the private wagons are placed; iron pipes and funnels make the transfer a very easy and quick operation."[97] But for longer hauls, steam tramways such as those employed in Stuttgart, Munich, Dresden, and Leipzig were advantageous. Pneumatic tubes were fashionable in Holland. These delivered a heavily diluted product effectively over comparatively level terrain to a central reservoir. The variability of seasonal conditions, most notably the impact of winter on the agricultural cycle, often necessitated a substantial storage capacity at either the supply or consumption end.

As in all things commercial, risks had to be acknowledged. Thus, as an agricultural input, urban sewage occasionally shared the fate of produce left too long without a buyer. Graz, Amsterdam, and Paris were forced to concede that "when no persons apply for the matter, it is thrown into the water."[98] Those embarking upon or determined to continue a course of putting excrement to use as an alternative to flushing were generally warned – in spite of Chadwick's optimistic projections – that market opportunities were largely confined to cost recovery. Only one profitable excrement operation appears in the record, a unique situation at the Baden barracks where the scheme succeeded "partly due . . . to the fact that the matter [was] carried in a fresh condition directly to the fields." Annual revenues from the initiative were estimated at sixty-two cents per capita.[99] Even Albert Lauder, despite his appreciation of manurial value, reluctantly acknowledged that, "Although many different methods of sewage disposal have been tried, up to the present no process has been introduced which combines the speedy clearance, and profitable utilization of the injurious matter."[100] A variant on retail manure sales also existed in the form of sewage farms. These, too, constituted an element of the conservancy alternative to water-borne waste disposal.

Spreading liquid sewage on agricultural lands was, of course, a widespread and long-standing practice around the world. Western advocates regularly cited Chinese practice and biblical sources alongside reference to

Prior to the widespread adoption of water-borne sewerage, human and organic wastes were collected and distributed on the land to support the next round of crops. Illustrated here is a sewage pumping system used in crop fertilizing. By the late nineteenth or early twentieth century, such practices was largely abandoned.

the fact that in Europe sewage had been used to irrigate private farms as early as 1559 at Bunzeau, Prussia. Milan's use of the Vettabia Canal to deliver sewage to an irrigation area of four thousand acres was another prominent forerunner of what came to be known as sewage farming. Edinburgh introduced sewage irrigation via the Foul Brook or Burn to the Craigentinny Meadows in the late seventeenth or early eighteenth centuries. By the late nineteenth century, the Edinburgh operation was dispersing about eight hundred cubic feet per minute in season across plots of land covering 236 acres.

The Edinburgh approach significantly influenced Chadwick's thinking on sewage management.[101] In the 1840s the health of towns commissioners reflected on the possible application of sewage water in agriculture, an opportunity – perhaps an obligation – inherent in the fact that "the sewer water of towns is unappropriated, and although all contribute to its production, none claim therein a right of property." To the commissioners, sewage was a valuable public property. Accordingly, they argued, "It becomes the duty of the guardians of the public interests to take steps in the first place for the investment in all sewer-water for the public interest, and therefore to provide legal facilities for rendering it practically available."[102]

Many initiatives were guided by a mid-nineteenth-century sewage inquiry that proclaimed, "The right way to dispose of town sewage is to apply it continuously to land and it is only by such application that the pollution of rivers can be avoided."[103] In its third report (1865), this commission encouragingly held out the prospect of "more or less considerable" profits from the application of sewage in agriculture, assuming favourable local conditions.[104] In parliamentary debate preceding legislation to permit municipal authorities to use sewage for agricultural purposes, Lord Ravensworth highlighted a further advantage: recycling urban wastes might deal with Britain's anomalous situation as a nation that "boasted of being the most civilized ever known, [but] wasted at our doors that which might enrich our soil, at the same time polluting our rivers and atmosphere, while we sent whole fleets to the ends of the earth to procure artificial elements of fertilization."[105] Nottingham took the message to heart. The city's sewage farm, extending over nine hundred acres, accommodated pigs, cattle, sheep, and horses, not to mention foreign observers intrigued by this bucolic resolution of flushing's increasingly voluminous conundrum.[106]

From a sanitary perspective, sewage farming achieved some successes, but, like the retail excrement market, it usually proved less economically attractive than forecast. Although the ingenious use of movable conduits, bituminous paper pipes, and steam pumps provided a good deal of flexibility, delivery was problematic.[107] When English chemists concluded that sewage fertilizer was less valuable than competing products and estimated its value for manurial purposes in the range of one to four pence per ton,[108] the material could hardly pass as a precious commodity. Thus, by the early 1900s the initiative had faltered in England. Sewage irrigation, ushered into the United States in 1872 by way of an installation at the State Insane Asylum at Augusta, Maine, shortly encountered skepticism when acreage requirements for economic viability were more systematically considered. It was recommended in 1880, for example, that Milwaukee should purchase five hundred acres for a sewage irrigation scheme.[109] Boston, having been advised that it was flushing $800 thousand worth of fertilizing matter into the sea each year, had no regrets, for it was estimated that the annual cost of recovering that value would amount to between $1.5 and $2 million.[110]

In arid regions, where some form of irrigation was required in any event, the water content alone enhanced sewage farming's appeal. This was particularly the case in the western United States, where sewage farming was

Before sewage systems removed waste and flushed it into rivers – with or without treatment – human waste was often recaptured for use as fertilizer. In combination with running water, the procedure contributed to irrigation as well. Shown here are the sewerage gardens at the London Asylum, ca. 1900.

under way as early as 1883 in Cheyenne, Wyoming. By the end of the century, such farms were in operation in twenty or more locations, including Salt Lake City and Los Angeles. In the latter city, a sewage irrigation system operated by the South Side Irrigation Company in the Vernon district was so successful that agricultural land values rose appreciably. In the wake of the ensuing residential boom, however, sewage irrigation – despite its original contribution to local prosperity – came to be seen as a nuisance and was ultimately discontinued.[111] Pasadena's sewage farm was renowned. The three-hundred-acre farm supported a hundred pigs, an extensive tract of English walnuts, and an alfalfa crop largely dedicated to the municipality's working horses, some of which, when no longer fit for firefighting and street work, continued to contribute to their own upkeep on the sewage farm itself. The operation generated a profit of $25,000 in the early 1900s, roughly $3 per resident.[112] At its peak, sewage irrigation was practised in over a hundred municipalities, notably in Texas, California, Arizona, Pennsylvania, and New Mexico.[113]

Over the course of the late nineteenth century, the dry-earth closet, the sewage farm, and other conservancy-oriented alternatives to the discharge of wastes into waterways proved no match for the convenience of flushing. In the context of self-purifying rivers and miasmatic theories of disease, the sewage question was essentially a matter of removing organic wastes whose decomposition was widely associated with the precipitating conditions of human illness. Cuthbert Johnson, author of *On Fertilizers*; Edwin Chadwick, public health pioneer; Joseph Bazalgette, renowned London engineer; and Colonel George Waring, American sanitarian, had failed in their efforts to demonstrate the value of human waste. Moreover, notwithstanding extensive sewerage to remove urban wastes, sewage treatment remained decidedly limited for decades.

The sewage from 20.4 million of the 28 million inhabitants of the United States who were connected to sewerage was discharged to fresh water around the turn of the century. A further 6.5 million flushed directly into the sea, leaving only 1.1 million connected with sewage purification works.[114] As of 1905, a mere 90 of the 1,100 sewered American communities had purification of any kind.[115] In 1909, 88 percent of the wastewater of the sewered population of the United States was still discharged to waterways with no treatment at all.[116] In Canada the situation was only marginally better. As of 1915, of the 279 sewer systems in Canadian cities, 204 discharged untreated sewage directly into neighbouring waters; only

75 systems included some form of treatment.[117] American observers perceived a more rapid adoption of sewage purification in Britain, something they attributed to population density, the comparatively smaller scale of inland waterways, and to pressure from river conservancy boards and the courts.[118] How would the nuisance created by municipal evacuation of waste by water, a new public source of advantage, be regarded by those who suffered the consequences and by the legal authorities to whom they appealed?

6

Learning to Live Downstream

THE INSTALLATION OF MUNICIPAL WATER supply systems during the nineteenth century represented an impressive physical accomplishment. Less visible, certainly, than railway lines, whose presence at ground level has helped them to stand as a symbolic achievement connecting communities with each other, waterworks also effected new forms of communication between communities – with unanticipated consequences. Sewage – the matter increasingly communicated from one community to the next as water-borne waste removal replaced the privy pit and the cesspool – would prove to be a somewhat unwelcome subject of inter-municipal intercourse.

Over the course of the nineteenth century, sewage provoked its fair share of complaints. Downstream residents adversely affected by the impact of urban flushing sought whatever legal redress might be obtainable. The concentration of population along waterways and the enthusiasm with which local governments embraced sewerage gave rise to a significant number of nuisance claims and other actions by riparian owners to safeguard their interests. Some of these contests were comparatively localized in nature, whereas others took on broad regional or national significance, sometimes even extending internationally when proponents and opponents of flushing sought legal precedent to support their positions. Insofar as the quality of waterways and the health of communities were involved in these struggles, participants and observers alike were compelled to reflect upon the comparative potential of private litigants and legislative determinations to resolve matters so frequently in contention.

Riparian Resistance and Statutory Accommodation

On the more modest scale of controversy, when public health legislation in Britain stimulated municipal sewerage after 1848, fishing interests were in the front ranks of the opposition. A proposal by Stratford-upon-Avon to construct a sewage outfall for the benefit of the town's five thousand inhabitants, for instance, was immediately challenged by the proprietors of a long-established fishery. The Oldaker family, seeking to defend a grant dating back to 1787, protested that the weak current below the outfall would allow sewage to accumulate, damaging their fishery and nets. The judges, basing their conclusion largely on the outfall's injurious effect on shoreline watering-places for cattle, agreed that an injunction was appropriate, for the local board of health, having failed to obtain the consent of persons with an interest in the stream, had exceeded its authority to carry out the sewerage enterprise. The decision was an important early indication that the vast new sewerage powers conferred on local boards were subject to legislative constraints – consent, compensation, and an obligation to prevent nuisance.[1]

The impact of controversy in larger communities, with Birmingham serving as the outstanding example, resonated widely indeed as the decision of a Chancery court in 1858 symbolized the potential of litigious individuals to safeguard environmental quality by protecting their own interests.[2] Charles Adderley (later Lord Norton), member of Parliament and owner of a family estate in Warwickshire, lived at mid-century in Hams Hall, close by the River Tame. Here, cattle were accustomed to drink, fish were plentiful, and sheep could readily be washed. Neighbouring proprietors drew water from the river for domestic and agricultural purposes as well as, it seems, for brewing. Meanwhile, several miles upstream on the River Rea (a foully contaminated tributary of the Tame), the remarkable rise of Birmingham as a nineteenth-century industrial centre was under way.

Adderley's estate remained comparatively unaffected by the presence of municipal effluent in the watershed until 1854. In that year, pursuant to the *Birmingham Improvement Act* of 1851, civic officials completed the consolidation of urban drainage into a formidable main sewer emptying directly into the Tame at Saltley. Adderley, joined in his complaint by Lords Bradford and Leigh, and Sir Robert Peel, among others immediately affected, protested the rapid deterioration of the waterway.

Birmingham's reassurances about improvement efforts alleviated the concerns of the landholders, at least temporarily. Council, it seemed, was earnestly in search of "a perfect drainage for the borough." In a manner reflecting the powerful mid-century remnants of the conservancy view that sewage should be reclaimed for its nutrient value, the municipality expressed its intent "to extract the manure from the sewage, and manufacture it into a highly concentrated fertilizer for agricultural purposes." Any residual fluid was expected then to enter the Tame "in a comparative pure state."[3] Four years later, with no signs of genuine progress, Adderley sought relief in court.

In the intervening years, he had presented Birmingham with a public park. Even in the summer of 1858, with the Great Stink dominating conversation in London, Adderley remained reluctant to press too heavily upon his fellow citizens. An interim injunction restraining the city from opening further public sewers, and an undertaking to deal effectively with those already in operation, was all he asked. Counsel for the municipality showed little inclination to compromise, arguing that the discharge was by no means a nuisance and even suggesting that the matter was properly a question for courts of common law rather than of equity. This proposition, offered while the integration of common law and equity courts was under discussion in England and not yet fully accomplished, invoked historic distinctions between the two jurisdictions: common law courts could not award so-called equitable remedies, including injunctions.[4]

Widening their scope beyond these customary judicial diversions, Birmingham's legal agents raised the spectre of an incalculable evil certain to result if the court ordered that drains be shut up. One ordinarily excuses the more outrageous excesses of lawyers with the thought that counsel do not necessarily believe their most preposterous utterances, and that able opponents will in any event make short work of posturing, sophistry, and bluster. Claims falling well outside the realm of legal rules, and advanced on behalf of important public authorities, invite closer scrutiny. "The entire sewage of the town will overflow," Birmingham's legal spokesmen wanted the court to appreciate. The city, they insisted, "will be converted into one vast cesspool." Given the elevation of the town (450 feet above sea level), lawyers argued, the area would naturally drain into the Tame as before, "only in a far more aggravated manner." Building to a crescendo, as counsel are wont to do, they predicted disaster: "The deluge of filth will cause a plague, which will not be confined to the 250,000 inhabitants of Birmingham, but will spread over the entire valley and become a national calamity." The town urged the court to adopt a particularly majoritarian perspective on this catastrophic prospect. "The increase of population, inseparable from the progress of a nation in industry and wealth, is attended of necessity by inconvenience to individuals against which it is vain to struggle. In such cases private interests must bend to those of the country at large. The safety of the public is the highest law."⁵ Birmingham's claim, had it been accepted, would have taken the idea of necessary inconveniences to a new level.

By the time Birmingham's diatribe got around to the democratic imperatives of municipal flushing in the national interest, Vice Chancellor Sir W. Page Wood, whose treatment of London's Metropolitan Board of Works, as we have seen, had already revealed him to be no particular admirer of municipal administrators, had lost patience. In reply to the spectre of plague and national calamity, he interrupted to say simply that it was not appropriate for a court to consider such allegations. Public safety, he bluntly declared, "is that which the Legislature has said is for the safety of the public, and no more."⁶ The court's function, he insisted, was not to serve as a committee for public safety with the power to prevent injury to the whole of England, but simply to determine what the legislature had authorized Birmingham to do in the course of implementing a program of municipal drainage. If Parliament had not removed Adderley's rights to enjoy the river, fish in its depths, and sustain his herds with its flow, those historic rights persisted, and Birmingham's immediate efforts to defend its actions would be at an end. And so, in 1858, they appeared to be.

If Birmingham insisted (as it did) that the city had already expended all its resources and utmost efforts to avoid contamination of the Tame and so must — unavoidably — override the rights of private parties, Sir Page Wood offered consolation in the form of an alternative course: civic officials should apply to Parliament for power to raise more money. Indeed, if Birmingham really could not, despite vigorous investigation of its options, drain the city without invading the private rights of the downstream owner, it must apply to Parliament to do so: "and if the case be one of such magnitude as it is represented to be, Parliament, no doubt, will take measures accordingly, and the Plaintiff will protect himself as best he may."⁷ This was not bad advice: by following it — with some variations, and by taking advantage of general legislation as it came along — Birmingham would manage to stave off the most severe consequences of the threatened injunction for the next half century.

Faced, not with "a plague," but with the more immediate prospect of a court-ordered injunction, Birmingham's public works committee soon proposed to establish a vast sewage farm in the Tame Valley. In keeping with the commercial spirit of the age and with due regard to the manurial value of Birmingham's outpourings, it was still hoped that a certain amount of raw material might be sold to farmers along the conduit route.

With a cost estimate of £275 thousand, these works faced severe challenge in the finance committee of a community whose civic improvement account was already well into overdraft.[8] A sewage filtration scheme successfully captured vast quantities of sludge, but the nutrient-rich effluent, when spread about to dry near the confines of the outlet, proved only modestly appealing to farmers.[9] In the absence of sales, the accumulating sludge built up to a depth of four feet on the poorly drained site.[10] National legislation inspired by Birmingham's plight allowed the city to acquire land outside its boundaries for a more extensive sewage farm. This proved to be no more satisfactory,[11] although with its 2,830 acres, it became the largest in England.[12]

Adderley, for his part, was in a comparatively good position to safeguard his own interests. Not only did he enjoy the support of Sir Robert Peel and other notable allies in the original complaint, but he was, as the litigation began, president of the Board of Health in the Derby government of 1858-59. A decade later, Adderley brought his unique personal experience to his appointment in 1869 as chair of the Royal Commission on Sanitary Laws. Adderley's sanitary commission ushered in significant public health reforms of the 1870s with a call for simplification and unification of the complicated and frustrating administrative framework. Hoping to ensure some level of overall guidance and direction, the commission called for a national policy on sanitation and health. It recommended as well the creation of a central authority, leaving the actual operation in the hands of local government.[13] Local government legislation in 1871 divided the country into sanitary districts, each of which would henceforth be required to appoint a medical officer and a nuisance inspector. Four years later the *Public Health Act* of 1875 set out a code of principles and responsibilities to direct the efforts of these new officials.[14]

While the sanitary inquiry was still under way, Adderley ran out of patience with the pace of progress in Birmingham. In 1870 he applied for a renewal and extension of the 1858 injunction. Residents in the immediate vicinity of the sewage outlet simultaneously launched litigation of their own. The prospect of further judicial action brought an end to the "leisurely maladministration" of Birmingham council,[15] as litigation elsewhere intermittently inspired local officials to address sewage concerns somewhat more systematically.[16]

Birmingham's municipal Sewage Inquiry Committee, consisting largely of Liberal associates of Joseph Chamberlain, the future mayor and prime minister, was struck to assess the sewage situation.[17] The ensuing report – a document of some three hundred pages – reviewed sewage treatment options in the context of a frank analysis of the desperate state of sanitation in Birmingham, where alarming mortality rates prevailed in certain wards. It seemed essential to reduce the discharge at the outfall; when council accepted the report in October 1871, some fourteen thousand open middens and ashpits draining directly into the sewers were the priority. Although water closets were not yet widely used in Birmingham, the committee creatively proposed to tax these fixtures "on the principle that persons who, by their means, introduced faecal matter into the sewers, may fairly be called upon to pay the cost of its necessary subsequent extraction."[18] The city declined to adopt this particular suggestion, but did explore a number of alternatives to alleviate the situation.

In broad outline, the basic nineteenth-century alternatives to discharging untreated wastes into local waters consisted of sewage or irrigation farms (as we have seen), chemical precipitation, and filtration beds. Attempts at "disinfection" were quickly discredited.

As a rule, precipitation systems consisted of two stages. Following the addition of some substance to sewage in order to produce a flocculent precipitate, a period of sedimentation encouraged the separation of the heavy sludge from the "supernatant" liquid.[19] Precipitation's initial appeal lay in the possibility of producing a valuable sludge consisting of a "portable and consequently marketable" form of the recovered constituents. Between 1856 and 1876, hopeful and industrious inventors took out over four hundred patents in England for various processes associated with the chemical precipitation of sewage, and soon pressed municipal officials to adopt proprietary methodologies. A good many of these may actually have been encouraged by the Birmingham litigation.

Techniques introduced in mid-century Leicester were actively promoted by their backers and studied for possible application elsewhere. In the early 1850s, the civil engineer Thomas Wicksteed claimed to have diverted much of the flow from one of Leicester's largest sewers, and "after separating all the noxious ingredients, *dissolved* as well as suspended in it, in the form of marketable manure, . . . returned the water to the river [Soar], in a state of as great purity as the water in the upper portions of the river before it is contaminated." In his entrepreneurial capacity, Wicksteed reported with delight that the Leicester town council had entered into a thirty-year contract for his company's services.[20] However, the undertaking, whose product fell short in the minds of farmers when compared with other fertilizers then available, failed to achieve profitability and was taken over by the city.[21] Alas, when economizing officials cut back on the quantity of lime required for effective precipitation, the Soar suffered the consequences and this particular initiative was somewhat discredited.[22]

Other common precipitants included charcoal, alum, and iron salts, but the results of precipitation experiments were frequently disappointing,[23] possibly because the ventures were more often driven by enthusiasm and convenience than by careful consideration of the alternatives and their implications. Inventors were accused of selecting cheap or worthless ingredients such as coal-ashes, soot, gypsum, and salt, and of combining them "without any definite notion of the part which they were separately and collectively to play." It was not unknown for promoters to specify alternative substances "whose action, if any, must be evidently quite dissimilar the one to the other."[24]

One of the more vigorously touted ventures was the Native Guano Company's ABC process, named after its Peruvian competition (guano) and its alphabet soup of ingredients. After the initial introduction of clay, carbon, or charcoal waste combined with "a little blood from slaughter houses" to attract suspended impurities, alum was added to settling tanks to promote precipitation. Proponents claimed the process resulted in a valuable sludge and an effluent of high quality, "fit to be charged into ordinary watercourses." Kingston-on-Thames, it was asserted, enjoyed "a ready sale" of sludge at £3 10s. per ton.[25] Skeptics and advocates of alternative approaches, especially land-based arrangements for sewage treatment, had cast doubts on ABC for some time.[26]

Where precipitation had been adopted, disposal of the sludge presented a further challenge, for the product was not only difficult to market but costly to eliminate. The rapidly growing community of Leyton in northeast London, for instance, had tried agitating a mixture of sewage and a sulphurous compound prior to precipitation in the presence of milk of lime. Following precipitation the water was drawn off, in their case, to the River Lee.[27] It was typical "to compact [the sludge] in filter presses and dispose of the press cake by burning it or burying it in the ground."[28] When spread across the local countryside, Leyton's press cakes were so unpopular that alternative arrangements were imperative. The community acquired a "Destructor" where the sludge could be burnt.[29] Although as many as 234 towns eventually used chemical treatment, only 30 generated income from fertilizer sales and none of these operations was judged profitable.[30] In the United States the comparatively high cost of labour and materials proved something of a deterrent to the adoption of precipitation techniques. One estimate suggested that an American city of fifty thousand might face annual costs of $37 thousand, substantially higher than the thirty-seven cents per person common for sewage treatment in England.[31]

The other contender as a basic treatment methodology, intermittent sand filtration, bore some resemblance to sewage farming but required much less land. Britain's Rivers Pollution Commission estimated that if a town of ten thousand inhabitants might require a hundred acres for sewage farming, "it would need but three acres of a porous medium six feet deep, worked as an intermittent filter, to oxidize and therefore purify the drainage water of such a town."[32] Contemporary American studies offered similarly hopeful predictions.

In 1872 Massachusetts legislators directed the State Board of Health to investigate "the disposition of the sewage of towns and cities," thereby eliciting a memorable series of reports from William Ripley Nichols and his associates. In an engineering report relating to the Mystic, Blackstone, and Charles Rivers in Massachusetts, land-based purification by means of filtration was explained: "If 20,000 gallons of sewage filter evenly through an acre of land, where the groundwater stands six feet below the surface, each gallon of sewage will be brought in contact with at least twenty-five times its bulk of air. The effects of this is to oxidize and change the organic impurities into harmless compounds, so that the effluent water, so far as can be determined by the senses and by chemical tests, is inoffensive and innocuous."[33] Septic systems, operating on an entirely different principle, were still several decades away as Birmingham confronted the practicalities.[34]

Using a combination of filtration, chemical precipitation, larger tanks, and a modest amount of additional land for sludge disposal, Birmingham satisfied the Court of Chancery in 1875 that the nuisance that so disturbed Adderley had been eliminated.[35] Professionals in the sanitary engineering community were less persuaded. Birmingham continued to struggle and experiment with precipitation tanks, and, as new ideas developed, with septic treatment, contact beds, and trickling beds.[36] Yet the city's problems on the flushing front only worsened as the century progressed. The 8,000 water closets, 35,000 midden privies, and 7,000 privy pails of 1875 gave way to 50,000 water closets by 1900.[37] All of them were flushed; none of them were taxed.

Chamberlain was deeply committed to principles of civic reform – or the "civic gospel," as the movement has also been described – including municipal ownership of such utilities as gas and water. Thus, in promoting Birmingham's acquisition of the Birmingham Water Works Company, he argued that regulated monopolies should be supervised by public representatives. For dramatic effect, Chamberlain added a principle that had been articulated shortly before by Adderley's Royal Commission on Sanitary Laws: the power of life and death ought not to be left in private hands. Yet Chamberlain then sharply distinguished water supply from gasworks in a manner that would profoundly influence the future of flushing. "Whereas there should be a profit made on the gas undertaking, the Water Works should never be a source of profit, as all profit should go in the reduction of the price of water."[38] Thus, Chamberlain joined the insurance industry as a leading advocate for generously supplied low-cost water, institutionally confirming Adam Smith's more thoughtful assessment of the substance as enormously useful, though exceptionally difficult to value, price, and ration. Chamberlain's rhetoric ran counter to the arguments of those who, for whatever reasons, were concerned to limit the waste of water, sometimes by pricing, sometimes by monitoring. Chamberlain, though not advocating free water, encouraged the shift of water from a private to a public good, thereby making it somewhat more difficult to invest in quality once it became evident that quality might need to be actively maintained rather than taken for granted.

On the sewage front, the Birmingham strategy (if the track record may be elevated to that status) had always employed delay. In this respect the city derived a tremendous though unforeseen advantage from the 1875 *Public Health Act,* which, by encouraging the formation of united districts to further sanitary objectives, led to the creation of the Birmingham, Tame, and Rea District Drainage Board. Joseph Chamberlain, as mayor of Birmingham en route to national prominence in the Gladstone and Salisbury governments, presided at the inaugural conference.[39] Thus in 1881, when Adderley sought – yet again – to enforce his hard-fought right to an injunction against the city of Birmingham, a rude shock awaited. The Birmingham, Tame, and Rea District Drainage Board was no longer the city, but an entirely new entity not subject to the legal liabilities of its predecessor. If Adderley, the chair of a royal sanitary commission whose recommendations were at least partially responsible for the new legislation, needed an injunction, he would have to begin again. He did in fact persevere, later obtaining a further injunction. This he employed graciously in negotiations with Birmingham. In 1892 the city contributed £5,000 to improvement of the Tame when Adderley finally abandoned his claim against it. Alderman Lawley Parker, for one, expressed pleasure at the outcome, which offered the city "splendid terms" although it was widely assumed that Adderley had been in a position to exercise the upper hand.[40] Adderley's personal decision to cede his rights to the city did not mark the end of court challenges against Birmingham's sewage and effluent flows, but by this point sewage had become a matter of public debate and a shift toward legislative solutions was well under way.

Flushing and the Courts

Had the common law effectively safeguarded waterways from municipal and industrial contamination, there would have been little need for royal commissions of inquiry, let alone proposals for statutory reform. The limitations of the common law, despite Adderley's original success against Birmingham, were the subject of an extensive critique presented in 1867 by a royal commission studying river pollution in Britain. Where abuse of a river affected private rights exclusively, it fell to each individual to protect himself, either by means of an action to recover damages at common law or by seeking an injunction. These options, argued the commissioners, were invidious, expensive, and uncertain. Invidiousness was inescapable, they asserted, whenever neighbour was set against neighbour, and whenever one manufacturer faced legal action for conduct that hundreds of others engaged in with impunity because they did not happen to offend a powerful neighbour. Litigation against one offending manufacturer seemed indefensibly expensive in comparison with the benefits that could be obtained by a comprehensive board or agency spending an equivalent amount on the safekeeping of an extensive stretch of river occupied by perhaps hundreds of factories. Moreover, the outcome of any legal proceeding was highly uncertain. Plaintiffs would generally have set out to establish that they had experienced injury from the pollution of the river and that the injury complained of was caused wholly or at least in part by the defendant. This would frequently be difficult or impossible to substantiate when, like the identified defendant, numerous other upstream manufacturers discharged their own liquid refuse, which so intermingled with everything else in the river that the offence of the defendant could not be readily distinguished.

Another major obstacle (at least occasionally) for private litigants was the possibility that polluters might successfully assert an entitlement to continue their operations. As long as the abuse fell short of a nuisance to the public generally, courts might recognize these claims where the practice had gone unchallenged for a sufficient length of time. Referring to such prescriptive rights as "privileged abuses," the rivers pollution commissioners lamented their devastating impact in a passionate account of the ultimate failure of the common law to protect water quality. A manufacturer exercising a prescriptive right to discharge solid or liquid refuse into a river might injure the waterway "to the extent of many times the money value of the right." The resulting loss to the public would be, contended the commissioners, too serious to be measured by a money standard. In their view, the problem compounded itself, for, "if some are permitted to pollute or obstruct the river, it is in vain that others abstain from abusing it." The common law's tolerance of exceptions was singled out for blame, since, by means of prescriptive exceptions, "the law discourages those who are well disposed, and renders ineffectual voluntary combination, even upon a large scale, amongst manufacturers, to preserve the river." It was even suggested that law offered a premium or advantage to those who abused waterways through pollution. A manufacturer might be "tempted to go on casting solid and liquid refuse into the river in order to establish a new right or in order to keep alive an old one."[41] These prescient remarks have seldom been matched as an indictment of the legal system's failure to secure the quality of the waterways. Similar sentiments supported recurring attempts to strengthen environmental safeguards.

Prescription offered no justification for pollution sufficiently widespread in its consequence to affect the rights to the public generally. But public nuisance was – and has long remained – a most malleable and equivocal term. As the commissioners viewed the problem, there was little likelihood in a manufacturing district of a manufacturer being indicted for causing a public nuisance merely for discharging wastes that rendered river water opaque, discoloured, unsightly, and quite unfit to drink, as long as there was no smell and no apparent danger to public health. The commissioners were equally of the view that in the vicinity of large towns, a river foul with sewage was perceived as inevitable: "Inhabitants are reluctant to come forward as witnesses to denounce that to which they have become long familiar, and in like manner jurymen are slow to find such things to constitute a public nuisance." It was extremely difficult to find anyone willing to prosecute sewage pollution. "For the principal offenders are the governing bodies of large towns. These do not prosecute one another for the reason that each is guilty of the same offence towards his neighbour, and that they are rarely prosecuted by private persons because few are willing to bear the expense and odium of acting as public prosecutors." Legal proceedings against a large town might well cost several thousand pounds and were sure to be contested. In view of the substantial costs involved, very few individuals would have sufficient personal interest in the stoppage of the nuisance. "Accordingly, inconvenience to the public, the nuisance, continues unabated. Rich and poor alike submit to it as a sort of destiny."[42] In this way acquiescence became endemic, and the ability of the legal regime to respond effectively to threats to water quality was severely compromised.

Even if these formidable obstacles were overcome, there was no guarantee that the river in question would enjoy greater protection. In the course of their deliberations, commissioners had become aware of several cases in which the defendant manufacturer had avoided liability not by ceasing to pollute, but simply by relocating the offending discharge to a point downstream from the complainant. The commissioners also lamented the structural weakness of the common law system, a deficiency that later generations of observers might describe as the absence of a watershed focus or lack of an ecological perspective: "The law, as it at present exists, is only applicable to local and individual cases. There is no power of general application. One town or one manufacturer may be proceeded against, but there is no authority having the means and the power to deal with nuisances throughout an entire drainage area."[43] To put the matter another way, the remedies available through private litigation could not guarantee improvements in water quality, and certainly not on a comprehensive basis.

Dissatisfaction with riparians for failing to defend the waterways was nevertheless so great that one commentary proposed to penalize them: "Suppose a case should arise of a manufacturer which could by no process really purify its refuse, and yet which had the prescriptive right by long usage to pour its filth into a stream. Towns may grow up below it, and may need clean water. In such a case, provision should exist, to enable the landowners below to inhibit the polluter upstream, giving him compensation for the loss of his acquired right to pollute. This right having been acquired by their supineness in former years, the burden of compensation should fall by rate upon the landowners below referred to."[44] This intriguing proposition appeared to assign to riparian property owners not merely a right but a duty or responsibility on behalf of the public to act where the condition of the river was threatened and to hold them accountable for failing to do so.

Through the mid-nineteenth century, the courts handled numerous cases in which plaintiff landowners, including riparians, challenged industrial operations that threatened them with property damage, economic losses, or inconvenience – or simply in order to prevent prescriptive rights from accruing to polluters, for that in itself was an injury recognized by the law. English courts clearly struggled to reconcile the concern of downstream interests not to be disturbed in the enjoyment of their property with acceptance of the idea that industrial activity, by and large, was the source of significant economic advantages to the wider community. On occasion, industrial interests explicitly challenged courts to confront the consequences of adverse decisions. One colliery, for example, defending itself against the threat that it might be forced to cease operations to prevent continuing pollution, invoked the spectre of massive unemployment and the loss of nearly £200 thousand of paid-up capital.[45] Such examples of the calculus of self-interest, though remarkably resilient, were generally resisted by the courts; the view that industrial activity so benefited society that its impacts would have to be accepted unless they resulted from severely negligent operational practices did not then prevail. Instead, the notion of reasonableness based on a fairly flexible range of factors continued to be helpful in resolving conflicts. At the same time, clear and rigorous safeguards against interference from industrial activity (something approximating natural conditions) were typically unsuccessful.[46]

But the idea of reasonableness was itself subject to close scrutiny. Thus, some years after a court had indicated in *Hole v. Barlow*, a controversy over air pollution, that industrial activity might be carried on if its location was "convenient, reasonable and proper" from the perspective of the industry itself, even if this resulted in a nuisance to neighbours,[47] an appeal court hearing a similar complaint in 1862 vigorously rebutted such a widely drawn proposition.[48] And all of the judges did so with reference to the unacceptable impacts on water quality that such a principle of law would sanction. Chief Baron Pollock, who was doubtful that nuisance law could ever be "capable of any legal definition which [would] be applicable to all cases and useful in deciding them" because of the infinite variety of circumstances in which such claims might arise,[49] also remarked that "water in the smallest degree corrupted is rendered useless for domestic purposes."[50] For his part, in rejecting the principle that an offensive trade might avoid nuisance claims by virtue of the fitness of the location in which it was carried on, Williams J. (in a judgment reflecting the opinion of a majority of the court) recoiled at the prospect of sheltering from suit "the transmission by a neighbour of water in a polluted condition." It would be inappropriate, he maintained, to advise juries to absolve polluters simply because it appeared to them that the manufacturer contributing to the contamination of the waterway was established "on a proper and convenient spot" and making a reasonable use of its own land.[51]

Baron Bramwell, dipping into an issue that would later immerse him, contemplated pollution from another perspective. The judge objected in principle to the suggestion that those who suffered a loss could be deprived of compensation simply on the grounds that some benefit accrued to the public as a consequence of the polluter's actions. The public, he insisted, consisted of many separate individuals, and an activity could not claim to benefit the public unless it was "productive of good to those individuals on the balance of loss and gain to all." Accordingly, "whenever a thing is for the public benefit, properly understood – the loss to the individuals of the public who lose will bear compensation out of the gains of those who gain." When industrial activity caused pollution, it might be difficult to assign a monetary value to the losses incurred by those affected, but, declared Bramwell, "it is equal to some number of pounds or pence . . . : unless the defendant's profits are enough to compensate this, I deny it is for the public benefit he should do what he has done; if they are, he ought to compensate."[52] Bramwell also noted that in the aftermath of *Hole v. Barlow* "claims [had] been made to poison and foul rivers . . . on the ground of public benefit."[53] He may have had in mind the assertion of a calico-printing business that, because its operations were suitably located alongside a waterway and conducted in a reasonable manner, its discharge of arsenic upstream from the Stockport waterworks would not constitute a nuisance in the eyes of the law.[54]

John McLaren, in one of the more elaborate recent assessments of the common law's capacity to respond to the environmental consequences of industrialization – at least in the British context – found no indication that the principles in themselves were incapable of providing a check on environmental degradation.[55] According to his assessment, there was, in effect, no "cardinal flaw" in common law theory; no doctrinal deficiency precluded litigation. That is not to say that common law rules were effective in discouraging industrial pollution, but simply to suggest that there was no obvious immunity either.

McLaren points to several factors that help account for the limited contribution of nuisance suits to the resolution of fundamental conflicts between the conservation of lands and waterways, on the one hand, and productive exploitation of resources and industrial opportunities, on the other. The cost of litigation and the problems of establishing causation in court were significant obstacles, even if the doctrine itself offered sufficiently robust protection from unwarranted environmental harm. Negotiated agreements resolving disputes may also have reduced the likelihood that nuisance complaints would ever reach court. In addition, McLaren identifies several social obstacles to legal action. With much of the population inclined to regard the proper functions of law and government restrictively, he suggests, willingness to resort to the law in order to mitigate the impacts of industrialization would have been limited. Moreover, there were economic inducements to accept adverse environmental impacts. If you were experiencing inconvenience as a result of nearby industrial activity, you were probably located in a district where your property was in demand for development. Selling, and perhaps with an attractive profit, might have been the preferred alternative to judicial wrangling.[56]

There are also indications that though riparian proprietors exercising their private rights might represent the best environmental protection available, they were equally capable of invoking those same private rights to monetize the inconvenience and withdraw from the front lines, as Charles Adderley eventually chose to do.

Best Practicable Means

Some early indications of general legislative concern were put in place shortly after Adderley's initial clash with Birmingham. In 1861, for example, municipalities were instructed to avoid "excrementitious" deposits that would deteriorate water quality. Locally applicable legislation had already introduced controls on certain designated streams.[57] Beginning in 1864, and for the better part of a decade thereafter, a commission on river pollution systematically investigated this matter, which by then was engaging the attention of industrialists, civic representatives, and professionals from the scientific and engineering communities. Legislative initiatives were soon under way. In 1865 Lord Robert Montagu introduced a measure to confer extensive authority on an inspectorate or board that would be empowered, he suggested, to check the continuing social and economic losses resulting from current levels of municipal and industrial contamination. Opponents successfully resisted the proposal for its likely tendency to impose high costs on manufacturing interests, undermining their undoubted contribution to the general well-being of the nation.[58] The issue had been engaged.

Reform advocates, sometimes vociferous, included members of the angling community, one of whom graphically recorded the destruction of fish in the Ribble when sudden flooding carried accumulated refuse from "its filthy tributaries," the Calder and the Darwen: "The edges of the river were lined with people picking up dead fish." This was not a case of poisoning but of suffocation, preventable, in the angler's opinion, only by a prohibition on the deposit of any solid matter in the rivers.[59] Some observers entertained less disruptive responses. For example, the chemist Edward Frankland of the Royal College of Chemistry, one of the commissioners looking into river pollution, acknowledged the scope of the challenge. In February 1875, in his "Discourse on River Pollution," he cited numerous examples of severe deterioration. The River Aire, even upstream from industrial Leeds, was fouled with the domestic drainage of a quarter of a million people combined with the effluent and refuse from

1,341 cloth and woollen factories, 26 tanneries, 35 dyeworks, 13 chemical plants, and numerous other mills and facilities manufacturing grease, glue, paper, and so on. Yet Frankland's message was hardly one of condemnation and lament, for he envisaged an achievable technical solution. Strict enforcement of comparatively mild legislation, might, Frankland proclaimed, restore many rivers to "nearly their original purity." Although he thought it unlikely, perhaps impossible, to make them suitable for drinking purposes, it was reasonable to imagine sufficient improvement that rivers would again "delight the eye" and that "the pestiferous and sickening exhalations which at present affect the multitudes of our population compelled to pass their lives on the bank of such rivers" could be eliminated.[60] In the end, even Frankland's olive branch was trampled by industrial opposition.

The royal commission's program for pollution prevention consisted of three principal elements: legal restrictions, scientific standards, and administrative supervision. There should be, in the commission's view, a comprehensive prohibition against depositing solid matter in rivers, minimum purity standards applicable to any liquid discharges, and authorization (under suitable conditions and subject to certain exceptions) of industrial discharges into municipal sewerage systems. The commissioners went so far as to recommend upper limits for suspended matter, organic carbon and nitrogen, metals, arsenic, chlorine, sulphur, acidity, alkalinity, and oil.[61]

After extensive investigations, the Rivers Pollution Commission endorsed precise standards. These were necessary, in the commissioners' view, because it would not be possible to prevent polluting liquids from entering the waterways "without first defining what is meant by noxious or *polluting* water." Since "absolutely pure water" was unknown in nature, it had to be recognized that springs, wells, lakes, rivers, and sewers actually constituted "a series gradually increasing in dirtiness." There was, the commissioners bluntly stated, "no definite line of demarcation separating the purest spring water from the filthiest sewage." Workable, efficient legislation therefore depended upon setting "an arbitrary line . . . between waters which are to be deemed polluting and inadmissible into streams, and such as may be considered innocuous, and, therefore admissible into river channels." With this distinction in place, incessant and uncertain litigation might be avoided, and it would become relatively straightforward to convict corporations or manufacturers that carelessly or recklessly contaminated Britain's streams. Standards of purity were also expected to ensure that industrialists in different manufacturing districts were treated equally. In the absence of such standards, "it is impossible to expect the same pressure to be exerted in a district where rivers are already almost hopelessly spoiled by an industry of enormous local importance, as in one where the evil is only growing out of its commencement, or upon a manufacturer who turns his drainage into a considerable stream, as upon one who at once succeeds in befouling a mere brook."[62]

Despite its apparent attractions, the standard-setting approach drew criticism. Some objections addressed controls on particular substances; others were more general in nature. To some the concept of prescribed standards was "wanting in elasticity." Other critics were concerned that effluent standards would "entirely disregard the proportion between the bulk of the polluting liquid and of the river into which it flows." It was even possible to ridicule effluent standards with reference to rivers whose regular flow already fell below the proposed standard, "so that a manufacturer who simply pumped the river water through his works without doing anything with it, would have infringed the standards of purity."[63] (Whether this constituted an objection to standards or an indictment of their absence might well be debated.) Protests arose over potential inequities, notably the possibility that upstream manufacturers would simply dilute the polluting liquid down to the proper point by pumping a little more pure river water through their works, and send the pollutant into the river with impunity. In locations where the water remained largely uncontaminated, it was suggested, the act should be most stringently applied.

Well before Parliament passed the *Rivers Pollution Prevention Act* in 1876, debate was closed on at least part of the reform agenda. Prescribed effluent standards were abandoned during legislative debate in the early 1870s on a series of public health and river protection bills. Manufacturing interests successfully advanced a preference for flexibility, buttressing their position with predictions of dire economic consequences if firm national standards were to be strictly enforced.[64] As enacted, the rivers pollution legislation simply established general prohibitions against refuse, sewage, and pollution from manufacturing and mining operations, but exempted in certain circumstances those able to demonstrate that they had employed the best practicable means (BPM) of rendering the offending substances harmless to the waterways. This legislative device, a mechanism mediating between environmental protection and industrial "practicalities," had been used in other circumstances where it assured those responsible for pollution that expectations of performance would not become commercially unreasonable.[65]

As applied to sewage – still almost universally understood to be a matter of local government responsibility – the BPM exemption was available only to facilities operating or under construction before the 1876 legislation. In the case of poisonous, noxious, or polluting liquids from manufacturing processes, the exemption was more liberal. In this context, the polluter's obligations were limited to *best practicable and reasonably available means*. The consequence, as assessed by one legal analyst, was that the court might "sanction means which are not the best available when the use of such would be unreasonable having regard to the source from which the liquid comes."[66] J.O. Taylor, who had a particular interest in Scottish rivers pollution legislation, concluded that the best practicable means tests left room for "an enormous latitude of public opinion," and perhaps even worked against adoption of methods to achieve the highest levels of purification. The latter problem arose, he suspected, because a higher standard of performance increased the prospect of penal consequences: "It is against the interest of manufacturers, so far as the statute is concerned, to seek for scientific improvements in means of purification, because each development diminishes their protection under the Act."[67]

To determine whether the best practicable means were in fact employed, a process of certification was established. Suitably qualified inspectors were authorized to attest that the means used for rendering harmless any sewage, or poisonous, noxious, or polluting solid or liquid matter discharged into any stream were the best or only practicable and available means. Such certificates, moreover, were considered conclusive evidence. Though applicants assumed the expenses relating to certification, the resulting immunity – and the degree of certainty associated with it – no doubt justified the outlay.[68]

The relaxation of prescribed performance standards to a "best practicable means" approach prompted concerns about who would actually enforce the new legislation. The common law required willing litigants; legislative alternatives depended upon enforcement mechanisms. Initially, proceedings against manufacturing and mining operations could be instituted only by a Municipal Sanitary Authority, and even then only with explicit authorization from the Local Government Board. That body was instructed to weigh the economic consequences of a successful prosecution. An official circular indicated that the decision to undertake prosecution should take into account both the industrial interests involved and the circumstances and requirements of the locality. Where the sanitary authority proposed to take action in any district that was "the seat of any manufacturing industry," the board would consent to prosecute only if satisfied that "means for rendering harmless the liquid refuse from the processes of such manufactures [were] reasonably practicable and available under all the circumstances of the case, and that no material injury [would] be inflicted by such proceedings on the interests of such industry."[69]

Understandably, fishing interests expressed disappointment over the new rivers pollution legislation. To the *Field,* a leading sportsman's journal, it appeared to be "a bill for the continuation of the pollution of rivers"; there were grounds to fear that the measure would actually "offer indemnity for the past and security for the future to the great body of manufacturers, miners and polluters generally; so that, when in the future they are compelled to give up poisoning the waters, they may be in a position to talk of confiscation of rights, and to ask for compensation."[70] The same periodical questioned the suitability of county courts to adjudicate alleged violations, since in small towns their sittings were infrequent and bargees accused of tipping rubbish in the rivers might be long gone before proceedings could be commenced. Local magistrates would be more effective in providing for prompt measures.[71]

England's *Rivers Pollution Prevention Act,* though fraught with procedural hurdles, did not entirely foreclose the possibility of enforcement. Prosecutions, including a number launched against municipalities responsible for sewage, averaged about nine per year in the immediate aftermath of the legislation in 1876.[72] Yet not all prosecutions were actively pursued to their conclusion, and even where orders were secured they were not always enforced.[73] Industrial polluters had some success in circumventing the statutory safeguards in the courts. One of the more egregious loopholes evidently allowed manufacturers to avoid legal interference by connecting their operations to a drain that had been functioning before 1876 rather than constructing a new outlet that would not have enjoyed immunity.[74]

Later observers of the 1876 rivers pollution regime sharply criticized institutional flaws embedded in its design. With particular attention to sewers as the primary threat to health, Frank Spence remarked in his 1893 "How to Stop River Pollution" that the legislation left the initiative "in the hands of men who have no interest in moving." The local sanitary authorities charged with remedial responsibilities often included among their members local manufacturers who were themselves river polluters. Other disincentives to action were deeply entrenched, as any city or county councillor sufficiently conscientious "to advocate spending money on precipitants before the bayonet is actually in the back of his committee, loses caste with his constituents." Any expenditure more likely to benefit downstream residents than the local taxpayer was hardly a priority. The consequences were predictable. "Is it any wonder," Spence asked rhetorically, "if there are sewerage committees which purchase a stock of chemicals and put a man in charge of the works, with strict injunctions to use the material only when the effluent is under inspection by powerful neighbours?" These same civic officials, he charged, "then plume themselves before the ratepayers on throwing so little money into the river for the benefit of towns, villages, and estates further downstream."[75]

Spence proposed an institutional reconfiguration – "imperial control of local authorities."[76] The principle already applied to numerous aspects of national life, from factories and mines to burial grounds and chemical works.[77] Spence believed that a national inspection system would transform the situation. An inspector with a national perspective, he predicted on the basis of his experience with alkali administration, would probably abandon requirements for purifying works conforming to official specifications in favour of devoting more attention to the results achieved. Local authorities would thus be free to adopt any treatment process, although they would be held firmly responsible for the resulting effluent.[78]

Others seeking adjustments to the regulatory regime included Major Lamorock Flower, sanitary engineer of the Lee Conservancy Board, who was an early advocate of addressing pollution on a watershed basis: "If we want to have our streams free from pollution, we must map out the country in watershed areas, and put each under the charge of an experienced man, backed by a good Board, who should legalize his acts."[79] Thus, as early as the 1880s, regulatory arrangements oriented around watersheds and designed to achieve water quality results, rather than to impose rules of operation, were under consideration. In a few instances, this perspective gained recognition.

Rivers Boards and Watersheds

Toward the end of the nineteenth century, British institutional reforms alleviated certain limitations in the 1876 rivers legislation. In 1888 county councils were permitted to pursue enforcement against river pollution on the same basis as local sanitary authorities.[80] More significantly, a suggestion dating back to Adderley's sanitary commission in 1871[81] was adopted, providing for the formation of joint committees along the lines of the Thames and Lee Conservancy Boards. The members of these bodies shared an interest in a particular watershed or tributary stream.[82] They sometimes even enjoyed the encouragement of manufacturers in the watershed who were anxious about the quality of their own water supplies or interested in ensuring even-handed enforcement. Several such joint committees soon came into existence: the River Ribble Joint Committee (1891), the Mersey and Irwell Joint Committee (1891), the West Riding of Yorkshire Joint Committee (1893), and the Tame Joint Committee (1894).[83]

Deplorable conditions were evident in each of these districts. Major Flower accorded the Irwell, where Messrs. Bealey and Shaw had contested water rights under riparian principles earlier in the century, the exalted status of being "the most foully used stream in the world," although there was no shortage of contenders for the title. Here England's staple trade had prospered, becoming indebted to the river for its success, yet "like many another faithful servant, its well-being [had] been ignored by those who [had] derived most advantage from its services." The neglect was long-standing, with all kinds of rubbish, manufacturing refuse, and sewage from a population of more than a million people entering the waterway with "scarcely an attempt at purification."[84]

Within a year or so of their creation, joint committees responsible for the Mersey and Irwell and for the West Riding of Yorkshire approached Parliament for statutory enhancement of their powers. The 1876 provision declaring that the Local Government Board would not authorize proceedings to be taken by a local authority "of any district which is the seat of any manufacturing industry unless they are satisfied after due inquiry . . . that no material injury will be inflicted by such proceedings on the interests of such industry" was eliminated.[85] The West Riding Rivers Board, as constituted in 1893, consisted of thirty members chosen from its major cities.[86] A staff of ten inspectors, with an annual budget of roughly £6,000, was soon actively engaged in investigation, negotiations, and education up and down the waterways.[87]

Charles Milnes Gaskell, chairman of the West Riding Rivers Board from 1893 to 1903, expressed considerable frustration with the state of the law and its functionaries. He protested against "all the dilatory pleas that the wit of clerks and lawyers can suggest." Excuses for inaction abounded. While one manufacturer resisted purification works on the grounds that he was too busy, another would protest that business was too slow for such facilities to be constructed. Some would claim to be willing to act as soon as upstream contamination was eliminated; others, when problems had been addressed below. Gaskell reported yet another case in which an industrial operator declined to treat the plant's refuse "because the stream into which he discharges is a very old one."[88] There were, moreover, institutional obstacles to be overcome: offending manufacturers, for example, could preside from the bench over the prosecution of a fellow culprit; the West Riding of Yorkshire Mill Owners' and Occupiers' Association was a body highly antagonistic to the board's ambition to restore the waterways.[89]

If Gaskell's view is representative, members of the West Riding Rivers Board would greatly have preferred further legislative support in their campaign to revive local waterways. But in the absence of such intervention, they sought to reinforce their authority through persistent persuasion and systematic use of law. An early foray into the courts must have been gratifying. In litigation against the Yorkshire West Riding County Council, Lord Justice Lindley expounded upon the importance of the *Rivers Pollution Prevention Act.* Citing the preamble, Lindley argued that, in the opinion of Parliament itself, prior legislation had not gone far enough to prevent river pollution. As a consequence, he suggested, restrictive interpretations of such previous legislation as the *Public Health Act* of 1875 would be of very little – if any – use in understanding the rivers pollution prevention measures, which were "deliberately intended to extend the previous legislation."[90]

The joint committees, certainly following the statutory reforms of the 1890s, brought about improvements even if continued industrial and municipal expansion severely complicated their efforts.[91] As chair of the West Riding Rivers Board, Gaskell naturally derived satisfaction from progress during the decade of his stewardship. At the outset, six non-county boroughs were without sewage works; by 1903 only one remained without such facilities, and these were about to be installed. The number of urban districts lacking sewage works had been reduced from eighty-one to thirty-nine. Of these, sixteen had the approval of the Local Government Board to proceed with proposed works; planning was under way at eleven others. To indicate advances on the industrial front, Gaskell recounted the results of a seventy-six-mile inspection of the River Wharfe. So active was installation, enlargement, or replacement of sewage works that Gaskell confidently allowed, "It may safely be asserted that the purity of the river will soon be secured."[92]

Overall, the volume of solid refuse being discharged from manufacturing processes had been dramatically reduced in a decade. Many manufacturers were claiming that "in place of tons a week of solid matters they [were] now using pounds." The result represented "a priceless inheritance to the thousands who delight in the scenery and surroundings of the loveliest of our Yorkshire streams."[93] After a decade of activity, the Mersey and Irwell Joint Committee, if not satisfied, was at least optimistic. It had taken twenty-one proceedings against polluters and applied for Local Government Board authorization in connection with a further thirty-nine.[94]

The advantages of a systematically enforced river-basin approach were evident. Where a river had been treated as a self-contained entity under the special authority, striking advances in water quality could be achieved.

Prime examples included certain river conservancies, as well as the water of the Leith, which was placed under a joint commission toward the end of the century.[95] Although the Leith, a short stream, was comparatively easy to deal with by means of sewerage draining directly to the nearby sea, longer rivers, especially those with extensive tributaries, presented greater challenges. Only at great expense could a complete system for carrying away waste by sewers be established; such arrangements "would also make a serious inroad upon the volume of the river." Thus, purification, as well as sewerage, would seem to be essential in such circumstances.[96]

As we have noted in following the story of sewerage projects in the United States, American observers frequented Europe to observe sanitary developments. The Birmingham situation was often instructive. Thus, C.F. Folsom, reporting to the Massachusetts State Board of Health on his examination of Birmingham, explained the apparent impossibility of the municipality's predicament: "The city, in accordance with an act of Parliament, built sewers, was compelled to introduce water-closets, and was advised to discharge its filth into the most convenient stream. It was then compelled by Chancery injunction to prevent a nuisance created in the very process of removing the former one." As Folsom wrote, in the mid-1870s, an initial parliamentary effort to secure land for sewage farming, as recommended by royal commissions, had failed. "It was difficult to know what to do."[97] When American observers confronted that very question, the ultimate lessons they derived from the Birmingham saga were equally uncertain.

They also noted institutional deficiencies. American critic George Fuller lamented the "intermittent or disconnected" character of municipal programs for sewage disposal, even in communities where other undertakings were carried out effectively. The explanation, he suggested, could be found in "the absence of comprehensive plans for intercepting sewage and delivering it to treatment works of a suitable type, size and location." There was a tendency for each successive administration "to take delight in changing the plans or ignoring the efforts of its predecessor." It was also problematic that sewage, in contrast with water supply, and despite hopeful experimentation, had no revenue-producing capacity. The fundamental difficulty was that waste treatment "does not add so obviously to the welfare of the community as a whole as do hospitals, parks, schools, streets, playgrounds and the like." In consequence, he reported, initiatives to eliminate stream pollution "are among the least, if not the least, popular projects for enlisting public funds."[98] Fuller found it even more disheartening to see "scores of sewage treatment plants . . . designed and built in a creditable manner, become practically abandoned through lack of funds or left largely or wholly to self-operation."[99]

The Massachusetts State Board of Health, charged with responsibility for developing legislation to address the threat of water pollution from the state's rapidly expanding industrial base, was somewhat dismissive of British initiatives. The 1876 act seemed the product of compromise in a community "not ready for more advanced measures."[100] The self-righteous tone was less evident very shortly thereafter, however, when the Massachusetts board, in the wake of a survey of the very industries whose operations were the subject of concern, reported that "it would be unwise to place too great restrictions upon manufacturers by setting up, for all cases, some arbitrary standard of purity, which must be always followed, but which could not be enforced." Regulation, rather than prohibition, would be the order of the day, although new pollutants were to be kept out of waterways within twenty miles of the source of any municipal water supply. Existing polluters, whether operating by prescriptive entitlement or legislative grant, were unaffected by the 1878 *Act Relative to the Pollution of Rivers, Streams and Ponds Used as Sources of Water Supply*. To its lack of retroactive application, the statute added certain exemptions, including, most remarkably, the Merrimack River, where intense industrial development and urban growth were notorious sources of contamination.[101] Yet there was nothing distinctive about the Massachusetts situation. Chicago was festering; New York Harbor was awash in sewage and manure; and Milwaukee's Menomenee valley as of 1887 had twenty years of accumulated sludge.

The challenge of sewage and wastewater disposal was addressed in clinical and practical terms during the nineteenth century. Clinically, sewage disposal processes were generalized as "methods for the conversion of the waste products of organic life and death into their oxidized and mineral forms."[102] The matter to be dealt with fell within three distinct categories: "excreta . . . ; slop-water, or the discharge from sinks, basins, baths, and the waste water of industrial processes; surface water due to rainfall."[103] The rather nondescript mention of "waste water of industrial processes" masked an enormous and growing variety of trade and manufacturing effluents. These continued to proliferate and often presented remarkably complex treatment challenges. Technical solutions depended for their eventual realization on a wider array of factors.

At the turn of the century, despite promising results from concerted remedial efforts, the overall condition of Britain's waterways was far from satisfactory. According to an assessment made in 1901, the *Rivers Pollution Prevention Act, 1876*, "has not resulted in the general purification of our rivers."[104] With reference to the situation in Scotland, Taylor examined evidence from the 1890s through to the mid-1920s. Inspections – where they had been undertaken – effectively stimulated remedial measures. But limitations in personnel meant that the burdens of diagnosis, prescription, and monitoring – not to mention the need to "negotiate with, conciliate, and perhaps cajole" the relevant interests and authorities – fell upon too few officials to accomplish widespread and sustained improvement. In the early 1920s, the Scottish Board of Health still dismally reported at least six hundred instances of "more or less constant pollution from sewage or industry." Adding their impact to those situations in which no effort at purification had been attempted were frequent indications that established treatment facilities were either neglected or wholly inadequate.[105] As Anthony S. Wohl, author of a comprehensive assessment of the nineteenth-century urban and industrial environment in Britain, concludes, "The prevention of river pollution must be viewed . . . as one of the least satisfactory chapters in the history of Victorian public health."[106]

Widespread recognition that wastewater and sewage treatment procedures were being developed, along with the evident intent of the law to discourage pollution, failed to lead to comprehensive remedial measures in the late nineteenth and early twentieth centuries. In fact, the abundance of alternatives was not entirely beneficial, for it confronted municipal officials – not yet all that comfortable or familiar with long-term projects and their associated financial commitments and potential legal implications – with complex choices. Moreover, those with overall financial involvement exercised considerable caution: "The general progress of sewage disposal in England has been seriously checked by the local government board, which enjoys extraordinary authority over any exercise of the borrowing power on the part of municipal corporations."[107]

Partisan political infighting over the prospect of substantial municipal expenditures – and the distribution of those expenditures – at least partially accounted for delays in the adoption of sewage treatment techniques. And the priority, as the comments of Joseph Chamberlain, among others, continued to highlight, remained focused on waterworks to provide generous supplies of water at low cost. Yet the extent of research activity and the positive results so often reported invite further reflection on the pace of implementation of sewage treatment procedures, particularly in light of the consequences. An early-twentieth-century commentator expressed serious reservations about the existing state of affairs: "It may be necessary for a municipality to send its sewage into a stream, but it is not necessary to pour it in without purification." The situation was all that more puzzling in light of his assessment of current practice: "The discharge of sewage by a municipality into a stream is in effect an appropriation of the bed of the stream as an open sewer."[108] The incidence of proceedings in which downstream residents took cities and towns to court over the discharge of raw or poorly treated sewage confirms that municipal reluctance to treat sewage was widespread.[109] Why were parties who might have been adversely affected not more successful in forcing the adoption of treatment techniques, either through litigation, private claims, or prosecutions?

Legal challenges to the practice of municipal sewerage using common law faced the kinds of obstacles McLaren and others have identified. And they were mounted in a context where it was widely understood that the municipal services in question were carried out for public purposes and rested on suitable scientific foundations. In the few situations where the river self-purification hypothesis and the coastal alternative of flushing into the boundless sea – two disposal models underpinning an earlier generation of design decisions – were supplanted by sewage treatment efforts, it is notable that these were designed primarily to address problems of local nuisance.[110] The emphasis, in other words, was on objectionable odour and visual degradation, rather than on efforts to avoid contamination of drinking-water supplies (except in the comparatively rare circumstance where treatment would help to safeguard a city's own water source). In regard to statutory controls, other limitations applied, including the reluctance of responsible authorities to prosecute, the lack of a cohesive strategy for any given watershed, and considerable difficulty in proving the source of pollution where so many discharged indiscriminately.

There were, however, more forceful detractors of the water closet and of the municipal sewerage arrangements this convenience had helped to encourage. As J.J. Rowley warned, "Of all our domestic institutions, the water-closet system is the most extravagant, the most wasteful, and the most dangerous to human life."[111] The first two charges, that vast quantities of water were being consumed needlessly and that the nutrient value of sewage was thoughtlessly being thrown away, found fewer and fewer supporters. The third, however, represented an emerging perspective. As the downstream impacts of flushing on public health became apparent and more reliably understood in the bacteriological era, would public health initiatives fare better against water pollution than their predecessors?

7

The Bacterial Assault on Local Government

THE WATER-BORNE SEWAGE REMOVAL systems that replaced land-based conservancy initially reflected essential elements of the miasma model of disease – that vapours originating in organic decomposition were somehow responsible for a variety of devastating illnesses. The persistent belief that running water would purify itself largely alleviated anxiety about the quality of the receiving waters. Efforts against pollution were thus significantly hampered by the widespread perception that waters fouled by sewage were simply matters of local annoyance or inconvenience accompanying vital municipal works. Nuisances of this nature were seen as the responsibility of individual riparian owners rather than as sources of incalculable harm to personal and public health demanding vigorous official intervention. Yet even if water-borne municipal evacuation substantially improved the living conditions of upstream residents, epidemics remained rampant as the nineteenth century drew to a close.[1] Early sanitary engineering had not eliminated public health concerns, and soon raised new questions about the location of suitable governmental responses.

Ground-breaking scientific advances, notably bacteriological insights derived from the work of Louis Pasteur and Robert Koch, conferred substantial authority on public health officials who were quick to challenge the assumptions of their legal and engineering counterparts. Health officials in the new bacteriological era eagerly set out to advance a water quality agenda consistent with the contagionist theory of the transmission of disease, even attempting to reshape the legal environment through new forms of regulation. Their efforts, by no means welcomed by municipal leaders, involved a renewed campaign to eliminate untreated discharges of civic wastes. The effort intermittently transferred debate from the local to the state/provincial or national level. By the First World War era, senior public health officials even spearheaded an ultimately unsuccessful international initiative to safeguard communities around the Great Lakes.

Death by Flushing

A few close observers of water quality expressed doubts about the capacity of rivers to purify themselves. In 1868, following a series of experiments designed to test the self-purification hypothesis, Edward Frankland, one of the rivers pollution commissioners, had forcefully asserted, "There is no river in the United Kingdom long enough to effect the destruction of sewage by oxidation."[2] A few years later, appearing in court as a scientific witness, Frankland reiterated, "No sewage can be admitted into a river without deteriorating the quality of the water." Although this might not matter from the perspective of manufacturing, Frankland conceded, he was adamant that sewage rendered water dangerous for drinking or cooking. Still more alarming was his caveat that chemical testing was an unreliable indicator: "the nature of the noxious ingredients which propagate small-pox, scarlet fever, typhoid fever, or cholera is unknown, and chemical analysis is therefore powerless to detect these ingredients."[3] Frankland's influential assessment was reflected in the work of the Rivers Pollution Commission, whose final report in June 1874 condemned as "altogether untrustworthy" the notion of the self-purifying power of streams that had experienced sewage contamination.[4]

One particularly prominent incident, coming not long after Frankland's critique, stimulated public outcry and inquiry. On the evening of 3 September 1878, a steel-hulled collier collided with the *Princess Alice,* a large saloon steamer transporting hundreds of London passengers home from a Sunday excursion to Gravesend.[5] Many of the more than six hundred resulting deaths were attributed to the foul condition of the Thames at the site of the crash.[6] The collision had occurred very close to the northern sewage outfall at Barking, a subject of intense local complaint, and in the vicinity of the Beckton gasworks.[7]

The *Princess Alice* disaster joined the list of grievances that inspired renewed inquiries into the London sewage question. In 1879, complaints by the Thames Conservancy about the impact of sewage on the river gave rise to an inconclusive arbitration,[8] soon to be followed by an elaborate public inquiry. Lord Bramwell, who was not unfamiliar with water contamination, and had put his views on compensation to downstream riparians on record,[9] brought considerable judicial experience to the task of chairing an investigation into metropolitan sewage disposal. Bramwell's inquiry reported twice in 1884. The initial findings, although complimentary with respect to the execution of London's sewerage works and their beneficial effects on the metropolis, were entirely frank as to their impact on the river. The effects of sewage discharge were "more or less apparent at all times" in some areas: due to the influence of the tides, discharged sewage oscillated back and forth along the river and was distributed upstream almost as high as the tide's reach at Teddington.[10]

The Bramwell reports ultimately made several recommendations: that it was neither necessary nor justified to discharge raw sewage into the Thames; that a precipitation process of some form should be installed immediately near the existing outfalls to separate out solids, which should be burned and deposited on land or dumped at sea; that if sufficient land could not be found for depositing precipitated sewage, it should be discharged further downriver; and that any future sewerage scheme should be designed to separate rainwater from sewage.[11] In 1887, amid mounting protests over the condition of the Thames near the outfalls, a precipitation process was introduced to separate solid from liquid sewage, the sludge then being conveyed by a dedicated fleet of tankers to the North Sea.[12] These vessels, each equipped with four oblong tanks with a combined capacity of a thousand tons, were soon dumping more than two million tons of sludge every year.[13] Concerned observers were assured, however, that organic matter was "rapidly consumed by the organic life in the sea water," and that microscopical examination and chemical analysis could detect no more than the "merest trace of the mineral portion of the sludge, either in dredgings from the bottom of the channels or on the surface of the sandbanks."[14]

The new arrangements, involving a disposal cost of about four pence per ton of sludge, though hardly ideal, offered London the virtue of economy while investigation of treatment options continued. Researchers in several centres – with London, Dublin, and Lawrence, Massachusetts, prominent – would through the 1890s and early 1900s identify new biological procedures for handling sewage with the use of bacteria and oxygen. Scientific monitoring was limited until the 1880s, but in 1893 the London County Council assigned to chemist William Joseph Dibdin the responsibility for regular monitoring of the entire tidal Thames.[15] Despite such major initiatives, as well as ongoing efforts to improve the quality of sewage discharges elsewhere, England experienced severe epidemics in the 1890s. In an 1893 typhoid outbreak, over 1,400 residents of the Worthing district were stricken, of whom 198 died; four years later at Maidstone, in the largest English epidemic of the age, 1,938 cases were recorded, including 132 deaths. Investigations exploring various causes implicated water supplies and sewage systems but, in the opinion of many observers, failed to convict these sources convincingly.[16]

By the late nineteenth century, popular and professional opinion in the Old World and the New had begun to associate water quality in some way with disease, and formal entities were established to provide guardianship over water. Yet, with the bacteriological transmission of disease still not well understood, linkages remained speculative. The efforts of newly empowered public health officials continued to focus on the vaguely characterized realm of "nuisance."

One Canadian official, for example, circulated a questionnaire concerning nuisances attributed to industrial activity in 1886. Dr. Peter H. Bryce, who had previously investigated the public health implications of the sawdust menace, now inquired into the number and extent of slaughterhouses, dairies, and cheese factories, as well as piggeries. Breweries and distilleries were also to be tallied up, and special attention paid to cattle byres in the vicinity of the distilleries. Did local officials, Bryce wanted to know, inspect slaughtered animals to ascertain their fitness for human food, or supervise dairies in any way? Another line of inquiry sought to inventory establishments such as knackeries, fat-rendering plants, bone-boiling operations, superphosphate works, oil refineries, and soap-boiling works, as well as tanneries and hide-storing premises. Bryce solicited a detailed account of the methods such industries employed to obviate or lessen nuisances. Moreover, Bryce asked, "Are any of your streams polluted by town or city sewage; and if so, what is the extent of this pollution?"[17] But he could offer no guidance as to any standard relevant to the assessment.

In the same year, a Massachusetts State Health Commission, having examined the condition of inland waters, advocated a permanent body to assume responsibility. The designated state guardians of inland waters would be expected to familiarize themselves with actual conditions bearing upon the relationship of water pollution and purity to public health. They were to address all remediable pollution and, through advice to cities, towns, and manufacturing concerns, to "use every means in their power to prevent further vitiation." In sum, the agency's function would be "to guard the public interest and the public health in its relation with water, whether pure or defiled." The ultimate goal, "which must never be abandoned," was that means might eventually be found to redeem and preserve all the waters of the state.[18] The shift away from the wishful thinking of such comparatively rudimentary investigation and exhortation came quickly in the wake of important discoveries regarding typhoid.

In 1885 an outbreak of typhoid traceable to one home upstream from Plymouth, Pennsylvania, had led to the deaths of 114 of the town's 8,000 inhabitants. Nearly 1,000 more experienced but survived the disease.[19] Newark, Jersey City, Louisville, Cincinnati, and Philadelphia were among other American cities to encounter first-hand the ravages of late-nineteenth-century typhoid, whose transmission was facilitated by sewerage and misunderstanding. Despite lessons from North Boston and other communities, Albany, for example – acting on the basis of a "childlike faith" and reassurances from an eminent chemist – located its water intake "in the cesspool of the Hudson." In the rueful words of Marshall Ora Leighton in a 1902 study, each city tended to learn the lesson "only when its group of mourners [had] become large enough to demand attention."[20] Countering this disheartening assessment was the possibility that, with rising numbers of victims, water-borne contamination might be sufficiently demonstrated and verified to stimulate preventive steps. For North Americans, such learning was derived from experience along the Merrimack River, one of the three major waterways that Massachusetts legislators had chosen to exempt from pollution control measures in the 1870s.

Contemporary professional opinion had supported the decision to exempt the Merrimack. Health officials calculated – or assumed – that, by virtue of dilution and distance, the river would purify itself between points of wastewater discharge and water intake sites. Indeed, they embraced an even more mischievous doctrine – the notion of beneficial contamination – whereby certain industrial wastes actually accelerated natural processes: "The sewage of Lowell is diluted with from 600 to 1,000 times its volume of water, and then flows a dozen miles to Lawrence, much of the refuse from the mills acting as a precipitant and disinfectant to it."[21] However comforting it must have been to imagine mill refuse as an antidote to the effects of sewage, beneficial contamination proved to be a mirage. In little more than a decade, the number of deaths from typhoid spiked dramatically in communities along the Merrimack. First in Lowell and shortly thereafter in Lawrence, whose water intake was nine miles downstream from the former community's sewage discharge, the toll of victims mounted. Investigation of the higher rates of illness and death along unprotected rivers produced "remarkably conclusive evidence of the river water supply being the direct cause of the epidemics."[22]

Only when Lowell's typhoid mortality rate increased to over 300 percent between 1889-90 and 1890-91, and the typhoid rate in Lawrence reached three times that of Boston,[23] did the members of the Lowell Water Board take action. They invited William Thompson Sedgwick, recently appointed as the first head of the newly formed Department of Biology at the Massachusetts Institute of Technology, to investigate.[24]

Sedgwick's meticulous observations established the point of origin for the outbreak in the privies of the neighbouring community of North Chelmsford and traced the passage of the typhoid bacillus "down the river and over the falls" along the Merrimack into Lowell's water supply.[25] His report on Lowell was unequivocal about the intimate connection between sewage practices and disease: "We shall look in vain for any adequate explanation of the constant excess of typhoid fever in Lowell and still more in Lawrence except to the fact that both these cities have constantly distributed to their citizens water, unpurified, drawn from a stream originally pure but now grossly polluted with the crude sewage of several large cities and towns."[26]

Sedgwick's condemnation of conditions at Lawrence must have been particularly disheartening, for here, half a decade before his inquiries into the typhoid tragedy began, Massachusetts had established an experimental facility for the study of sewage disposal. As "Chemist in Charge" at the Lawrence Experimental Station, the nineteen-year-old Allen Hazen had begun to establish the foundations of a profoundly influential career that took him across the United States, into Canada, and as far afield as Australia and Japan. In one Lawrence study, ten circular cypress tanks seventeen feet in diameter and six feet deep were filled with various materials including sand, gravel, peat, river silt, loam, garden soil, and clay, to serve as experimental filters through which doses of city sewage were passed at twenty-four-hour intervals. Although peat and garden soil proved excessively impervious and subject to clogging, the other filtering materials converted sewage into a clear and completely nitrified effluent. The results of purification were comparable, it had been confidently asserted, to well waters used in the city of Lawrence.[27]

The Lawrence experiments helped to reveal that sewage purification resulted from bacterial oxidation, and demonstrated the role of intermittent application in maintaining the necessary oxygen levels.[28] The essential conditions, as explained to the Massachusetts State Board of Health, involved "very slow motion of very thin films of liquid over the surface of particles having spaces between them sufficient to allow air to be continually in contact with the films of liquid." Here, bacteria did their work, with the consequence that during an experiment conducted over several months, the intermittent filtration process over gravel stones removed 97 percent of the organic nitrogenous matter – a large part of which was in solution – as well as 99 percent of the bacteria. These organic matters were oxidized or burned, so that the resulting effluent contained only 3 percent of the decomposable organic matter of the sewage.[29] By the early 1900s, twenty-three Massachusetts towns and cities operated such intermittent filtration sewage treatment plants to encourage bacterial decomposition.[30]

The importance of supplementing traditional chemical analysis of water with bacterial research had been firmly placed on the agenda by the pioneering bacteriological inquiries of Louis Pasteur and the subsequent investigations of Robert Koch. The former's microbial studies in the context of beer, wine, and vinegar production were soon followed in 1883 by Koch's investigation of a possible linkage between a distinctive "comma shaped" organism and the spread of cholera.[31] Very shortly thereafter, the etiology of typhoid and its relationship to sewage in waterways was more clearly understood: coliform bacteria, prevalent in human and animal feces, though not ordinarily found in water, signalled fecal pollution and indicated the possible presence of pathogenic organisms.[32] The biological revolution suggested from the 1880s that germs, rather than noxious smells, putrefaction, or miasmas were responsible for many diseases. Nonetheless, miasmas and their cousin, "sewer gas," retained their status as treacherous foes for many years. Earlier theories, often deeply embedded in patterns of thought and even in institutions, proved extremely difficult to dislodge. Even as late as 1900, health departments were still discouraging gas and water companies from excavating city streets during warm weather, for fear that potentially dangerous miasmas might be released. Sewer gas was widely regarded as a source of disease until the First World War.[33]

Even where the new contagionist principles were acknowledged, the implications encountered resistance. A number of US courts had shown a singular reluctance to impose preventive obligations on companies engaged in water supply. In 1891, for example, Pennsylvania water companies were relieved of the obligation to respond to new knowledge, even though the court recognized that typhoid fever was "produced by a specific typhoid germ existing in the excreta of a person sick with that disease, which, being deposited in a stream, multiplies so that it contaminates the body of the water and reproduces the disease in the persons who drink it." A few years earlier, another Pennsylvania court had sharply lowered the performance bar: "Even comparatively pure water is hard to be obtained in large quantities, for in populous sections of the country where waterworks are most needed, neither rivers nor small streams can be kept

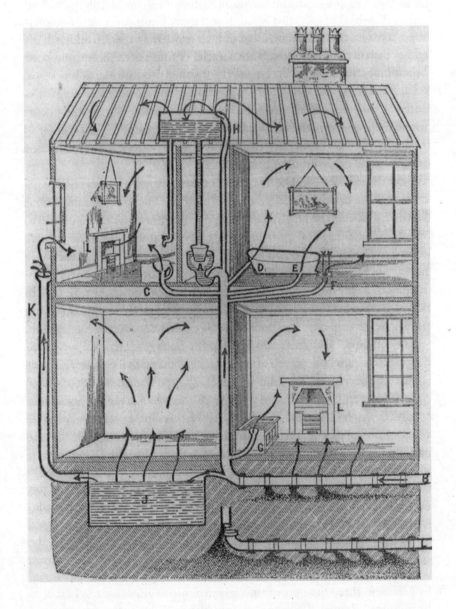

One of the risks believed to be associated with sewerage and plumbing was the release, in improperly constructed systems, of gas resulting from decomposing wastes, which could poison household residents.

entirely free of sewage." To the court, these were matters of "common observation" requiring no substantiation from experts; "purity" would thus be interpreted pragmatically to mean "wholesome, ordinarily pure." To put its dismissal of specialist opinion still more bluntly, the court emphasized, "We must use [the word purity] as it is used by the world at large and not in the abstract or chemical sense." In no other way would it be possible to attain the court's chosen outcome of ensuring that a water company charter would remain "economically valuable."[34]

State Policy, Local Funding, and Someone Else's Health

A few North American cities – first New York (1866) in the wake of a severe cholera epidemic, and then Chicago (1867) – established boards of health at the municipal or metropolitan level. Yet when local officials proved disinclined to address such public health problems as the contamination of water supplies by sewage, solutions were sought at the state or provincial level. As one early-twentieth-century summary of governmental responsibilities for waste management commented, "The interest of the city is to get rid of its waste; the state sees to it that one municipality does not commit a nuisance upon others."[35] Following Massachusetts's example, state and provincial boards of health dominated by medical practitioners and public health professionals began to emerge elsewhere.[36] National health organizations sprang up: the American Public Health Association in 1872, followed within a decade by the short-lived US National Board of Health, a federal government response to yellow fever devastating Memphis, New Orleans, and the Mississippi Valley. In Britain, the *Public Health Act* of 1875, with its requirements for local boards of health, represented a pivotal accomplishment. The public health movement also gained recognition through research initiatives and specialized training programs in centres such as Trinity College, Dublin, in 1870 and at Cambridge. Equivalent institutions appeared in Canada.

Yet in both Britain and North America, controversy persisted about the appropriate location of responsibility for water quality. In 1882 a legislative committee in Ontario determined that the water supply was being polluted by privies in three-quarters of the eighty municipalities that responded to its inquiries. Remedial efforts were virtually non-existent and disease was widespread. The committee called for a provincially appointed board of health.[37] This body repeatedly encountered municipal penny-pinching in its efforts to persuade local councils to act systematically and methodically in dealing with the sewage of their burgeoning populations. In Toronto, where health officials had responsibility for sanitary conditions affecting nearly a hundred thousand people, civic leaders allocated a mere $500 to the local board of health.[38] The situation was not unlike that in Massachusetts a few years earlier, when nuisances such as polluted water and contaminated food were also accepted as the responsibility of town and city governments whose commitment was at best uneven.[39]

As health officials assumed the role of the public "conscience" for water quality, their efforts remained grounded in the common law of nuisance. Law offered a few gratifying successes, but was equally a source of frustration, thanks to its preoccupation with property rights and procedural preconditions. Legislation greatly extended – indeed, it formally created – the authority of health officials. Simultaneously, it constrained that authority within broader norms.

Ontario health officials were initially encouraged by successes in the courts. In 1884 they reported enthusiastically, "The reading of the law has been so clear that verdicts against offenders have been obtained and remedies have been effected."[40] But if early successes, typically involving privy complaints, were celebrated, satisfaction was short-lived. Waves of individual offenders were dishearteningly common in certain communities, and public health professionals soon grew skeptical of legal process. It seemed excruciatingly difficult to establish nuisance at trial – "not only whether this or that condition is *injurious to the public health*, but whether it is *materially offensive to the senses*, or interferes with the *enjoyment of life and property*." Health officials lost confidence in the capacity of juries to reach decisions that they – as experts – would consider appropriate: "To make the question of whether a man with senses rendered obtuse is or is not nauseated by a smell a criterion of the existence or absence of a nuisance is as crude as was trial by fire in old Saxon times, since the guilt or innocence of the accused was tested by his power to endure pain."[41]

Pollution, including sewage contamination, extended well outside urban centres. Several years after his survey of industrial nuisances in Ontario, public health secretary Dr. P. H. Bryce investigated sanitary conditions at the health resorts of the fabled Muskoka Lakes. Shocked by his findings, Bryce launched a remedial campaign. He intended not only to curtail the spread of typhoid and similar diseases but also to safeguard "the further aesthetic sense which must be satisfied if the celebrity which the District already enjoys is to be maintained."[42] Five years later, though, when Bryce's colleagues conducted a follow-up examination of a wider portion of the cottage districts, they found very few dry-earth closets in the tradition of the Reverend Henry Moule and no arrangements for sewage treatment. "Many of the summer residents draw their water from the same bay or locality into which the closets drained, and the kitchen slops ultimately found their way."[43] Regular examinations by health officials continued to reveal alarming conditions such as a widespread tendency to install pit privies with "rarely any pit" on the smooth rocky slopes of the Canadian Shield. Frequently these faced neighbouring homes. "Rain runs down this rocky slope, carrying the soakage under the sawdust in the ravines to their outlet, or through crevices in the rock, from one ravine to another, until the river is quickly reached."[44]

Systematic and determined inspection throughout the jurisdiction began producing results. By the 1920s, the Muskoka situation was more encouraging. Sanitary conditions were good, with the water generally well protected from contamination, even if occasional recommendations asked cottagers to lengthen their water intakes or to relocate them away from bathing beaches. Much of that progress, however, depended on the nature of emerging regulatory norms regarding inland waters and on indications that officials would actively enforce them.

In reviewing American law on inland water pollution for the United States Geological Survey in 1905, Edwin B. Goodell found that despite uneven levels of "public enlightenment as to the deleterious effects of water pollution," there was no shortage of statutory initiatives to alleviate the problem.[45] These he presented in three rough categories. In the first, represented by seventeen states, Goodell could ascertain "no sense of the general desirability of pure natural waters, but only a desire to prevent certain acts recognized as criminal in intent or as likely to injure special groups of persons whom the legislature desires to protect."[46] These jurisdictions had simply enacted prohibitions, albeit often accompanied by the threat of imprisonment, for wrongs related to the offence of knowingly or wilfully depositing noxious, poisonous, or offensive matter in or near water supplies, springs, wells, or reservoirs. If we may judge from the frequency with which specific prohibitions appear, dead animals were particularly adept at finding their way into the sources of water supply.

A further twenty states had made somewhat more progress in protecting their water supplies. In this second grouping, prohibitions similar to those in Goodell's first category were often supplemented by greater detail – concern about contamination of ice supplies, for example. In addition, a number of these states had conferred modest regulatory authority over water pollution on boards of health, occasionally funding enforcement actions or the operation of laboratory facilities. A few states required permits for the discharge of wastewater, some even insisting that sewage be treated before effluent could be released.

Eight states – Connecticut, Massachusetts, New Hampshire, New York, New Jersey, Minnesota, Vermont, and Pennsylvania – constituted Goodell's third category. These jurisdictions, mostly in the east, he credited with "stringent methods to enforce the right of their citizens to unpolluted natural waters."[47] Their enactments, he anticipated, would control pollution so as "eventually to prevent all danger to public health." Refinements adopted in these states served to encourage regular water quality investigation and reporting. Authorizations to enter premises subject to public health regulations or considered possible sources of pollution were also commonly granted. Some states, New Jersey, for example, provided for sewerage districts or boards with supervisory responsibility over permits, treatment facilities, and means of financing the costs of infrastructure. Remedial measures and the prevention of pollution were also addressed.[48]

Legislative measures to bolster the formal authority of health officials, refining prohibitions earlier embedded in local municipal waterworks statutes, were also welcomed in Canada.[49] Toronto's 1872 legislation, for example, had already made it an offence to "bathe or wash or cleanse any cloth, wool, leather, skin or animals, or place any nuisance or offensive thing" within three miles from any source of supply for the city's waterworks. Citizens had also been discouraged from conveying "any filth, dirt, dead carcasses or other noisome or offensive things" into sources of water supply and enjoined not to "cause, permit or suffer the water of any sink, sewer or drain" to find its way into whatever lake, river, pond, or other source the city selected to supply its waterworks. Offenders were liable for a penalty of up to twenty dollars, of which ten would go to the waterworks themselves and ten to whoever initiated the proceedings; the presiding justice of the peace had the discretion to impose up to a month in jail "with or without hard labour."[50]

In 1906 a more sternly worded general prohibition was promulgated on a province-wide basis: "No garbage, excreta, manure, vegetable or animal matter or filth shall be discharged into or be deposited in any of the lakes, rivers, streams or other waters in Ontario, or on the shores or banks thereof."[51] In a later revision of the *Public Health Act* (*PHA*) that further fortified the public health arsenal, officials were empowered to develop regulations for preventing pollution in the province's lakes, rivers, streams, and other inland waters.[52] Perhaps most significantly, for purposes of the *PHA*, "nuisance" was redefined to pertain to more than inconvenience or aesthetic sensibilities: "any condition . . . which is or may become injurious to health or prevent or hinder in any manner the suppression of disease."[53]

Regulatory action against those whose acts threatened public welfare was undoubtedly an important alternative to the procedural and financial pitfalls of private litigation – or to formal criminal prosecutions in which technical and evidentiary requirements might prove insurmountable. But, as public health officials increasingly recognized, prohibitions against pollution were no more self-enforcing than the Ten Commandments. Indeed, the paradoxical coexistence of permissive regulations alongside prohibitions risked undermining the authority of the latter. As judges and other officials considered prohibitions and regulations on the front lines in local communities, the practical and symbolic significance of legislative measures was publicly tested, and anomalies in enforcement exposed.

Other incidents more positively indicated the potential of a determined environmental and public health bureaucracy to achieve its objectives – when supported by the judiciary and the legislature. In 1912, Dr. John W.S. McCullough put Ontario residents on notice that anyone contravening the pollution provisions of the *PHA* would be "prosecuted to the full extent of the law."[54] An opportunity to test the strength of administrative authority under the province's statutory nuisance provisions arose in 1918. The Waterloo Board of Health sought to enforce a local abatement order against a trio of offenders – the city of Kitchener, a business provocatively operating as the Riverside Garbage Disposal Company, and a landowner, Campbell, who lent his name to the case.[55] When these three parties disregarded the Waterloo board's order of abatement and removal concerning garbage from Kitchener that had been deposited by Riverside near a creek on Campbell's property draining into the Grand River, local officials called upon the Provincial Board of Health (PBH) to investigate. The presence of ashes, metal- and enamel ware, and kitchen wastes along the creek bank was confirmed, as was that of decomposing animal and vegetable material. The board declared the situation "a most insanitary one, a serious nuisance, and extremely dangerous to the public."[56]

The local Waterloo board then applied for a court order to remove and abate the nuisance pursuant to a provision authorizing a judge to impose such a requirement "upon the report of the Provincial Board or upon such further evidence as he may deem meet." In upholding a court order issued against the city of Kitchener, Mr. Campbell, and Riverside Garbage Disposal, the appeal court unanimously declined to open up the question of the existence of a nuisance. "The Act," stated Chief Justice Mulock, "confers jurisdiction upon the Provincial Board of Health to determine that question, and there is no jurisdiction in the Court to try that question of fact."[57] Encouragement of this nature may have reinforced the confidence of public health administrators, even emboldening them in some respects to contemplate further intervention.

Frederick A. Dallyn, sanitary engineer for the PBH, believed the time was ripe for the province to suggest collaborative ways for municipalities to handle their sewage as well as improve their water supply. Smaller municipalities were "keenly concerned" about the situation, he urged, but, as they lacked local engineers, could take no steps to assess the practicality of remedial alternatives. Assuming that the province would take some initiative, Dallyn outlined further issues to be considered. Would the PBH be content to discuss generalities and ultimately to generate a little business for consulting engineers, or would it wish to furnish each municipality with a plan and a general cost estimate, either at no charge or on the basis of some formula for cost recovery? Given provincial support, Dallyn argued, the engineering department might (without waiting for civic initiatives) collaborate with local health officers to campaign for improved sewers, treatment facilities, the extension of water supply systems, and water purification processes – especially in smaller municipalities.[58]

As an example of the difficulties that had to be overcome, Dallyn outlined the situation at Kincardine on Lake Huron's eastern shore, where the mayor was anxious to install sewerage facilities. "Their desire," he explained, "is to sewer one little section of the town and drain the same into a septic tank, allowing the effluent to discharge directly into the river." On a recent visit he had observed "very little flow of water in the river and in some places it was practically dry." In such circumstances, Dallyn cautioned against permitting the town to discharge untreated effluent, and warned against a partial or patchwork solution. A previous attempt to implement a comprehensive scheme had foundered, he added, "principally owing to the fact that [the ratepayers] had not consulted the provincial board of health."[59]

Unwillingness to address the challenge of treating sewage was not confined to smaller communities. Many major centres had a less than sterling record when it came to dealing responsibly with residential, commercial, and industrial wastes. Nor was it entirely clear that the public health administration actually could influence or accomplish sewage treatment to the degree that sanitary officials might have wished. Such vignettes from small communities were rapidly compounded on a scale that affected entire North American watersheds.

Wastes Unlimited in Boundary Waters

A survey of major watersheds revealed the difficulties facing downstream consumers of water as the twentieth century got under way. On the roughly eight thousand square miles of drainage area above Philadelphia's intake point on the Delaware, for example, lived 360,994 people; the city's other supply source, the smaller Schuylkill basin, was inhabited by a further 257,415 citizens, densely packed at 134 per square mile. St. Louis took its water from the Mississippi below the discharged waste from over five and a half million upstream urban residents. Further along the same river, New Orleans drew water at a point downstream from twelve and a half million people. Louisville water came from the Ohio below a drainage basin inhabited by over three million urban dwellers, and so on.[60] Though distances and densities varied, such statistics provided, in the surprisingly unhurried expression of the United States Geological Survey, "some indication of the problem which must be met in the near future."[61] The same report, again revealing no particular sense of urgency, described the challenge of treating discharges in the Boston area alone as "a problem in applied chemistry of no mean magnitude."[62] Although Bostonians would defer to the ocean for some time, the question of whether to address sewage treatment – in addition to sewerage – in the interests of public health and water supply was engaged elsewhere.

But the magnitude of the public health and environmental challenge to international watersheds was not yet widely recognized. Allen Hazen, an experienced student of treatment engineering, dismissed sewage treatment in 1914 as a viable contributor to public health: "The Great Lakes are so large, and the dilution and time intervals and exposure to sun and air are so great that there is no chance of infection being carried from one of the great cities to another."[63] The sewage of Detroit, he categorically insisted, was harmless to Cleveland, and sewage from Cleveland posed no threat whatsoever to Buffalo. Perhaps Hazen was unaware of the extent to which the Detroit and Niagara Rivers were being exploited for waste disposal purposes in the late nineteenth century. Perhaps he had not heard of the barges of municipal waste that were being towed out by the Detroit Sanitation Company for dumping near Amherstburg, Ontario, emboldening Canadian customs officials to arrest the perpetrators. With a similar approach to waste management by Buffalo meeting the same fate,[64] mounting expressions of concern from both sides of the border forced the United States and Canadian governments to address water supply and sewage treatment on a bilateral basis.

The challenge was evident enough in comparative typhoid mortality rates. These exposed a sharp contrast between the overall incidence in Canada and the United States (35.5 and 46.0 per 100,000 respectively) and the vastly more satisfactory results then being achieved in much of Europe. Even the worst European experiences, in Hungary (28.3) and Italy (35.2), were better than the North American record. In the assessment of George Chandler Whipple, author of *The Microscopy of Drinking Water* (1899), the overall situation in the United States as of 1907 was that in cities with "reasonably good water supplies" the typhoid fever death rate was around twenty per hundred thousand. In communities whose supplies were "more or less contaminated," the rate rose up to forty or sixty deaths per hundred thousand.[65] A good many communities around the Great Lakes suffered substantially higher rates.

Powerful voices were being raised against the flood of sewage. Charles Evans Hughes, New York State's influential governor, had risen to prominence through his exposure of malpractice in gas utilities and insurance companies. Turning his energies to water quality, Hughes proclaimed in 1909 that the state could "no longer afford to permit the sewage of our cities and our industrial wastes to be poured into our watercourses." In 1910, former president Theodore Roosevelt emphasized before a Buffalo audience the importance of protecting the quality of the Great Lakes. He took the unarguable position that "civilized people should be able to dispose of sewage in a better way than by putting it into drinking water."[66] In a series of articles that appeared in the *Toronto Globe,* T. Aird Murray, a Canadian civil engineer, joined a number of fellow citizens in calling attention to the extent of the public health crisis attributable to contaminated water supplies.[67] A prominent American counterpart similarly questioned presumptions about the security of water supplies in the bacteriological era: "He who says that a polluted river will purify itself in the course of several miles reckons with an unknown force which will probably fail him at the critical time."[68]

In the aftermath of the 1909 Boundary Waters Treaty, Canada and the United States took advantage of the newly created International Joint Commission (IJC) to put water pollution on the international agenda. The neighbouring countries specifically asked the commission to investigate the location, extent, and causes of boundary water pollution that was injurious to public health and rendered the affected waters unfit for domestic or other uses. Remedies were requested, whether involving the construction and operation of suitable drainage canals or treatment plants. The inquiry also encompassed potential preventive measures to make the waters of the

Great Lakes sanitary and suitable for use, so as to fulfill treaty obligations. The parties agreed that boundary waters and waters flowing across the boundary would not be polluted on either side to the injury of health or property on the other. This was a tall order whose scale was perhaps not fully realized even after the eventual completion of the inquiry's work in 1918.

A preparatory conference in Buffalo, 17 December 1912, brought together officials from Canada, Ontario, and Quebec as well as the United States and several individual states most directly affected. Next summer, follow-up hearings in Buffalo attracted municipal representatives as well as persons connected with business associations and other parties. The Buffalo gathering identified a research agenda, which Dr. Allan J. McLaughlin of the US Public Health Service (PHS) would oversee as chief sanitary expert and director of fieldwork. Among the Canadian participants, Dr. J.W.S. McCullough and Dr. John A. Amyot of the Ontario board of health were named as consultants to the undertaking. By September 1913, the scope of the investigation had been determined, and arrangements formulated to examine the Niagara River, the Detroit River and connecting waterways from Lake Huron to Lake Erie, St. Mary's River, the St. Lawrence River from Lake Ontario to a point where it departed from the boundary, and a portion of the St. John River.[69]

At this point, more than seven million people lived along the boundary waters from Lake of the Woods, separating Ontario and Minnesota in the west, to the St. John River, flowing between New Brunswick and Maine in the east. Extensive pollution, signalled by the presence of certain microorganisms in water samples, was common in centres of population.[70] Although the nature of the IJC and the terms of reference suggested that the inquiry would be confined to pollution on one side of the boundary which affected waters on the other, researchers were naturally anxious to incorporate a number of purely domestic considerations into their operational mandate.

The research program involved analysis of about eighteen thousand samples taken from fifteen hundred locations around the Great Lakes, and reviews of the historic incidence of certain diseases, accompanied by an extensive program of interviews and correspondence. In concluding what they described as the most extensive investigation and bacteriological examination ever made in the world, the commissioners presented their preliminary findings in 1914. In the absence of comprehensive information establishing historic baselines, the report's authors made comparative references to conditions in other jurisdictions. However, the use of these horizontal benchmarks – perhaps the best or most persuasive indicators that might have been obtained – had the effect of establishing standards already far removed from pre-industrial conditions on the lakes. Pollution was therefore being defined against a baseline or norm that appeared already to take for granted a significant level of contamination from human activity.

Addressing the effects of pollution on public health, commissioners indicated that – with the exception of public water supplies – the sanitary and climatic conditions of cities and towns around the Great Lakes were much better than national averages, and infinitely better than those pertaining in the filthy, overcrowded, and often impoverished cities of Europe. Yet despite such advantages, excessive rates of typhoid fever had been documented for years in Great Lakes communities. The explosive epidemics sometimes seen were said to be without parallel in the European context. Death rates attributed to typhoid fever averaged less than five per hundred thousand in the large cities of Northern Europe where water supplies – often underground – had been secured, but the Great Lakes inquiry revealed disturbingly high impacts in many North American communities. Between 1910 and 1912, the rate of death per hundred thousand ranged from fifteen in Detroit to well over a hundred in many other cities, and skyrocketed to over three hundred in Ashland, Wisconsin.[71]

The IJC's advisors advanced a straightforward explanation directly implicating untreated sewage in the public health crisis: "The greatest single factor in this avoidable and remediable pollution is the sewage discharged without restriction or treatment of any kind by the municipalities situated on the boundary waters."[72] The Niagara River situation illustrated this principal finding. On the American side, a population of roughly 615 thousand (including 100 thousand rural and more than 500 thousand in the cities of Lackawanna, Buffalo, Tonawanda, North Tonawanda, and Niagara Falls) and occupying approximately two thousand square miles, with one exception discharged raw sewage directly into the river above the falls.[73] The waters below the falls were dangerously polluted, affecting municipalities on both the Canadian and American sides. Buffalo, a city of 460 thousand people, was the most important contributor to Niagara River pollution, for the city discharged all its sewage in an untreated state into the river above the intakes of the public water supplies of all the cities below.[74] The researchers' analysis of the tourist mecca was clear in its assessment of the implications: they rejected the popular impression that the action of the falls purified sewage. "It simply mixes it more thoroughly with the water; it does not remove it or its danger. The pollution below the Falls is gross."[75]

The Canadian situation had yet to be addressed in detail by sanitary experts, but at Niagara on the Lake, the researchers reported, "the injurious effects of the pollution from the upper cities on the river have been seriously felt." These findings were in marked contrast to the assessment gratuitously offered by British engineer James Mansergh in his 1896 report on Toronto's water supply. Mansergh had quoted Dr. Edward Frankland on the quality of the Niagara River at the entrance to Lake Ontario: "The water of the Niagara River as it enters Lake Ontario is of excellent quality, for, although it has received the sewage of Buffalo and other places, the immense volume of water with which this is mixed renders its effect upon the chemical, as distinguished from the bacteriological character of the water, inappreciable." Yet Mansergh had added a point that was taking some time to register with municipal officials around the Great Lakes: "The bacteriological condition of the water intended for dietetic purposes is probably of greater importance than its chemical composition."[76]

There was no particular reason apart from size to single out individual municipalities for critical comment, since the investigators had quite categorically concluded, "Every municipality, without exception, in the area investigated of the Great Lakes and their connecting rivers, avails itself of the opportunity to discharge its sewage untreated into these international waterways. This is the largest factor in their pollution."[77] Although these observations might seem to provide a powerful foundation for a vigorous assertion of the need for sewage treatment, that was hardly the result.

The IJC offered up success stories as examples for other communities to follow: Cleveland saw its typhoid death rate fall to single digits of seven per hundred thousand after 1912 and Erie, Pennsylvania, reported equally positive improvements. The report then advanced a finding with immense and continuing significance for future water management: rather than treating sewage – the outflow – communities such as Erie and Cleveland had taken advantage of new chemical or mechanical procedures to treat water prior to consumption.[78] These means would increasingly expose a gulf between the protection of human health and the preservation of the natural environment.

Commission member Dr. McLaughlin underlined the futility of a sewage-oriented campaign: "The source of Detroit's water supply is polluted," he declared, "and the attempt to purify is ineffectual."[79] The Tonawandas and Lockport suffered a still more intemperate dressing down: Tonawanda and North Tonawanda "still drink sewage-polluted water, expending their energies in a fruitless effort to improve sewerage conditions in the Upper Niagara River instead of protecting themselves by treating their own water supplies." The town of Lockport drew water from the same source, and, "in spite of repeated warnings and advice," followed the same course as the Tonawandas. In forty-eight hours and at a cost of under $1,000, Dr. McLaughlin claimed, a plant could be installed to treat the water supply with hypochlorite of lime. Treatment costs would be less than fifty cents per million gallons. Even if it were later decided to construct a filtration plant, temporary arrangements of this nature were vital to save lives in the interim: "There is no excuse for delay in making the temporary installation."[80]

The priority accorded human health in these circumstances is by no means surprising, and where modest expenditures would allow civic officials to deal quickly and effectively with the threat of typhoid – after several promptings in the case of the Tonawandas – McLaughlin's rebuke was well deserved. But emphasis on remedial water treatment in the immediate interests of human health rather than on a comprehensive preventive alternative also signalled official acknowledgment within much of the medical community that, despite Teddy Roosevelt's vision of civilization, the flushing of untreated municipal wastes would not readily be curtailed at the start of the twentieth century.

A notable exception to the investigation's recommendation to forsake sewage discharge controls for water treatment – and a direct rejoinder to Allen Hazen's broad assurance that the water supplies of Great Lakes communities were entirely secure from each other's sewage – concerned vessels plying the Great Lakes. The scale of the phenomenon and its potential contribution to contamination of the Great Lakes was apparent from the fact that in 1912 alone twenty-six thousand vessels passed through the Detroit River. These and other vessels navigating the Great Lakes and connecting waterways annually transported a population of at least fifteen million. The sewage these vessels discharged indiscriminately along their routes – and in harbours – contributed materially to pollution in both countries.[81] Even ballast was problematic, since some vessels took on water ballast before leaving port and discharged it just before entering the port of destination. There was therefore a danger of polluted water being discharged near the intake of a city water supply in an otherwise uncontaminated harbour.[82] And, of course, passengers themselves were at risk, for even though lake vessels were supposed to fill their drinking tanks in midlake – ostensibly far removed from sources of pollution – the distance that pollution travelled from shore made it difficult to find unpolluted areas. "There is excellent evidence," the commissioners noted, "to show that vessels frequently fill their tanks from polluted sources."[83] Well before the IJC inquiry, though, shipping on waterways such as the Hudson had been identified as a source of contamination in itself: "If all city sewage were excluded from the river, the *ejecta* from passenger steamers passing up and down would make the raw water unsafe as a domestic supply, especially when such matter was discharged near a supply intake."[84]

Officials on the American side conducted a survey of lake vessels to determine whether any were equipped with holding tanks or other retaining devices. Without exception they were not so equipped; the sewage outlet pipes from these lake vessels discharged directly into the water. However, in a comment suggesting a cooperative approach, the vice president and general manager of the Detroit and Cleveland Navigation Company added that in response to a request from the Buffalo health department new procedures had been instituted: "Our rules require all toilets to be locked and remain locked, and have no sewage discharged from the time our boats leave the mouth of the Buffalo River until they are two miles beyond the intake pipe of the Buffalo Water System."[85] The remark was interesting for its indication not only that some measures were already under way, but for the fact that a local or municipally based initiative was involved; for, though local solutions had dominated the introduction of water and sewerage arrangements, the linkage between bacteriological contamination and disease was encouraging a shift toward more comprehensive responses.

As chief sanitary officer of the 1912 international investigation, Dr. McLaughlin had perhaps not entirely abandoned sewage treatment as a public health measure, but rather recognized that it would probably not come about through the ad hoc initiatives of individual communities. He argued, as British and Canadian authorities were also inclined to observe, that a more encompassing authority was essential: "The problem of pollution of interstate and international waters is so broad and affects so many interests that it necessitates for its equitable and efficient handling a central directing authority independent of local influences and prejudices."[86]

It was becoming increasingly urgent on both sides of the Canada-US border to ascertain which level of government was most suited to respond effectively to the distinctive and growing challenges of controlling water pollution. The existing options – local, national, or the intermediate-level jurisdictions of state or provincial governments – each had plausible claims. Local governments have always claimed a degree of responsiveness to community sentiment greater than state, regional, or provincial jurisdictions might offer, and all of these have tended to insist that the remoteness of national institutions disqualifies them from involvement in activities and services intimately associated with the preferences and well-being of individual communities. From the perspective of effectiveness, however, the calibre of personnel and access to financial resources – purse-string politics – have sometimes favoured national-level initiatives.

Whereas in the United Kingdom tension between local and national institutions was perpetuated in philosophical considerations and deeply rooted traditions, in the North American federations the potential for inter-jurisdictional controversy and uncertainty was embedded in constitutional documents. As a new generation of public health issues came to prominence, national governments were forced to reflect on their potential contributions and responsibilities, and on the jurisdictional basis for any actions they might contemplate. When pollution concerns of the Progressive era coincided with the formation of the US Public Health Service (PHS) in 1912, Congress authorized the new institution to study the problem. The PHS established a Center for Pollution Studies in Cincinnati, although its mandate was congressionally confined to navigable waters. Despite the absence of any powers to compel abatement, the PHS enjoyed considerable success in persuading state and local authorities to adopt water treatment along uniform standards.[87] Thus it came about that interstate transport provided the leverage on which federal drinking-water quality regulation oriented around bacterial standards was introduced in the United States.[88]

Navigable waters and transportation also grounded federal government water quality initiatives in Canada, where the public health implications of sewage discharges were widely apparent. Within a decade of the first US federal drinking-water quality standards, the Canadian government followed suit with bacterial quality standards for drinking water and water used for culinary purposes on vessels engaged in interprovincial and international transport.[89] Yet authority over navigable waters was ultimately too limited a basis on which to proceed against the ever-increasing volume of sewage and industrial waste pouring into Canadian waterways.

National Default and International Failure

Even within the confines of Canada's capital city, the Ottawa River, an industrial thoroughfare for the lumber trade of the nineteenth century,[90] was mistakenly – perhaps perversely – presumed to be pure as the new century got under way; the linkage between health and contaminated water supplies was not readily embraced.[91] In October 1910, Allen Hazen, whose consulting services extended internationally, arrived in Ottawa to advise city councillors. He informed them that the Ottawa River was "a yellow water, and yellow waters are hard to treat."[92] He recommended a dual process: chemical treatment to deal with bacterial contamination from sewage pollution, and mechanical filters to reduce turbidity and objectionable colouring. But to meet the long-term needs for a high-quality supply, Hazen proposed a connection to the McGregor Lake system in Quebec's Gatineau Hills, a network of lakes encompassing a catchment area of ninety-four square miles. McGregor Lake water appeared "extremely cheap" in relation to expenditures made by American and European cities for equivalent sources of supply, but in comparison with water pumped from the Ottawa River the Gatineau option seem expensive. In any other circumstances, Hazen insisted, "the opportunity of securing so large a supply of excellent water at a sufficient elevation, within a distance of 15 miles of the centre of the city, would be a most remarkable opportunity not to be allowed to pass."[93]

Electoral opposition to the opportunity presented by the Gatineau Hills, combined with judicial challenges on the part of ratepayers, delayed adoption of the scheme. In 1914, with a new city council in place, the Ottawa electorate formally expressed a continued preference for water drawn from the Ottawa River and filtered locally, over the more costly scheme to deliver uncontaminated water from Quebec. But when the actual plans for the less costly river water reached provincial health officials for approval, the Provincial Board of Health unanimously rejected the scheme, observing that the Ottawa River was "beyond any question, a polluted source of supply at all points in the vicinity of the city of Ottawa." Accordingly, the board concluded, it would not be consistent with its duty to the citizens of Ottawa or to visitors to the national capital to "countenance the use of water which, after mechanical filtration, constantly require[d] chlorination, when a pure and adequate supply, requiring no treatment whatever, [might] be readily procured."[94]

This particular skirmish was not an isolated clash between local politicians concerned with the practicalities of municipal finance and a remote provincial agency intent on imposing abstract and artificially high standards. The need to address drinking-water quality in the nation's capital had been pressing for some time, and had become particularly acute after the deaths of 174 people in successive typhoid epidemics in 1911 and 1912. With the outbreaks of disease attributed to pollution from the untreated sewage of the community of Hintonberg flowing down Cave Creek to Nepean Bay, where it entered the city's faulty supply pipe, the situation was urgent. Just when health officials once more seemed close to securing acceptance of their preferred source, a reassessment of the city's water volume requirements for firefighting purposes conclusively removed the Gatineau plan from contention.⁹⁵

Meanwhile, against this unfolding local backdrop and frustrated like so many other thoughtful observers by the dearth of municipal initiative, the distinguished francophone senator Napoléon Belcourt embarked upon a lengthy campaign for national action. As an Ottawa resident, Belcourt was thoroughly familiar with the national capital's dilatory approach to water supply; he had first called for amendment of Canada's *Navigable Waters Protection Act* in the interests of public health as early as 1910. The successful earlier ban on sawdust, he had suggested, might usefully be supplemented by a further prohibition. He called on Parliament to join him in declaring that "our noble rivers shall no longer be made the receptacles of the raw sewage of the country."⁹⁶ His proposal was diverted to the newly created Commission of Conservation, which recommended a modified version. Although passed by the Senate, the measure was not considered in the Commons because of the unexpected dissolution of Parliament.

In 1911, with renewed determination in light of another major typhoid outbreak, Senator Belcourt introduced draft legislation developed by the water pollution committee of a national public health conference. The proposal was in essence a prohibition against contaminating navigable water in Canada as a whole, subject to exemptions authorized by regulation or specifically permitted. Directed at surface waters throughout the country, the proposed protections were wider in scope than most provincial efforts to safeguard sources of water supply.⁹⁷

Belcourt's advocacy of anti-pollution measures was vigorous and wide-ranging. To demonstrate the severity of the underlying public health crisis, he quoted newspaper reports from a number of Canadian communities; he cited devastating typhoid statistics from Chicago where the Lake Michigan city had been polluting its own water supply; and he recounted at some length a local family's tragic experience with the water of the Ottawa River. With one of their twelve children already dead and five more hospitalized with typhoid, the Gravelles of nearby Deschenes, Quebec, illustrated the senator's point that water quality was the most important public issue of the day.

Belcourt turned to history, and to Roman law in particular, for the principle that water "is a natural commodity provided by the law of creation for the use of man." This foundation was not unlike the point of departure that had led Justice Joseph Story and Chancellor James Kent to the doctrine of reasonable use so widely credited with eroding the natural flow principle in the early nineteenth century. But for Belcourt, creation's gift for the use of man had other implications. "Consequently," he argued, putting his claim on a very high plane, "the individual and the public as well, have an inalienable and indefeasible right to pure water."[98] But the senator's proposition never became a rallying cry.[99] Belcourt furnished evidence that stringent legislative provisions had been implemented to this effect in European jurisdictions. And, lest apprehension about the practical challenges might deter action, he offered a brief inventory of successful – ostensibly even profitable – sewage treatment procedures.[100]

Outside Parliament, T. Aird Murray, the civil engineer whose views in the *Globe* we have noted, was among those who hoped for federal government action along the lines of the Belcourt initiative. At an October 1910 conference of the Public Health Committee of the Commission of Conservation in Ottawa, he discerned a general consensus along these lines within the ranks of provincial health authorities. Murray voiced the concern that isolated provincial actions would never achieve more than localized responses based on local interests, and that many aspects of the pollution problem would be ignored: "For example the province of Ontario may have the most stringent laws relative to water pollution, and after putting its house in order would be yet dependent upon the action taken by the province of Quebec relative to the pollution of the Ottawa river whose banks are interprovincial."[101] Similarly, in the United States, he wrote, "while one state may have drastic laws with reference to river pollution, the adjoining state may have none."[102] Action at the national level, on the other hand, offered attractions: standardized information could be assembled, neglected problems of interprovincial pollution could be addressed, and the array of questions associated with Canada-US boundary waters could effectively be confronted.[103]

In the debate, Belcourt had on his side, or so he evidently believed, important work by his own colleagues – an inquiry into water pollution by a Senate committee on public health chaired by Senator L. George De Veber. The report, from which Belcourt quoted extensively, fully supported his assertions on the severity of the water quality problem he hoped to address by amending federal legislation to prohibit pollution of navigable waters on a national scale. However, by the time Senator De Veber rose to speak, hope of proceeding expeditiously with the proposed amendment was already lost.

Senator McSweeney had voiced the first reservation, inquiring on behalf of the city of Moncton, New Brunswick, whether that community would be put to great expense by the far-ranging proposal. Moncton was by then well accustomed to discharging its sewage into the Petitcodiac River, confident in the capacity of this tidal waterway to flush municipal waste thirty miles into the Bay of Fundy, whence it would then be swept into the ocean.[104] Exemptions were available under the proposed amendment, Belcourt assured his senatorial colleagues, but the expression of doubt was under way. Other senators queried the constitutional authority of the federal government to enact the proposed measure, imagining it to fall more appropriately within provincial jurisdiction. The suggestion was made that the criminal law, statutorily codified in Canada in 1892, was a more suitable location for a prohibition of the sort envisaged. But it was the measure's practical implications that occasioned the most doubt.

Perhaps established communities could be spared, one senator reflected, if the proposed measure could be confined to new localities. Having satisfied himself that Montreal could not possibly prevent its own sewage from accumulating along the St. Lawrence waterfront, Montreal-based senator Casgrain uttered the self-interested proposition that "it would be a great improvement if in all the new places being constantly established above Montreal such a system were adopted."[105] The *Sanitary Review* was thus fully vindicated in its assessment of Montreal as "a hygienic disgrace to civilization."[106]

A former Canadian prime minister, Sir Mackenzie Bowell, advanced the opinion that the amendment as drafted was too wide in its implications to be carried out. By way of example, he described the circumstances of his own community on the Moira River, which flowed through the city en route to Lake Ontario's Bay of Quinte. The Moira, he explained, extended some hundreds of miles to the north along a course into which twenty or more villages of various sizes emptied their sewage. Numerous other communities along tributary creeks and branches similarly discharged wastes which descended the Moira to the very navigable Bay of Quinte. The proposed legislation, Bowell protested, "provides that if a dead horse is thrown into the river a hundred miles north of its outlet, or sewage from any of the towns or villages upstream is deposited in the waters running into the Bay of Quinte, then the operation of this law could be invoked, because the River Moira empties into the Bay of Quinte."[107]

The former prime minister's line of thought is difficult to discern, yet he emerges ultimately as a stalwart defender of the right to throw dead horses and to discharge sewage into rivers against the inalienable right to pure water championed by Belcourt. To this end, Bowell invoked "scientific treatises" purportedly establishing "very clearly that once sewage is emptied into a running stream, after it has travelled a certain distance it purifies itself."[108]

Belcourt had at least one strong ally in the person of Senator James Lougheed, who seems to have appreciated both the promise and the limitations of his colleague's proposal. Although by no means a panacea, Belcourt's measure struck Lougheed as an initiative that had "set public opinion in motion." Something useful might result to address a tragic state of affairs, he said, for, "It goes without saying that all our public streams, provincial and inter-provincial are becoming practically the great sewers of the Dominion." Municipalities, he observed, find it cheaper in their attempts to avoid indebtedness, "to empty their sewers into the streams which run by their doors, than to adopt some scientific method which possibly will cost more, for the purpose of cremating or otherwise destroying the sewage of that community." Concerted national action, Lougheed asserted, was essential to confront the intolerable situation that had developed: "We seem to have concluded that nature has placed those streams by our doors to carry off our sewage, and notwithstanding the fact that the community requires pure water, yet we will reject the best methods of purification and take the consequences. It seems to me we have reached that stage."[109]

Lougheed's intervention, however prescient, came too late to restore senatorial momentum for Belcourt's initiative. Senator De Veber's remarks had already sealed the fate of the amendment. He had personally undermined his colleague by asserting that Belcourt's liberal references to his own public health committee report had failed to acknowledge the difficulties the committee identified of implementing an integrated sewage prohibition such as the one under debate. A face-saving consensus eventually formed around the idea that Belcourt's draft bill and De Veber's committee report should both be subjected to further study.

Napoléon Belcourt's renewed campaign against pollution of navigable waters during the 1912-15 period met with the same kinds of opposition that had checked his original effort to keep municipal sewage out of Canada's lakes and rivers, much of it rooted in the expenditure-conscious attitudes of civic leaders whose ponderous responses to the water quality challenge imperilled their citizens. Municipal critics, notably in coastal communities blessed with the apparent infinity of the undrinkable oceans, rejected national sewage treatment measures as irrelevant to their circumstances, and found amenable parliamentary allies. Constitutional reservations persisted, while the boldness of the flushing constituency reached new heights: Montreal senator Henry Cloran maintained that Canada's geography provided "rivers and lakes large enough to contain all the refuse that the inhabitants of the country could discharge into them, without danger of contagion to the people."[110] A blessed country indeed.

Although Belcourt's proposal had actually secured a sufficient number of allies, even including government supporters, to sustain active interest in Canadian navigable waters pollution legislation, that pressure dissipated with anticipation of the IJC's final report on boundary water pollution. The desire to avoid inconsistent action, the virtues of being more fully informed, and the significance of simple courtesy or respect for the joint commission's efforts all counselled delay. Unfortunately, the IJC did not report in 1915 as expected, nor in 1916, nor the year after that. Only in September 1918 did the product of over half a decade of scientific and engineering research, public consultations, and vigorous deliberations emerge from the IJC.[111]

Whether a sincere interest in receiving the findings of the IJC's work, as opposed to an exploitation of a convenient source of delay, had caused Canadian authorities to set aside the Belcourt initiative is of some interest. Evidently the commission itself considered the delay unnecessary or merely opportunistic, for it cited the disinclination of governmental authorities at all levels to take responsible action as grounds for endowing yet another jurisdiction with authority over Great Lakes waters and effluent quality.[112]

In relatively short order – that is, by March 1919 – the two national governments agreed to call upon the IJC to formulate a convention or to draft concurrent legislation for the purpose of conferring such authority as would be necessary to remedy existing pollution problems. With this supplementary assignment completed, commissioners lamented the absence of preventive mechanisms that would allow them to "maintain boundary waters in as healthful a condition as practicable" rather than having to wait for pollution contravening the treaty before taking action. Ontario officials, however, saw some signs of interest in preventive action on the part of communities bordering international waters and expressed the hope that if these municipalities initiated sewage treatment (at least in the form of sedimentation), the impulse to do so might reach inland.[113]

Intermittent international negotiations throughout the twenties finally lapsed completely in 1929, to some degree in consequence of other preoccupations triggered by the Great Depression. At the end of the decade, though, health officials still imagined that some basic treatment standard would be adopted for international waters and that the example would inspire communities elsewhere to do the right thing. Only a collective initiative, it was now assumed, could overcome the natural inclination of communities to defer significant local measures in the interests of a wider constituency until they were confident that their efforts would be reciprocated. "So many of our municipalities are located on international waters where similar conditions exist on both sides that there is a distinct tendency to make no move until assurance is given that other offenders will follow the same course." A committee representing communities bordering on international waters had already been formed. It had agreed upon sedimentation as a minimum treatment at such time as treatment might be considered necessary. Ideally, such an example "should have an excellent effect on inland centres where conditions are generally more acute by lack of sufficient dilution water."[114]

This forecast proved overly optimistic. The most persistent and creative champion of the dilutionary impulse turned out to be the lakeside city of Chicago, where a unique engineering response to public health concerns laid the foundations for decades of political and legal controversy.

8

The Dilutionary Impulse at Chicago

CHICAGO'S DEEPLY ENTRENCHED COMMITMENT to dilution as a response to sewage contamination and urban wastewater dramatically demonstrated the obstacles that public health, engineering, and legal professionals faced in their attempts to combat the perils released by flushing. Here on the shores of Lake Michigan, civic preferences and local priorities held firm against regional, national, and even international concerns.

CONTINENTAL INVERSIONS

Until some four thousand years ago, Lake Michigan drained southward via the Illinois River. The natural reversal, a continental divide that ultimately made streams on the western shore flow into – rather than away from – the Great Lakes – St. Lawrence system, was never pronounced. As a result, when the city of Chicago began depositing sewage and other pollutants into Lake Michigan, the lake's current was insufficient to disperse such contaminants – or even to keep them away from the city's offshore water supply. Chicago's poorly drained and boggy terrain was prone to flooding, and mosquitoes abounded, contributing to a malaria problem that persisted, along with cholera and typhoid, until the early 1900s.[1] The Chicago River, whose north and south branches cross the city in their course toward the harbour, was so lethargic that in 1848 an observer had condemned it as "a sluggish, slimy stream, too lazy to clean itself."[2]

With such drawbacks, nineteenth-century Chicago appeared ill-suited to accommodate massive population increases brought about by its successive waves of immigrants – Irish, German, and Scandinavian. Many newcomers had been recruited as labourers for the Illinois and Michigan Canal, authorized by Congress in the 1820s and constructed pursuant to Illinois legislation of 1836. The ninety-six-mile canal, opened in 1848 after twelve years fraught with land speculation, funding shortages, and work stoppages, connected the Illinois River to Lake Michigan via five and a half miles of the Chicago River. With commerce to the south facilitated, Chicago, the gateway to the state of Illinois, fast became the hub of the American Midwest, prospering as a commercial centre and eventually gaining prominence over its rival, St. Louis, Missouri, whose access to Mississippi shipping was severely curtailed during the Civil War.[3]

After the centralization of Chicago's livestock operations in 1864, the new Union Stock Yard south of the existing city limits provided temporary accommodation for seventy-five thousand hogs, twenty-two thousand sheep, and twenty-one thousand head of cattle in some two thousand pens spread across a hundred acres. The visible infrastructure afforded access to an elaborate network of railway lines; beneath the surface, thirty miles of drainage pipes fed two massive sewers that transported water and stockyard offal to the "sluggish," "slimy," and "lazy" South Branch of the Chicago River.[4] And of course, the burgeoning human population of what was by now "The Great Bovine City of the World" was already making its own great demands on the local waterways.

The official response to Chicago's epidemic conditions and sanitary challenges combined engineering and public health dimensions. Municipal officials ordained the elevation of city streets – sometimes by as much as fourteen feet – to improve drainage,[5] and the city's Board of Sewerage commissioners obtained the assistance of Boston city engineer Ellis Sylvester Chesbrough to design suitable infrastructure. Chesbrough, who had begun his career in hydraulics in 1846 as chief engineer of the Boston Water Works, became commissioner of the Boston water system in 1850 and city engineer in 1851. Half a decade later, he took up the same position in Chicago, which in 1854 had lost over 5 percent of its inhabitants to cholera. Here, until his resignation in 1879, Chesbrough spearheaded a vast program of urban infrastructure designed to drain municipal sewage more effectively via the Chicago River into Lake Michigan. The implementation of Chesbrough's recommendation to use the Chicago River as the city's de facto sewer involved widening, deepening, and straightening its natural channel flowing into the lake.

By the time Chesbrough assumed the position of chief engineer of Chicago's Board of Public Works, he was forced to confront navigational concerns posed by the build-up of wastes entering the Chicago River. He had to face the dilemma presented by the city's simultaneous use of Lake Michigan as a sewer and as the primary source of municipal water supply. The navigational concerns he dismissed as "groundless." To those who objected about possible impacts of his scheme on public health, Chesbrough confidently cited dilutionary theory: health hazards would be eliminated "by pouring into the river from the lake a sufficient body of pure water into the North and South Branches to prevent offensive or injurious exhalations."[6] Yet the typhoid death rate continued to be alarming, averaging 65 per 100,000 between 1860 and 1900, and peaking in 1891 at 174 per 100,000.

The next effort at a solution, implemented during the 1860s, was to rebuild the Illinois and Michigan Canal, lowering its summit level to below that of Lake Michigan. Between 1866 and 1871, at the city's request, the canal trustees operated pumps at Bridgeport, near the mouth of the Chicago River, to flush sewage from the river into the canal and away from the lake. "Quite by accident," as Chicago historian Louis P. Cain recounted the episode, the flow of the Chicago River was reversed.[7] During those years, with the exception of periods of excessive rainfall, the Chicago flowed south, away from the lake. In this first engineered reversal of the Chicago River, the total annual diversion from the lake amounted to about 125 cubic feet per second (cfs).[8] In the following decade, additional pumping stations were added to deliver water directly from Lake Michigan to the canal, eventually increasing the annual diversion to about five hundred cfs.[9] Although they may have helped the quality of Chicago's drinking water, these measures did not noticeably alleviate pollution in the Chicago River.

Even with its dramatic growth to a population of some 325 thousand by 1871, Chicago might have proceeded to address water supply and sewage treatment requirements using land disposal or other techniques like those being developed elsewhere. But on 8-9 October 1871, a conflagration known thereafter as the Great Fire consumed virtually everything in a pathway some two-thirds of a mile wide and stretching four miles. With three hundred of its citizens dead, a further hundred thousand homeless, and $200 million worth of property destroyed, Chicago set out to recover from what historian William Cronon describes as a "key mythic event" in its history.[10] Inspired by leaders determined to overcome the setback, the resurrection and reconstruction of Chicago accelerated urban and industrial development. Louis Sullivan, Frank Lloyd Wright, and others of the Chicago School of architecture envisaged skyscrapers of iron, brick, stone, and glass raising the city to new heights,[11] while suburban growth rapidly extended the lateral reach of the metropolis.

As communities expanded, the question of pollution of inland rivers intermittently engaged the attention of the Illinois State Board of Health. A modest initial expenditure in the board's first year of operations, from 1877 to 1878, of $19.20 to collect and analyze water samples from the Chicago River provided no real indication of what lay ahead.[12] After investigating Chicago's pollution of the Illinois River in 1879, the board turned its attention to stream pollution at Chicago, Peoria, Springfield, Quincy, Rock Island, and Rockford. In 1885 the legislature set up a contingency fund to examine water supplies and stream pollution, and by 1888 analysts had demonstrated that a number of larger rivers in Illinois were polluted to varying degrees. A decade or so later, specific investigations were launched into contamination of the Mississippi River by sewage from the Southern Illinois Penitentiary at Chester.[13] Cases of typhoid were mounting toward the state of emergency declared at their peak in 1891. Once again, something had to be done.

The Chicago Steal

Vociferous complaints from downstream residents of Illinois and continuing public health concerns induced the civic leaders of Chicago – by now a city of 700 thousand – to seek a further solution to its problems. In 1887 they recruited Rudolph Hering, a New York civil engineer with considerable experience and knowledge of European practice, to the position of chief engineer for a water supply and drainage commission. This body examined such alternatives as a lakeshore interceptor system and land disposal using intermittent filtration on a sewage farm, perhaps extending over as much as fifteen thousand acres. For reasons of economy, however, it ultimately recommended the excavation of a new canal, which would be filled with water taken from Lake Michigan. The purpose of the canal was to extend the Chicago River westward, out of the Great Lakes – St. Lawrence drainage system. Beginning at the river's South Branch, near the shore of Lake Michigan, the canal would cross the continental divide, some twelve miles west of the lake, to flow into the Des Plaines River, which drained south into the Illinois River. The reconfigured waterway would transport sewage directly downstream to the Illinois, an arrangement that would replace the previous system of pumping into the Illinois and Michigan Canal.[14] To complete the diversion and to safeguard the city's water supply, a dam would be built on the Chicago River, to ensure that it could no longer enter Lake Michigan. The scheme, some proponents argued, offered the additional attraction of providing a new source of waterpower and "a magnificent waterway" linking Chicago and the Mississippi River.[15] This "new river" – in contrast with Sir Hugh Myddleton's London venture of that name almost three centuries earlier – was designed for elimination rather than delivery.

In order to dispose of Chicago's sewage and that of the surrounding territory through this common drainage outlet, Illinois in 1889 established sanitary districts, including the Sanitary District of Chicago (SDC). Thus was conceived one of the main protagonists in Chicago's controversial relationship with the North American environment. The SDC's mandate, conferred – or ostensibly conferred – by the state and initially extending over 185 square miles, transformed Chicago's relationship with the Great Lakes. The SDC would subsequently figure centrally in clashes between the interests of Chicago and those of its upstream and downstream neighbours, including jurisdictions bordering the Great Lakes as well as states along the Mississippi watershed, with which the city was to be linked by means of modern engineering.

Working from preliminary expert advice, the state legislature endorsed a dilution formula for the oxidation of sewage in the canal of twenty thousand cubic feet of water per minute (333 1/3 cfs) for every hundred thousand residents of the Sanitary District. Those later responsible for implementing and defending Chicago's unilateral appropriation of Lake Michigan water repeatedly invoked the experts who had originally determined the formula. Accepting the ratio as necessary to render the sewage inoffensive and unobjectionable when it reached Illinois Valley towns, the citizens of Chicago anticipated an expenditure of $50 or $60 million to comply. "In the interest of health, in the interest of the lives of our people," it would be said, "no sum of money is too great for us to expend in order to comply with the opinion of the experts."[16]

The advice on which this momentous decision was founded should have been followed by a final report setting out the findings in substantial detail. Unfortunately, Chicago declined to allocate sufficient funding for the report, Isham Randolph, chief engineer of the SDC, later explained, "so that we do not know what it would have contained."[17] Surprisingly, despite an expressed willingness to spend whatever it took to implement expert opinion, Chicago refused to provide enough resources to assess the

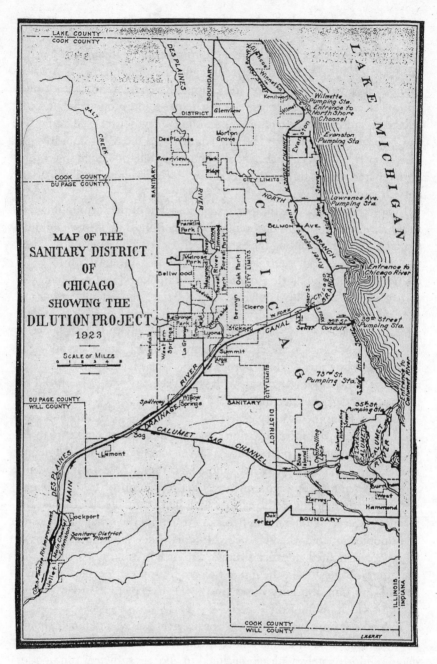

Map of the Sanitary District of Chicago, 1923. To safeguard Lake Michigan as a source of municipal water, the Sanitary District of Chicago redirected the flow of the Chicago River southwards, away from the lake and towards the Des Plaines River. While other communities moved to develop and install sewage treatment facilities, Chicago officials defended a statutory scheme that endorsed dilution as a waste management process.

reliability of that expert opinion. As a consequence, the database on which Illinois enshrined dilution theory in legislation remained somewhat lacking.

Shovel Day, 3 September 1892, brought the commencement of work on the new canal, a moment celebrated, as such occasions tend to be, in the speeches of dignitaries and officials. Standing on the continental divide, Lyman E. Cooley, then a trustee of the Sanitary District of Chicago, staked out the ancestral grounds of his city's entitlement to the anticipated flow. Only "strange mischance," he daringly suggested, had caused nature to favour the St. Lawrence with the outflow, with the result that "we lost our heritage." The re-engineering was therefore undertaken "by right of primogeniture," whereby Chicago claimed "all that which we should possess and have the energy and purpose to acquire." Nature's caprice would be remedied by man's creative intelligence. Not only would the ancient outlet be restored, the project would "extend through the continent from fog bank, to tropic breeze as though it were the sea, joining coast, lake and river systems in one whole as is not possible elsewhere on earth." It would then, in the psalmist's words, indeed be possible to "go down to the sea in ships."[18]

The Drainage Canal, otherwise referred to as the Chicago Sanitary and Ship Canal (or simply the Main Channel), was officially said to be capable of accommodating some ten thousand cfs, a volume that would, according to the Illinois formula, meet the needs of a city whose population was projected to reach two and a half million.[19] In fact, the channel had been designed to accommodate as much as fourteen thousand cfs, a rate of diversion that planners had prepared for in anticipation of still higher population levels.[20] The city's 1892 *Standard Guide to Chicago* offered a glimpse into Chicago's grand ambitions for this new infrastructure: "The one great object of this ship canal . . . is to dispose of Chicago sewage . . . plans are being perfected for a channel which will answer the double purpose of disposing of the city's sewage and establishing a navigable waterway for the interchange of commerce between Lake Michigan and the Mississippi River."[21] In short, a navigable sewer.

This was the eve of the World's Columbian Exposition, 1893, an extraordinary opportunity for ambitious Chicago boosters to enhance their renewed city's stature on the world stage. No details were overlooked, as even waterworks and sewage were called into service. For the exposition itself, two water plants with a total capacity of more than sixty million gallons were installed. Claiming to have taken account of all relevant considerations, promoters championed drainage arrangements while insisting that "Perfect sewerage, too, is planned." More than six thousand water closets, constructed for the event at a cost of some $450 thousand, along with cafés and kitchens throughout the fairgrounds, would contribute their wastes to a series of injectors. From there, with the assistance of compressed air, the sewage would travel through underground pipes to four huge tanks for chemical treatment that was promised to render it "entirely inoffensive."[22]

Almost a hundred years after the event, the Chicago fair was still described as a "sanitary wonder" representing the "apotheosis of nineteenth-century urban sanitary engineering."[23] But it was not new. The treatment system installed at the World's Columbian Exposition closely resembled the Dortmund tank designed by Carl Kinebühler, by then already extensively used in Germany to treat industrial waste.[24] The adoption of chemical treatment at the Chicago fair was therefore hardly innovative. Still, it would be another three decades before Chicago began to process municipal wastes, acting then only under compulsion.

The divergence between fair organizers' aspirations for perfect sewerage and the SDC's statutory commitment to flushing on a continental scale is apparent. Momentum certainly played a part in maintaining the gulf between promise and practice, for the state legislature's endorsement of dilution theory and local civic commitment to redirecting the Chicago River occurred only slightly ahead of a further round of late-nineteenth-century sanitary initiatives. Just as fair organizers were laying the foundations for the exposition, William Thompson Sedgwick was reconstructing the passage of typhoid bacteria along the Merrimack at Lowell (see Chapter 7) in an investigation that greatly increased popular and professional awareness of germ theory. Once the Sanitary and Ship Canal was under way at Chicago, however, there was little inclination to put alternatives to dilution back on the agenda. By 1900, when the canal was finished, the direction of the Chicago River had been permanently reversed.

General public attitudes, influenced in part by the post-fire aspirations of Chicago's civic and commercial leaders, must also have been significant. A sense of confidence bordering on invincibility inspired the city, "Nature's chosen tabernacle," to pursue its chosen course, drawing upon vast regional resources, including an infinite supply of water. As the 1892 *Standard Guide to Chicago* boasted, the city's water supply came from the "inexhaustible and magnificent reservoir of pure water" of Lake Michigan.[25] To support this claim, the authors offered statistics on the Great Lakes – St. Lawrence system overall. Considered by many of its other residents to be part of their own natural heritage, the Great Lakes were here presumed by Chicago publicists to constitute their city's own inexhaustible reservoir, notwithstanding the fact that the Great Lakes – St. Lawrence system flowed eastward to the Atlantic.

The president of the Chicago Board of Trade gave other indications of the importance of the water system to the city. Speaking to the International Waterways Commission in 1906, Mr. B.A. Eckhart assured the panel that his association, given its extensive shipping concerns, was as interested as any commercial organization in maintaining lake levels. On the other hand, he incongruously maintained, "no community, city or state or large municipality can long exist without a supply of pure water. It is more essential to the health and welfare of a community than even pure food."[26] Members of the Chicago Industrial Club shared this sense of priorities: "If the lake level should be affected, the humanitarian question is the great question, and ... the health of the people is paramount."[27] Yet these questions of paramountcy and priority had always been matters of law, and the relative ranking of fish, sewage, ships, swimmers, and so on would soon be tested again – repeatedly.

Since 1890, federal rivers and harbours legislation in the United States had contained measures against the obstruction of navigable waters. Designed to deal with threats to navigation ranging from construction through refuse dumping, the progeny of these provisions, the *Refuse Act* of 1899, established a durable framework for federal permits applicable to wharves, bridges, bulkheads, piers, and "any refuse matter of any kind or description whatever other than that flowing from streets and sewers and passing therefrom in a liquid state" into navigable waterways.[28] Acknowledging the importance of federal approval for the Sanitary and Ship Canal, Sanitary District officials eventually contacted the US secretary of war in 1896 to request a permit to proceed with the Chicago diversion scheme.

By way of reply, the (federal) chief of engineers stated that no objection could be taken to the physical work itself, every aspect of which would improve the navigation channel of the Chicago River. More problematic, however, he responded, were issues relating to the actual flow of water through the channel. Would the Chicago River's new channel be capable of carrying 300 thousand cubic feet of water per minute (five thousand cfs) without undermining navigability on the river, and would Chicago even be permitted to withdraw such quantities of water from the Great Lakes, thereby inevitably lowering levels? As these matters were not within the scope of his immediate investigation, the chief of engineers scrupulously cautioned that the authority granted should "not be interpreted as approval of the plans of the Sanitary District of Chicago to introduce a current into [the] Chicago River." This proposition would have to be separately considered at a later date.[29] This unfortunate case of administrative incrementalism did nothing to deter the local enthusiasm.

With construction nearing completion, Chicago officials learned in December 1899 that St. Louis was taking steps to prevent the opening of the canal. The Mississippi River city – perhaps, like Chicago, subscribing to the view that the health of the people was paramount – had become aroused by the pollution threat from the Chicago dilution process.[30] A contest of no small importance was under way.

On 2 January 1900, Lake Michigan water entered the new canal via the Chicago River, and a two-week process of filling the channel began. By 16 January, with representatives of Missouri preparing to appear before the US Supreme Court, SDC officials scrambled to secure permission from Governor Tanner of Illinois to open the channel. On 17 January, having obtained that authorization, the SDC lowered the Bear Trap Dam at the south end of the channel and water began to flow from Lake Michigan to the Gulf of Mexico.[31] When Chicago sanitary officials learned that the St. Louis injunction had formally been filed, they had already settled contentedly into a celebratory lunch.[32] In the disdainful words of the *New York Times*, "The ceremonies marking the consummation of the great enterprise were brief and marked with something like undignified haste."[33]

A marked improvement in sanitary conditions at Chicago elicited more favourable notice on the United States east coast. Observing that the Chicago River was becoming clear, the *New York Times* noted that "the impossible" was under way. "The stream which . . . has been known rather as a solid than a liquid body, which polluted the city water supply for a quarter of a century . . . is actually approaching translucency and can be seen to move!" The *Times* added in an astute denial of one of flushing's major premises, "Of course the filth of fifty years is still at the bottom of the river, and the clearer water now seen is merely that of the lake flowing over the surface of the stream." It could be anticipated, though, that when the canal water began to run through the Illinois River en route to the Mississippi the flow from the lake would cleanse the muddy bed and sides of the river. At that point, the city would at last have a stream of which it would "not be ashamed." In due course, the greasy black water, which was gradually taking on a pale green colour, would turn "a blue like that of Lake Michigan."[34]

Residents benefited enormously from the re-engineering of the Chicago River, which led to a significant public health claim: "In general the result has been to make the death rate of this city lower than that of any other important municipality of the United States."[35] That was indeed a major shift, for in 1894, as determined by the US commissioner of labor, sanitary conditions in Chicago had lagged behind those of major eastern cities.[36]

Those facing the downstream consequences were much less appreciative of Illinois' accomplishment. Missouri estimated the volume of filth and sewage flowing in its direction – created by Chicago's one and a half million residents, the stockyards, slaughtering establishments, rendering plants, distilleries, and other riparian enterprises – at about fifteen hundred tons per day. Novelist Upton Sinclair graphically described Bubbly Creek, an arm of the Chicago River, as "a great open sewer a hundred or two feet wide" serving the Packingtown district. One long blind arm of the creek trapped filth "forever and a day." Here grease and chemicals, undergoing "strange transformations," gave the creek its name, for it was "constantly in motion, as if huge fish were feeding in it, or great leviathans disporting themselves in its depths. Bubbles of carbonic acid gas [would] rise to the surface and burst, and make rings two or three feet wide." In places where grease and filth had caked solid, Bubbly Creek resembled "a bed of lava" that chickens could walk upon, feeding. Long disregarded, the creek caught on fire now and again, burning furiously until the fire department put it out.[37]

Vast quantities of rubbish, together with less savoury accumulations already lining the bed and banks of the river system, were expected to descend upon Missouri residents, who depended heavily on the Mississippi for drinking water. Speaking at Anna, Illinois, on 9 October 1902, Illinois governor Richard Yates welcomed the release of a 1901 State Board of Health report titled *Sanitary Investigations of the Illinois River and Tributaries*. The findings, Yates asserted, demonstrated that the Illinois River, charged with the removal of over three-quarters of the sewage of Chicago "purifies itself through natural causes." In Yates's view, the board had not only "demonstrated to scientists the self-purification of running streams and thus vindicated the wisdom of the people of Chicago" in undertaking the diversion project but had also "prevented years of litigation."[38]

Governor Yates may have contributed to the durability of the dilution and self-purification models, but his triumphal announcement that the Illinois State Board of Health study had forestalled years of litigation proved somewhat apocryphal. Nor would the self-purification claim – a central tenet of the flushing doctrine – go unchallenged. In January 1906, the next round of the Missouri-Illinois litigation reached the Supreme Court of the United States. By this time Missouri had a comparative record of typhoid deaths before and after the opening of the Chicago Drainage Canal to put before the judges. In 1890, 140 people had died in St. Louis from the disease. The typhoid death rate had surged dramatically in 1892, but had generally declined thereafter – until 1900, when, coincident with the Chicago diversion, a steady increase began.³⁹ Yet the aggrieved downstream litigants still faced the challenge of demonstrating an actual linkage between Chicago's effluent discharges and the increased incidence of typhoid deaths in St. Louis. Scientific opinion, as presented by the two sides, was sharply divided.

Justice Oliver Wendell Holmes, newly appointed to the Supreme Court of the United States after two decades on the Massachusetts bench, now encountered Chicago's dilution-based approach to sewage for the first, but by no means the last, time. The relevant principle, he remarked, must be approached with caution, for the issues involved were of primary importance. "It is a question of the first magnitude," Holmes believed, "whether the destiny of the great rivers is to be the sewers of the cities along their banks or to be protected against everything which threatens their purity." Thus it would be premature to decide the whole matter "at one blow by an irrevocable fiat." Rather than addressing the question he had formulated, Holmes considered ways to avoid it. In particular, he took note of general practice along the Mississippi, concluding that "the discharge of sewage into the Mississippi by cities and towns [was] to be expected." This was, simply put, part of "the practical course of events" which courts would naturally consider. Adverse inferences resulted for St. Louis and Missouri, for the latter, he pointed out, "deliberately permits discharges similar to those of which it complains." That, in Holmes's view, offered a performance standard to which Illinois might reasonably appeal. Moreover, as some of Missouri's own discharges were actually upstream from the St. Louis intake, Illinois was entitled to demand the "strictest proof" that Missouri's injuries were not self-inflicted.⁴⁰

Holmes forthrightly acknowledged the almost hopeless position of the downstream protest in relation to the legal doctrine of an earlier era. Fifty years earlier, Missouri's claim would almost certainly have failed the simple test of nuisance familiar to the older common law: "There is nothing which can be detected by the unassisted senses – no visible increase of filth, no new smell." These were precisely the limitations embedded in the common law that universally frustrated the current generation of public health professionals. Yet Missouri's situation was even more difficult, for, as Holmes carefully noted, the introduction of pure water from Lake Michigan had noticeably improved the condition of the Illinois River. "Formerly it was sluggish and ill-smelling. Now it is a comparatively clear stream to which edible fish have returned." Holmes was even impressed by the results of an exercise in aqua-roulette engaged in by local fishermen who drank from the river, "it is said, without evil results." The legal obstacles

Missouri faced were formidable indeed, for its case against Illinois' sewage, diluted by the Lake Michigan diversion, depended upon what Holmes described as "an inference of the unseen." The inference, he explained, depended on two propositions: first, that an increase in typhoid could not be attributed to any other cause, and second, that the typhoid bacillus could, indeed, survive the downstream journey from the sewers of Chicago to St. Louis' water intake pipe.[41] The court was unpersuaded.

Although Holmes's decision had local and immediate significance, it was also notable for establishing a high standard applicable to similar conflicts. The law would intervene only in cases of "serious magnitude, clearly and fully proved," and only where the court was prepared "deliberately to maintain against all considerations on the other side" the relevant principle of liability.[42] These guidelines significantly strengthened the flushing position. US courts would not easily be stampeded into protecting rivers – or even the health of downstream residents.

Negotiating with the Great Lakes Neighbours

With the expansion of such manufacturing enterprises as Illinois Steel, the Pullman Industries, and International Harvester, as well as accompanying residential districts extending along to the south of the city, SDC officials concluded that Chicago's Calumet River must also be reversed to safeguard the city's water supply.[43] Like the Chicago River, the Calumet, which flowed east into Lake Michigan, would be re-engineered to flow west, joining the Main Channel via a new canal. This proposal came to the attention of the International Waterways Commission (IWC), then engaged in an investigation of issues relating to the preservation of Niagara Falls. The IWC was immediately curious about the possible implications of an additional diversion from Lake Michigan at Chicago for water volumes passing through the Great Lakes system. Might navigation and downstream power production be adversely affected?

As Isham Randolph, chief engineer for the SDC, explained what came to be known as the Cal-Sag diversion, "It is proposed to take through the Calumet channel, when built, about 240,000 cubic feet of water per minute [4,000 cfs]; that is about as little as we could get along with, and also as much as we could take care of through our main channel."[44] Randolph explained that from an engineering perspective the anticipated ultimate limit of the Lake Michigan diversion at Chicago would be about fourteen thousand cubic feet per second, the capacity for which the original engineering had been conceived, albeit without the approval of federal authorities.

One forward-thinking member of the commission wondered what might happen after that limit was reached. This inquiry elicited only proverbial and fatalistic assurances: "Sufficient unto the day is the evil thereof. I hope somebody will develop something by that time to help us out."[45] But in the meantime, or so a good many representatives of Chicago's interests continued to insist, the health of the residents of that city remained their paramount consideration.

At the International Waterways hearings, Canadians – both witnesses and panel members – were less willing than some of their American counterparts to accept at face value Chicago's claim that the existing level of diversion, not to mention contemplated increases, was indeed essential to the health of the city's population. A representative of Canada's Dominion Marine Association challenged the proposition that "Chicago must have all the water that may be necessary for sanitation purposes without reference to the extent of that amount."[46] Other Great Lakes cities also expressed reservations about the open-ended nature of Chicago's watertaking: a spokesman for the Buffalo Chamber of Commerce imagined Chicago's population reaching six million and wondered whether other lakeshore centres – Milwaukee, Cleveland, and Toledo, for example – might not embark upon similar engineering schemes.[47] In navigational terms, the stakes were enormous, although they had hardly been factored into the least-cost solutions that were invariably favoured according to the municipal calculus operating at the Chicago end of the Great Lakes system.

IWC commissioner Gibbons inquired pointedly at the Buffalo hearings "whether any other means could be devised for the preservation of the health of Chicago, at an expenditure of money, or whether this is the only one?" Randolph replied simply, "I do not know of any other means within the financial ability of the city."[48] When the IWC reconvened in Chicago later in the year, Gibbons came back to the question of alternatives. Positing that Chicago was entitled to ten thousand cfs, Gibbons asked whether it could content itself with that volume and seek out some other system "such as is used in other great cities" to avoid expanding the diversion. But it took another brief exchange, this time between General Ernst, an American member of the IWC, and Randolph, to expose the full extent of Chicago's unwillingness to look ahead, or even to keep up to date with its planning. When Ernst incisively inquired whether any study of alternative methods of disposing of Calumet-area sewage had been made since the 1887 commissioners' preliminary report, Randolph, who had been in charge of the work for thirteen years, answered "No, sir." Ernst also requested in Chicago, as he had in Buffalo and by mail on several occasions, whatever report existed on studies to support the new plans for the Calumet diversion. Again Randolph confessed, "There is no report covering those studies," and simply referred the general to the proceedings of the Board of Trustees. Birmingham, in the Adderley case, had not been able to get away with this in the House of Lords; but Chicago, not having attended to the completion of the original SDC dilution studies, remained undeterred in its commitment to further diversion. Apparently oblivious to the irony created by his chief engineer's acknowledgment that alternatives were simply not being studied, the SDC's president rose shortly thereafter to repeat, "The sanitary district of Chicago and the municipal corporation have taxed themselves over fifty million dollars, and are taxing themselves, not with the idea of getting any profit out of it, but because the sanitary district does not know, in its legal and constitutional capacity, any other way of disposing of its sewage."[49] Evidently none of the lessons that might have been learned about perfect sewage treatment from the Columbian Exposition over a decade earlier had suggested a course of action to those responsible for the city's sewage.

Randolph acknowledged the relationship between Chicago's appropriation and lake levels elsewhere, while insisting, quite properly, on the complexity of the relationship. "I do not see how you could take water out of a vessel without affecting its level," he admitted, but observed that paradoxically, after six years of withdrawals, the lake was actually higher than before. This might be explained by greater rainfall, he suggested, or in severe winters ice might have choked off the discharge into the St. Clair River, forcing water to accumulate higher up the system.[50]

The Sanitary District's ambitions and agenda were clearly set out by its president, who underlined the expenditure record – $50 million already invested with a further $12 to follow. "And then what?" It was alarming, to say the least, even to contemplate "that the whole thing may be wiped out, the canal closed, and the great, growing prospects of the West left floating in the air, without a thing to grasp or light on." The people of the West, the IWC was firmly informed, were "very much in earnest upon this matter." They hoped to see a treaty ratified between Great Britain and the United States governing the Great Lakes, and, of course, preserving Niagara Falls. And it was equally important to encourage the commission to avoid making any recommendation that would "cause the failure of any trade whatsoever."[51] Some decades before negotiation theorists popularized "win-win" strategies for successful dispute resolution, Chicago revealed its own vision of the win-win paradigm: first you set out to win everything you need, and then you try to win some more.

Before the hearings concluded, representatives of various smaller downstream Illinois communities had dutifully paraded out to endorse the diversion: "We do not object to the City of Chicago sending sewage down the river, provided that they give us sufficient water to dilute the sewage, so that it will not affect the health of our people." The high-water mark of obsequiousness may then have been reached by the mayor of Ottawa, Illinois, some eighty-four miles away: "We are glad to have Chicago's sewage pass before us if properly diluted and will stand for it, being ready at all times to further the interests of our metropolitan city." Yet this acceptance of a seemingly inevitable fate was combined with the clearest possible insistence that residents of the Illinois Valley were entitled "to have all that has been guaranteed . . . by the laws of the State of Illinois." For the governments of the United States and Canada now to limit the diversion to ten thousand cfs, the mayor goaded, "we would regard as an outrage and a violation of confidence."[52]

Tough talk from Illinois was indeed the order of the day, constituting the hard-edged backup, if repeated invocation of the health of Chicago and the future of the West failed to secure acquiescence. When Governor Charles S. Deneen, unable due to important matters to attend the IWC hearings in Chicago, delegated Randolph to address the commissioners, there was no hint of apology or backing down. "Speaking for the Governor of this State," Randolph categorically asserted, "representing its sovereignty, we declare that it was a mistake on the part of the Government of the United States to permit any inquiry into the uses of waters wholly within the confines of the United States." He insisted that, as the waters of Lake Michigan were never mentioned in any treaty between Great Britain and the United States until 1871, the state had "rights in Lake Michigan which are inherent and which even the Government of the United States cannot deprive us of."[53] These were forceful, provocative remarks. For the chief engineer to have uttered them in the name of the Illinois governor might have appeared curious to some. Randolph's bluntness was no doubt strengthened by his own personal acquaintance with Governor Deneen, a Cook County state attorney and attorney for the Sanitary District of Chicago before his election in 1904.

If splashing water on the nose of the alligator is the aquatic equivalent of waving a red flag at a bull, Chicago now seemed more than willing to arouse the United States government, which had been sunning itself on the banks of a growing controversy, or to call Washington's bluff over the importance of national rivers and harbours legislation.

Ensuring the Navigability of Sewers

Chicago's claim to ten thousand cfs was not seriously questioned at the IWC hearings, although even this rate of withdrawal – well short of Randolph's projected diversion level – had never received federal approval. From the outset, US officials had held significant reservations about the impact of the sanitary diversion on Great Lakes water levels and shipping. One early estimate anticipated that an abstraction of between 300 thousand and 600 thousand cubic feet per minute (five to ten thousand cfs) would permanently lower Lakes Michigan, Huron, and Erie between three and eight inches, with a consequent reduction in the carrying capacity of commercial lake vessels.[54] There was also concern about conditions of navigability within the canal itself. Thus, a permit of May 1899 from Secretary of War R.A. Alger, in connection with the opening of the Main Channel, was conditional. The secretary of war intended to submit matters concerning the Sanitary District of Chicago to Congress "for consideration and final action." In the event that the current in the Chicago River resulting from the drainage scheme became "unreasonably obstructive to navigation or injurious to property," Alger specifically reserved the federal right to close off or modify the discharge through the channel.[55] The original permit for 5,000 cfs was modified in December 1901 and lowered to allow for 4,167 cfs.[56]

The engineering project that Chicago officials had hastily and unceremoniously ushered into operation early in 1900 may have been called the Sanitary and Ship Canal, but its primary function was clearly understood. Even three decades after the authors of the *Standard Guide* had bluntly declared in 1892 that "the one great object of this ship canal [was] to dispose of Chicago sewage," SDC planners unabashedly presented the overall diversionary work as the Dilution Project. Notwithstanding condemnation of these arrangements as "a simple but barbaric method of turning the city sewage from its own waterfront to a river flowing into and through neighbouring states,"[57] it would prove remarkably difficult to islodge the mindset that had envisaged waterways as waste-removal channels. Just how difficult soon became apparent in further rounds of protracted litigation.

In 1906, the International Waterways Commission more or less conceded ten thousand cfs to the Chicago diversion.[58] However, the updated consulting work of Rudolph Hering, who had originally advised Chicago in the 1880s, led the commission to conclude that continued reliance on the dilution approach to sewage was neither suitable nor necessary. It specifically refused to allocate an additional four thousand cfs for the proposed

Calumet-Sag channel. Despite Hering's change of heart – and even in the face of a clear denial of its permit application by William Howard Taft, Theodore Roosevelt's secretary of war – the SDC went ahead with the new project. The SDC's chief engineer showed an unbounded capacity to disregard the interests of other Great Lakes communities by insisting that "when you skim off the top of the lake the bottom will not drop out."[59] Perhaps Chicago officials ought to have shown more consideration for their neighbours, who were eventually pushed to take legal action to quell the city's unquenchable thirst for diversionary flushing. They might also well have demonstrated more respect for Taft's firmly held opinion that this was a matter for the consideration of Congress, for two decades later they were to encounter Taft once more. He was then chief justice of the United States.

When, in March 1908, the US federal government sued the SDC to stop construction of the Calumet-Sag channel, the SDC fumed that federal constraints were forcing Chicago to contravene Illinois law on the dilution ratio and threatened to waste their substantial investment. Chicago's continuing argument for dilution was that other treatment approaches were not "tested and proven" for large-scale applications,[60] an assertion whose credibility suffered somewhat from the fact that the SDC had not troubled itself to conduct any new studies on Calumet River disposal options since 1887.[61] Nevertheless, subject to a clear understanding that it would not exceed the total previously authorized volume of water, a connection through the Calumet River involving half the water originally requested by Chicago (i.e., two thousand cfs) received the secretary of war's approval on 30 June 1910.[62]

With Chicago's population already at two and a half million and increasing, the SDC sought permission in February 1912 to enlarge the authorized diversion to ten thousand cfs in order to dispose of Chicago sewage.[63] Sanitary District officials evidently felt it unnecessary to disclose that the existing diversion greatly exceeded the volume authorized: by some reports it was already very close to ten thousand cfs.[64] Objections to an increased withdrawal of Lake Michigan waters came from six states, twenty-three cities, a number of private navigational interests, and Canada, whose officials had been monitoring the Chicago situation for some time.

As early as August 1895, J.L.P. O'Hanly of Canada's Department of Marine and Fisheries had attempted to determine the probable effects of the Chicago Drainage Canal on the water levels of the Great Lakes and connecting waterways.[65] This hydraulic problem, O'Hanly reported to his minister (with considerable professional satisfaction, one suspects), was complex, intricate, and unique. To illustrate his assessment, he invited his minister to visualize the course taken by a hypothetical gallon of water as it left Lake Michigan. To the two existing avenues of evaporation or discharge through the Straits of MacKinac into Lake Huron, O'Hanly added an imaginary third – a natural subterranean passage extending out from Lake Michigan and equal in capacity to the proposed Chicago drainage ditch. The passage, he further suggested in language conjuring up Jules Verne, was fitted with automatic valves and sluices, such that, under similar heads of pressure, identical volumes would discharge from either outlet. To determine, on the basis of present knowledge of the physics and hydraulics of the Great Lakes, whether the gallon had left Lake Michigan via evaporation, discharge, or the underground passage would require "a century of minute, elaborate scientific research." And even after its century of research, science might fail to solve the mystery. Nonetheless, the fact would still remain "that Lake Michigan [would be] bled during every second of that century to the tune of 12,500 cubic feet, with the doctors still diagnosing the patient."[66] That the discharge would have been enormous, there could be no doubt. Imagining that his estimate of forty trillion (39,446,161,250,000) cubic feet would not readily be comprehended, O'Hanly offered a more tangible conception of the loss. It was, he suggested, equivalent to the mass of water occupying a hollow prism of 283 miles long, 50 miles wide, and 100 feet deep.[67]

O'Hanly eventually had the benefit of a US study that was launched only days before his own inquiry got under way. A report by Brigadier General O.M. Poe of the Corps of Engineers – an official who had criticized the US federal government's late-blooming inclination to safeguard navigation from the perils of sawdust – emphasized that "actual measurements only" could be relied upon to assess the impact of the Chicago diversion.[68] He estimated, however, that declines of six inches or so might readily be anticipated.[69] An impact of that order of magnitude was generally acknowledged by Chicago officials. Together, the Poe and O'Hanly studies from the mid-1890s cast some doubt on Randolph's claim that Illinois sanitary district legislation had been enacted on the basis of "full cognizance of the effect it would have upon the level of the lakes and the natural flow of the waters."[70]

Several distinguished Canadian engineers and sanitary officials spoke up in criticism of the Chicago scheme. To Thomas C. Keefer, Montreal's philosopher of waterworks, the proposal was unquestionably an abstraction of water from the Great Lakes system and by no means a diversion within the watershed. To a Quebec City engineer, the removal of about one-thirtieth the outflow at Niagara was an invasion of the riparian rights of Canadians. To Ontario's public health secretary, Dr. Peter H. Bryce, the public health implications of "the good people of Chicago, who, getting tired of drinking their own sewage, proposed to supply it to all the dwellers along the Father of Waters down to its mouth," could not be overlooked.[71]

Some years later, Canada's Department of Marine and Fisheries, the Department of Public Works, and the Commission of Conservation jointly recounted Chicago's disregard of US federal controls, noting, "Although the federal authorities have never given permission to divert a greater amount than 4,167 cubic feet per second, it is a matter of public notoriety that at least 8,000 cubic feet per second, and probably nearly nine thousand cubic feet per second, are now being diverted through the canal." The Canadian officials observed that the capacity of the canal between Sag and Lockport was fourteen thousand cfs and that the SDC's intention to divert that amount if it could be done without interfering with navigation on the Chicago River was clear.[72] Indeed, Canadian public servants eventually learned of a 1911 SDC report in which that organization's president had asserted that Chicago's needs alone should govern the diversion level. He described the attempt to limit the diversion to 10,000 cubic feet per second, or 600,000 cubic feet per minute, as "gratuitous and mischievous." "I believe," he continued, "that we should have the volume requisite to our needs as they appear and are justified."[73]

Canadian officials maintained that the country might ultimately lose the benefit of an expenditure of more than $200 million on canals and navigational improvements along the St. Lawrence waterway if the existing diversion (8,000+ cfs) continued. In some respects, the Illinois actions even appeared inconsistent with a British-American treaty of 1871 providing for freedom of navigation on the St. Lawrence system.[74] Through the cumbersome diplomatic channels then available to it, the Canadian government advised Washington that expansion of Chicago's diversionary capacity would be "highly detrimental." The State Department's reply to His Majesty's ambassador, Christmas Eve 1912, was curt – but hardly reassuring: "The Government of Canada has been misinformed in the matter."[75]

Only a couple of weeks after the State Department's unconvincing reassurances, US secretary of war Henry L. Stimson released his decision in relation to the SDC application for ten thousand cfs. Canada and other critics of the Chicago solution now stood to fare much better, for Stimson declared with absolute clarity the unwillingness of the United States government to sacrifice the general interests of the Great Lakes community to Chicago's murky vision of perfect municipal sewerage. With the viability of alternative approaches to sewage treatment more widely established, prolongation of the dilutionary interlude appeared in question.

Stimson recalled the terms and conditions of previous authorizations, as well as the decision of his predecessor, Taft, to deny permission to reverse the Calumet in 1909 at least in part on the ground that "this question of capital and national importance" should be submitted to Congress. The secretary of war made it plain that Chicago's sanitary canal had "never received the direct sanction of Congress." It was built, he emphasized, solely under state authority. Numerous expressions of opposition to Chicago's diversion plan had demonstrated a serious conflict between "the vital interest of a single community on the one side and the broad interest of the commerce of the nation on the other."[76]

Every drop of water taken out of Lake Michigan, Stimson calculated, undermined the benefit of costly improvements made with congressional approval throughout the Great Lakes. Withdrawal on the scale proposed by Chicago would nullify the value of millions of dollars of such expenditures and impose still greater losses on navigational interests. None of Chicago's self-interested assertions seemed persuasive: If a deep waterway linked to the Mississippi was wanted, a mere thousand cfs should suffice. For every unit of horsepower generated by the diversion at Lockport, four could be produced at Niagara. Not even the sanitary situation – assuming this was a relevant consideration for the secretary of war – required the volume of water that Chicago proposed to withdraw. The secretary of war demonstrated a clear-eyed appreciation of incentives when he added, "So long as the city is permitted to increase the amount of water which it may take from the Lakes, there will be a very strong temptation placed upon it to postpone a more scientific and possibly more expensive method of disposing of its sewage."[77]

Stimson also reflected on the nature of his power under the relevant sections of the 1899 *Rivers and Harbors Act*, legislation that made the secretary of war "the guardian of the commercial interests of the nation represented in their waterways." This responsibility might sometimes call for decisions reconciling the interests of one class of transportation with the competing interests of another, or indeed establishing priorities. Yet Stimson doubted whether Congress intended that substantial navigational interests should ever be subordinated to entirely different claims such as municipal sanitation. He did not, of course, intend in any way to "minimize the importance of preserving the health of the great city of Chicago," yet he was clearly of the view that "when a method of doing this is proposed which will materially injure a most important class of commerce of the nation and which will also seriously affect the interests of a foreign power, it should not be done without the deliberate consideration and authority of the representatives of the entire nation." He noted, as had others previously, that those asserting the interests of an ever-growing Chicago were "quite unwilling to put any final limit to the demand which may be made upon the waters of Lake Michigan for its sanitation under the system now in use." The obvious consequence of pursuing the dilution model would be to "materially change this great natural watercourse now existing through the Lakes."[78]

The competing interests were not easily reconciled. It was hardly a straightforward, merely technical, matter to weigh sanitary concerns – and perhaps the health of one community – against national commercial interests. Relations with Canada, including international obligations concerning waterways, only compounded the complexity of the decision. The matters at issue constituted "broad questions of national policy" and were sharply different from matters ordinarily considered under the governing legislation, the height or width of a drawbridge over a navigable stream, for instance.[79]

Stimson's decision, issued very shortly after Washington's formulaic assurances to Canada about effective federal diversion controls, finally set the stage for direct conflict between the congressional commitment to safeguard navigable waters and Chicago's open-ended sewage diversion program. The intent had been evident as early as 1890, when Congress enacted the *Rivers and Harbors* legislation to prohibit obstructions "not affirmatively authorized by law." In 1899, the prohibition against any obstruction to navigation "not affirmatively authorized by Congress" was confirmed and supplemented with a firm declaration: "It shall not be lawful to excavate or fill, or in any manner to alter or modify the course, location, condition or capacity of, any port, roadstead, haven, harbor, canal, lake, harbor of refuge, or inclosure within the limits of any breakwater, or of the channel of any navigable water of the United States, unless the work has been recommended by the Chief of Engineers and authorized by the Secretary of War prior to beginning the same."[80]

To uphold the integrity and credibility of its rulings, the US War Department was compelled to challenge the SDC for exceeding the permitted withdrawal. Yet Chicago continued to assert the importance of Illinois' dilution statute, and some particularly imaginative interpretations were circulated. One argument had it that the federal permit constituted a restriction on the volume of diversion through the Chicago River, not on the total removal of water from Lake Michigan. Since some of the flow actually entered the river by way of a lakeside pumping station and conduit along Thirty-ninth Street, this hopeful sophistry suggested that the Sanitary District could still manage to comply simultaneously with its federal permit and with Illinois law.[81]

The difference of opinion between the Illinois agency and federal authorities eventually moved off to the courts, where it was consolidated with a claim, launched in 1908, to prevent construction of the Calumet channel itself. Judicial proceedings advanced at a glacial pace. When the matter finally reached the US Supreme Court after more than a decade, Justice Holmes sardonically noted "some delay in concluding the case."[82] In Illinois, Judge Kenesaw Mountain Landis retained the file without explanation or excuse for about six years before delivering an oral opinion in favour of the federal government on 19 June 1920. A motion for reconsideration produced no further action until Landis resigned in 1922 to pursue a higher calling as commissioner of organized baseball.[83] In 1923, another judge issued an order limiting the Chicago diversion to 4,167 cfs, and deferring the effective date of the decree by six months to permit the defendant to present the record to the Supreme Court.[84]

In delivering the judgment of the court in *Sanitary District of Chicago v. The United States*, Oliver Wendell Holmes reviewed Chicago's position. The SDC claimed to be acting under the Illinois statute of 29 May 1889. This legislation called for the construction of a channel large enough to take care of the sewage and drainage of Chicago as the increase of population might require. Sufficient capacity was needed to allow for an ultimate flow of not less than 600 thousand cubic feet of water per minute (10,000 cfs) and a continuous flow of not less than 20,000 cubic feet for each 100,000 residents of the Sanitary District. Holmes noted that the SDC denied that it was already abstracting between 400 thousand and 600 thousand feet per minute (6,666 to 10,000 cfs), before proceeding to discredit that claim: "As it alleges the great evils that would ensue if the flow were limited to the amount fixed by the Secretary of War, or to any amount materially less than that required by the state Act of May 29, 1889, and as it admits present conditions to be good, the denial cannot be taken very seriously." The SDC also denied that an abstraction of water substantially in excess of 250 thousand cubic feet per minute would lower the levels of the lakes and rivers concerned, or impede the navigability of those waters, and, of course, continued to emphasize the expense incurred to flush sewage through the Chicago River and away from the city's water supply. Then, as Holmes put the matter, the defence "finally takes the bull by the horns and denies the right of the United States to determine the amount of water that should flow through the channel, or the manner of the flow."[85] Holmes emphasized that it was not the object of the United States in launching this litigation to eliminate the channel, but only to limit the amount of water taken through it from Lake Michigan. He saw little merit in the sanitary authority's legal position, and on 5 January 1925 confirmed the injunction previously awarded by lower courts limiting the diversion to 4,167 cfs, but without prejudice to any future permit that the secretary of war might lawfully issue.

Such a permit was indeed issued later in the same year. As a condition of the new permit, issued in 1925 and more than doubling the authorized diversion to ten thousand cfs, the SDC was expected to implement a program of artificial sewage treatment capable of handling the sewage of at least 1.2 million people. Moreover, if regulatory or compensatory works were required to restore levels elsewhere on the Great Lakes, the SDC would have to pay its share.[86] For several Great Lakes states, the increased authorization in 1925 was entirely too much. Wisconsin, Ohio, Minnesota, and Pennsylvania – later joined by Michigan and New York – immediately challenged the validity of the expanded diversion.[87] Their position, in essence, constituted a defence of natural lake levels, with navigation as the highest priority use. Arguably, the claim even amounted to an attack on the constitutionality of trans-basin diversions.[88] Overtures followed to bring Canadian critics on board.[89]

With the new dispute pending before it, the United States Supreme Court appointed the Honourable Charles Evans Hughes as a special master to conduct hearings at which the parties would make submissions that he would condense and submit as findings to the court for its deliberations. Hughes, a former governor of New York, had gone on record in 1909 against flushing untreated sewage into waterways.[90]

Despite the public impression that Chicago's abstraction rate might be curtailed to 4,167 cubic feet per second by 1935 from the then current level of 9,700 cfs, this was merely a recommendation of the US engineers and not a condition incorporated into the official permit. For its part, Chicago indicated at the hearings that it had no plan or program in place to reduce the consumption level. Its intention to develop sewage treatment facilities would do no more than keep pace with increased requirements, such that the current dilution procedure would need to be maintained.

Both Illinois and the Sanitary District of Chicago vigorously resisted the Wisconsin-led challenge, maintaining that the permit – and consequently the diversion – was entirely lawful. To support this contention, Illinois invoked the opinion of James M. Beck, acting attorney general of the

United States. In that capacity, Beck had previously advised the secretary of war on the nature of his approval powers as they affected US interests in navigation.[91] Beck specifically addressed the congressional prohibition against impairment of navigable waters by any obstruction "not affirmatively authorized by Congress," as well as the exception applicable where "the work had been recommended by the Chief of Engineers and authorized by the Secretary of War." Congress, Beck indicated at the outset, enjoyed paramount jurisdiction over navigable waters. Accordingly, no state action could authorize impairment of navigation. Nor could federal assent be implied on the basis of federal inaction. What sense could be made, therefore, of the statutory foundation of Chicago's dilution program, which enjoyed the affirmation of the secretary of war, but certainly lacked explicit congressional approval? In other words, where, in the absence of approval by Congress itself, could affirmative authorization be found?

In Beck's view, the legislation indicated that Congress intended to delegate administrative authority to the chief of engineers and the secretary of war. If those officials concluded that the proposed use of navigable waters did not impair navigability to a degree that required prohibition, then the construction was "affirmatively authorized by Congress." This result followed "because the administrative agency, to which Congress had delegated the ascertainment of the facts, had found the fact to be that such use was not, for the time being, an impairment of navigable capacity, such as Congress intended to prohibit."[92]

In the Chicago case, Beck distinguished between the original authorization and the subsequent supplementary approval. In connection with his assessment of the first volume of diversion (up to 4,167 cfs), Beck drew a distinction that might not have been so evident to earlier observers: projects seeking federal authorization might impair navigation, or they might *unduly* impair it. Only the latter degree of impairment gave rise to a prohibition, Beck reasoned, since the War Department had in fact issued permits authorizing Chicago to divert up to 4,167 cfs from Lake Michigan and "theoretically, the diversion of so great a volume of water affected the navigable capacity of the Great Lakes." The authorizations could be explained, therefore, only by the fine distinction between impairment and undue impairment which Beck creatively described.[93]

The additional diversion, that is, of a volume of water greater than 4,167 cfs, required even more intricate legal acrobatics. Beck rose valiantly to the challenge. Admitting that the current level of diversion exceeded the amount permitted and was not authorized by Congress, he suggested that it would be, "if and when the Chief of Engineers and the Secretary of

War determine that the interests of navigation – having in mind all the complicated circumstances – would not be unfairly prejudiced by a temporary use of the waters to the extent that the War Department, in its sound discretion and having in mind the general policy of Congress, sees fit to permit."[94] In the case at hand, Beck prescribed a course of action largely dictated by the status quo. He advised that in exercising political discretion, the War Department should do what it believed Congress would wish in the circumstances. Those circumstances, as Beck spelled them out, indicated that "a great city, with over three millions of inhabitants and with mighty industries situated within its boundaries, has built important works for the diversion of water from Lake Michigan, and is in fact diverting such waters, to an amount in excess of the amount permitted." Such a situation, he advised the secretary of war, "cannot be changed in a day."[95] Navigation, notwithstanding *Magna Carta,* appeared to be slipping down the hierarchy. Flushing continued to ascend.

Having thus established, to its own satisfaction at least, the legitimacy of the extended sanitary diversion under federal legislation as being not inconsistent with the general interest in navigation, Chicago now went even further. The entire diversion program, Chicago's defenders insisted, was no longer really about sewerage at all but rather about navigation. This extraordinary claim had three aspects.[96] First, the SDC asserted, the permit provided the water from Lake Michigan required, according to Illinois legislation, to operate the canal. This channel constituted one of the essential linkages between Lake Michigan and the Gulf of Mexico, a waterway in which Illinois had been interested since it became a state of the union. This waterway, the SDC insisted, enjoyed congressional authorization dating from the 1820s. The second aspect of the SDC's navigation claim emphasized that the diversion facilitated navigation by providing oxygen for the oxidation of sewage and drainage in the Des Plaines and Illinois Rivers, thereby preventing a nuisance and inconvenience to navigation. So, let us say, if you were going to use navigable waterways as sewers, and you wanted those waterways to remain navigable, you would want a large volume of water to maintain them for that purpose. Similarly, by keeping Chicago's sewage out of Lake Michigan, the diversion actually benefited navigation on the lake itself. Finally, added the SDC, in a forlorn effort to put itself on the side of the angels, the diversion increased depth and helped to enhance the navigable capacity of the Illinois River.[97]

Other objections of a technical or procedural nature were not unimportant, but the SDC's position that its sewerage arrangements now actually *facilitated* navigation was the centrepiece of a defence designed to establish the validity of the 1925 permit. In other words, having denied for over a quarter century that the secretary of war had any authority over the diversion, Illinois now sought shelter beneath that same authority in a manner not dissimilar to Justice Best's attempt to establish a right to sea bathing with the thought that swimmers save ships.

Chief Justice Taft was unmoved. Two decades after he had personally rejected the Calumet-Sag sewerage application in his former capacity as secretary of war, Taft found no reason to disagree with himself. The validity of Wisconsin's protest against the sanitary diversion was eventually acknowledged in the plainest terms. The secretary of war "could not make mere local sanitation a basis for a continuing diversion." Thus, as Taft explained, the 1925 permit was temporary in nature and conditional upon the adoption of alternative means of sewage disposal.[98]

Taft called for the calculation of a diversion volume that would be consistent with the maintenance of Great Lakes navigation and sought advice on a timetable for reductions to that level. The decree, as issued in April 1930,[99] set the maximum volume of withdrawal at 6,500 cfs until 31 December 1935, when a reduction to 5,000 cfs should occur, with a further decline scheduled to take place three years after that.[100] The step-down decree, which, as one commentator stated, "takes control of the Chicago diversion from the Secretary of War and places it under the court"[101] had a two-dimensional rationale. The ultimate volume limitation was derived from estimates about navigational needs, and the timetabling was driven by assumptions about the SDC's capacity to make progress toward a comprehensive sewage treatment program.

Although an SDC president had once described any limit on the diversion volume as mischievous, sanitary officials had grudgingly begun to pursue treatment alternatives in the aftermath of the IWC's denunciation of the Calumet-Sag scheme. By 1909 it had been apparent to Isham Randolph that dilution could not meet the growing demands of the expanding metropolis. The chief engineer's report for 1911 addressed the challenge of dealing with a population above the three million maximum that had been the basis for the diversion program. That threshold would be crossed by 1922, adding a burden to the infrastructure quite apart from that provided by chemical and industrial wastes. Proponents of sanitary sewerage for Chicago had appreciated the burden of commercial and manufacturing wastes no more than their counterparts and contemporaries elsewhere. These had never been factored into the equation and were well beyond the capacity of the SDC system to handle.[102]

As early as 1910, the SDC had actually begun to contemplate sewage treatment by means of settling basins and sprinkling filters, with a target date of 1930 for putting such arrangements in place. Considerable attention was later devoted to alternatives, and by 1919 projections were in place for a twenty-five-year schedule of construction that would see substantial reductions in the amount of sewage passing through the Drainage Canal. A treatment facility beside the Des Plaines River began operations in 1922, and additional plants were established elsewhere around the city over a twenty-year period. When completed in 1939, the Southwest Sewage Treatment Works was the largest facility of its kind in the world.[103] By all accounts, it was fortunate that the program ever reached its conclusion.

On 5 February 1932, the trial in criminal court of several former trustees and employees of the Sanitary District of Chicago on conspiracy charges came to an end. More than seven hundred witnesses from all walks of life had testified before Justice Harry M. Fisher, each "telling the same monotonous tale." For nearly a year and a half, "without rendering one dollar's worth of service," they had regularly drawn upon public funds. Furthermore, not satisfied with receiving their cheques, they had developed a scheme "by which expense money was paid by the district running to vast sums." Justice Fisher readily concluded that "hideous corruption" in the form of an extensive plot to rob the public treasury had been in place for some time. Hundreds of "employees" were paid for services unrendered; investigative trips never taken on behalf of the Sanitary District generated extensive expense claims; dummy corporations enjoyed generous overcharges. Even members of the Illinois legislature were implicated. Many had "received from the coffers of the municipality vast sums of money by themselves going on the payroll or by members of their families or favorite friends being put upon it," with no expectation of service being rendered.[104]

Chicago, clearly no island unto itself, finally did address the treatment issue. But not before the city's devotion to its dilution project had generated some of the twentieth century's more creative justifications for sewering rivers and streams. The water lost to the Great Lakes through Chicago's navigable sewer also provoked at least one compensatory fantasy. C. Lorne Campbell, a Toronto engineer, responded to "the Chicago steal" with an eye to practicalities. In a transparent application of the principle "if you can't beat them, join them," Campbell proposed a countervailing restoration of Great Lakes water levels in the most direct manner possible – filling them up with more water, diverted from the north. Indeed, Campbell envisaged the creation of a sixth great lake. After years of exploration in the vast Albany watershed, extending several hundred miles north of Lake Superior, Campbell went on record about "how purposeless, how

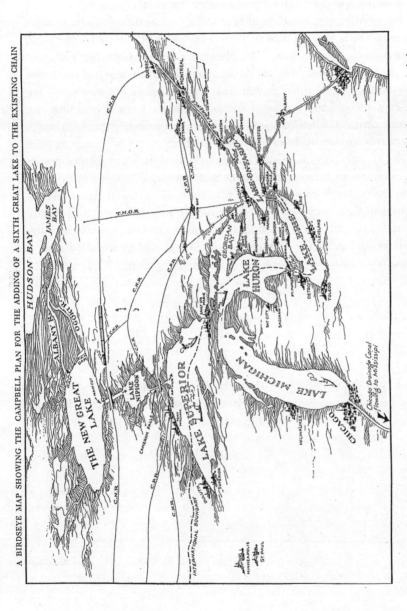

When Chicago redirected large volumes of water away from the Great Lakes, other residents of the basin worried about the impact of declining water levels. Lorne Campbell, a Canadian engineer, even imagined the creation of a "Sixth Great Lake" to compensate for the losses from the Chicago diversion.

useless, was the flow of this great volume of fresh water into the salt waters of Hudson Bay." The remedy was at hand, however, in the form of a massive flooding and diversion scheme to store the waters of the Albany River and its tributaries in an artificial lake some fifteen thousand square miles in extent. From there, a flow of 18,000 cfs could be directed southwards through a cut channel into Lake Nipigon and over Cameron Falls into Lake Superior. This was roughly double the volume Chicago had attempted to secure and several times greater than the authorized withdrawal.[105] Yet, as General Poe had forecast, Chicago's impulse toward diversion was brought under US federal supervision: even under the guise of health imperatives and the artifice of navigational improvement, sewage disposal had not entirely trumped Great Lakes navigation. It finally seemed preferable to treat sewage rather than to navigate sewers, although this situation proved to be temporary. Soon the condition of lakes and rivers suffered renewed neglect, attributable in large measure to the emerging proposition that using chemical or mechanical means to treat municipal water supplies could avoid the responsibility of safeguarding natural waters.

9

Separating Water from the Waterways

Safeguarding Municipal Water Sources

CHICAGO'S PROLONGED AND DETERMINED campaign to flush untreated municipal wastes away from Lake Michigan was inspired originally by the city's wish to safeguard its own water supply from contamination. As other growing communities endeavoured to secure water for their residents around the turn of the century, they too encountered threats to the quality of their supplies, notably those arising from municipal wastewater. In various settings, municipal water sources might be protected, either by preserving their isolation from human activity or by prohibitions against contamination including the use of modern sewage treatment. In some fashion, each of these strategies linked population health with environmental quality. That connection would eventually be ruptured by a new round of innovation in water supply technology that emphasized water treatment over the protection of sources.

SAFEGUARDING MUNICIPAL WATER SOURCES

Waterworks systems proliferated in the United States, Britain, and Canada during the late nineteenth and early twentieth centuries, but rising demand and the threat of contamination placed considerable pressure on supplies.[1] To secure the quality of water sources needed to meet the rising demand, several American municipalities incurred substantial expenditures. Newark, New Jersey, anxiously struggled to safeguard the Passaic River and even bought up numerous properties in its watershed in order to reduce the resident population. By 1889, however, the city had abandoned all pretense of preserving the Passaic for use, even for industrial purposes, and secured a new water source from a private supplier on an upstream tributary.[2] New York City, having invested in sewage purification facilities for several communities along its Croton supply corridor, subsequently had to battle to prevent the establishment of a facility for the insane in the watershed.[3] Pennsylvania, after a typhoid outbreak, approved legislation in 1905 to safeguard water quality in the interests of public health.

When forced to reconsider water supply arrangements, many urban centres – wherever located – contemplated new or remote sources to ensure quality. Alternative sources of water supply for London had almost continuously been under consideration. Schemes to draw water from the Lake District or from the Severn watershed in Wales had surfaced during the 1860s and attracted the attention of a royal commission inquiry into the metropolitan water supply set up on the heels of a cholera outbreak in the summer of 1866.[4] Through to the mid-1880s, when further thought was given to London's ever-increasing needs, Welsh water still had some advocates. The engineer John Burns, MP, among others, urged an expenditure of £20 million to link the capital with Wales so as to bring "from the mountain and the mist our water supply to London, by gravitation and without pumping." One consequence of the failure to tap into "cloudland," as Burns described it, was continued reliance on the Thames, such that "along the banks of one of the most beautiful rivers . . . you see enormous tracts of land scheduled for use as reservoirs." Burns was no fan of these, for no matter how they were designed, reservoirs "are never so beautiful as natural, undulating scenery." The Thames, he concluded, had been badly marred, and, moreover, the alternative source was "running to waste in the most extravagant way in the West Country."[5]

Outside London, industrial contamination frequently undermined water quality in precisely the areas where alternative sources were needed. Where underground sources were inadequate, local authorities extended their horizons to more remote sources of supply and, where necessary, impounded water behind dams in mountain valleys. Consultants looked for areas with particular features: "Only open mountain pastures, unploughed and unmanured, and with practically no population, are considered good enough as collecting grounds for town supplies." If a river had to be tapped, the risks and implications would have to be carefully considered.[6]

As improved construction techniques enabled greater storage capacity, the quest for suitable water supplies took on a more regional aspect, and legislative arrangements were made to establish regional gathering grounds.

The first such monumental structure was the massive Vyrnwy Dam in Wales, constructed between 1881 and 1892. Behind its 144-foot-high wall extending over a fifth of a mile, twelve billion gallons of water accumulated to form Europe's largest artificial lake. Manchester and Birmingham soon embarked upon similar storage schemes, the former by means of the Thirlmere Dam in the Lake District and the latter by means of a series of masonry works along the Elan Valley in Radnorshire.

North American municipalities also faced renewed water supply challenges. Although New York, having completed development of the New Croton Aqueduct between 1883 and 1893, appeared amply supplied for the foreseeable future, neighbouring Brooklyn, still independent, was rapidly depleting Long Island's groundwater reserves and had begun to experience shortages. The water imbalance figured prominently in political debate over consolidation leading in 1898 to Greater New York.[7] Alternative sources of additional supply were immediately on the civic agenda.

In 1899, the city's water commissioner proposed a forty-year contract with the Ramapo Water Company for 200 million gallons daily at $70 per million gallons, a high cost that "startled the city and convulsed the engineering profession." A contemporary review of alternatives conducted under Rudolph Hering's leadership identified the Hudson River upstream from Poughkeepsie as the preferred source for a daily supply of one and a half billion gallons. The Hudson proposal, complete with reservoirs in the Adirondacks to maintain the river's flow during dry seasons and filtration to overcome the effects of upstream sewage, would have provided up to 500 million gallons at a cost of roughly half the Ramapo rate. Yet the Hudson's condition was questionable. Estimates suggested that if New York adopted the Hudson as a water source, "the polluted condition of the river would necessitate an annual expenditure of at least $2 million in addition to the initial cost of establishing so gigantic a filtration system."[8]

The Catskill Mountains were eventually selected by the state Board of Water Supply as a more appropriate solution than either the badly deteriorated Hudson or the Ramapo proposition. Between 1907 and 1917, an aqueduct over a hundred miles long, supported by storage reservoirs that had necessitated the displacement of several thousand rural residents, was constructed to supply over 500 million gallons per day to the New York metropolis.[9] Another water supply crisis was barely averted.

The Toronto water supply saga typified the pattern of indecision and neglect that so often characterized the response of civic leaders. By the 1880s, the city's waterworks system was showing its age. In addition, the number of households served had more than tripled between 1877 and

1883, while quantities consumed also increased as residents found more uses for water.[10] These changes put the water supply question back on the agenda as officials began to imagine either alternative sources, renovations to existing works, or both.[11] Toronto's options included local rivers and continued reliance on Lake Ontario water. The former were repeatedly rejected on the grounds that costs to acquire rights from mill owners and other riparians would be prohibitive. Gravitational supply schemes, such as the use of Lake Simcoe, fifty miles to the north, were also considered intermittently.

Until Christmas Day 1892, Toronto's interest in gravitational schemes had been relatively dormant for several decades. On this holy morning, the intake pipe rose unceremoniously from the bottom of Lake Ontario. Residents of Toronto Island crossing over to the city were greeted by the disconcerting spectacle of the municipal supply pipe, broken in at least four places and caught in the ice. Weeds had choked off the intake valve, despite the efforts of one Captain Goodwin, who was responsible for keeping the system from clogging up (but who might have been more attentive); as a result, the shoreline pumping station (possibly not monitored as closely as might have been desired during the holiday festivities) had drained the entire length of the line (perhaps not as well anchored as might have been wished), causing it to float to the surface. It took a day for news of the accident to become known to the public, the relaxed pace of the notification process adding to the litany of municipal shortcomings.

With supply temporarily restored and the location of actual breaks carefully studied, residents were encouragingly, if somewhat mystifyingly advised that the water was "at least two-thirds pure and that the impure portion [came] not from the bay near the sewer outlets, but from the point furthest from such pollution."[12] Thoughtful residents no doubt deliberated whether to drink from the top or bottom two-thirds of the glass.

As the finger pointing died down, a more positive line of inquiry led Alderman Shaw to visit London in September 1894. He called upon James Mansergh, an experienced British engineer whose credits included Birmingham's water supply arrangements. By the time Mansergh arrived in Toronto a year later, the intake pipe had found its way to the harbour surface a second time, an incident giving rise to "considerable excitement in the City in consequence of the fouling of the water."[13]

Mansergh inspected several possible alternative sources of supply, including the Briar Hill area and the Don and Rouge Rivers, and "unhesitatingly" turned them down. He then directed his attention to the central remaining issue: should Toronto's future water supply be provided by pumping from Lake Ontario or by gravitation from Lake Simcoe?

Mansergh observed that the water of Lake Ontario, with filtering, and the water of Lake Simcoe, with a service reservoir, would offer "high-class waters of unimpeachable character, with practically nothing to choose between them."[14] Yet the Lake Ontario option presented incontestable financial advantages, as the Simcoe scheme would require a vast initial construction program involving a high capital outlay and immediate extensive repayment obligations. The Lake Ontario plan, despite higher operating expenditures for pumping facilities, could be developed incrementally to track actual demand levels more closely at a lower cost. These considerations dictated his recommendation to retain Lake Ontario as the source of supply.

Although exceeding his mandate, Mansergh volunteered caustic observations on Toronto's existing sewage arrangements, a state of affairs no one attempted to justify beyond the thought that remedial measures would be costly. "To discharge all the sewage of 175,000 people in its crude state into a tideless and practically stagnant harbour is obviously a very wrong thing to do, and every rational man must condemn it." Until this "blot" was wiped out, Toronto could not reasonably aspire to be held in high esteem as a residential city, for people were attentive to "the importance of safe sanitary surroundings" and increasingly inclined to ask, "Where does the sewage go to, and where does the water come from?"[15]

These remarks on sewage were not entirely gratuitous, for there were clear linkages between the water supply question on which Mansergh had been asked to advise and Toronto's continued discharge of untreated sewage. The British consultant was less concerned about contamination, however, than about the impact of the untreated waste on the lakeshore and the bay. As he categorically asserted, "There is no risk whatever of harmful pollution of the water to be supplied . . . The offence arising from the stirring up of the foul mud in front of the wharves by the steamboats in hot weather, is . . . very great, and the discomfort caused to the people carrying on their business on the waterside, must at times be almost intolerable, not to speak of injury to their health." These matters, Mansergh assumed, were beyond dispute, and only costs stood in the way of a response; but that response would be delayed, in fact "relegated to the dim future," if Toronto chose "to indulge in the luxury of Simcoe water."[16]

In a somewhat surprising twist, Mansergh identified several ways in which continued reliance on Lake Ontario as a source of supply would offer incentives to address the sewage issue. He argued that Toronto would be in a better position to afford necessary sewage work as a result of the relatively lower costs of water supply inherent in expanding its existing arrangements. Moreover, the desirability of controlling waste would become apparent, both to save money to devote to sewage treatment and to reduce the cost of providing that treatment by reducing the volume of sewage. The final incentive to treat sewage flowed naturally enough from the "desire to remove entirely the last trace of uneasiness with regard to the intake."[17]

Regardless of whether sewage treatment was yet at a stage to remove "the last trace of uneasiness," innovations were widespread and rapid advances were under way.

Good Clean Sludge and a Fine Clear Liquid

For nearly two decades, from 1898 to 1915, a British royal commission diligently investigated sewage treatment and disposal. The precipitating factor in the formation of the commission lay in local authorities' resentment toward the Local Government Board (LGB). Fervently devoted to land-based sewage treatment and exercising authority over municipal finance, the board dominated decision making. Yet as towns grew, land for sewage treatment, even if available, became costly, and any community that chose to pursue attractive alternative arrangements faced formidable financial obstacles. Some pursued independent action: Salford, for example, secured its own funding for a precipitation and filtration plant whose costs were estimated at roughly half those of land-based treatment.[18] But most communities were effectively precluded from adopting such innovative treatments as the artificial filters pioneered in the United States to accelerate aerobic decomposition or other less land-intensive technology such as septic tanks.

Following prolonged deliberations, the royal commission produced nine separate reports supported by numerous appendices and auxiliary studies. This work has been credited with an influential and long-lasting consensus over standards for sewage treatment. Its effect was to restore and secure the legitimacy of municipal flushing on generally acceptable foundations, thereby signalling the end of the long-cherished (if widely honoured in the breach) tradition of using sewage for fertilization.

What came to be known as primary and secondary treatment, the former involving the removal of solids and the latter oriented toward the decomposition of organic matter, emerged from numerous proposals for a two-stage approach to sewage that were presented to the commissioners. On the primary side, the options included sedimentation tanks, chemical precipitation, and septic tanks. Depending on local circumstances, secondary treatment could be carried out using traditional land treatment systems, contact beds, or various types of biological processes that were then emerging. The most significant of the new secondary treatments came to be known as activated sludge. Its inventor – or so he claimed – was Dr. Gilbert John Fowler, an English sewage treatment specialist.

In another round of trans-Atlantic exchange, Fowler journeyed in 1912 to consult with American counterparts inquiring into the condition of New York Harbor. From there he ventured to the pioneering experiment station in Lawrence, Massachusetts, a research facility he regarded as the mecca of sewage purification. Fowler later attributed an "illuminating idea"[19] of his own to this tour.

Returning to Britain, Dr. Fowler, a professor at the University of Manchester and consulting chemist for the city's Rivers Committee,[20] initiated a series of experiments at Manchester's Davyhulme sewage works in January 1913. After sedimentation, sewage was aerated in the presence of a bacterial sediment or sludge.[21] A remarkably clear effluent emerged, the result, it was speculated, of a process of organic conversion.[22] To pursue the scientific inquiries further, Fowler consulted a local manufacturing concern where additional aeration equipment for the bacterial treatment of sewage was designed and constructed. The new equipment – paid for by the city of Manchester – became operational between 20 and 26 September 1913.[23]

Meanwhile, Fowler, though failing for some time to advise Manchester's Rivers Committee of his new arrangements, agreed to serve as a consulting chemist for the sewage purification firm of Jones and Attwood.[24] For his part, the inventive Walter Jones had already addressed the intriguing practicalities of circulating sewage solids in the context of increased aerobic activity. He filed a provisional British patent specification on 11 October 1913 and three months later applied again in connection with further improvements to the apparatus. As Jones explained in seeking to protect his continuing refinements, it was advantageous to "bring and keep certain forms of bacteria and other life found in or added to the finely divided particles of sewage sludge into intimate contact with and equally dispersed throughout the whole of the liquid portion of the sewage and in the presence of suitable chemicals and of ample atmospheric oxygen." Jones' invention provided a means of maintaining the process by preventing the settlement of heavier solid matter and "keeping the whole in motion for any desired period of time, and for aerating it until the process [was] completed."[25] Thus emerged – or so it appeared – a biological treatment process known as activated sludge, consisting in essence of "an aeration tank, a final settling tank and a pump for return of sludge."[26]

The addition of over two hundred titles to the corpus of activated-sludge literature between 1914 and 1916, alongside a flurry of new patent applications, signalled the intensity of activity.[27] Activated-sludge plants were soon installed in England: Salford, 1914; Davyhulme (Manchester), 1915; Worcester, 1916; Sheffield, 1916; Withington, 1917; Stamford, 1917; and Bury, 1921. They were also immediately popular in the United States, where one early historian of the phenomenon was even inspired to assert that it was "in keeping with the tempo of life in America that there should be more activated sludge plants in the United States than anywhere else in the world." The vast scale of these plants was in keeping with "the American reverence for bigness."[28] But surprises lay ahead: as we will see, whether the Fowler-Jones narrative of the invention of this popular technology could hold up against rival claimants to a hugely valuable piece of intellectual property would later be questioned.

The keen interest of the British sewage disposal commission in new technological advances such as activated sludge was closely associated with the inquiry's concern for the standard of purity that an effluent would somehow have to achieve in order to obviate the risk of nuisance. When the nuisance-producing power of normal sewage or effluent was found to be broadly proportional to its de-oxygenating effect upon the water of a stream, that relationship became a central focus. The commission proposed an enduring test, known as the Biochemical Oxygen Demand (BOD) test, to ascertain the rate at which oxygen was removed from water as a result of bacteriological decomposition.[29]

Taking BOD as one measure of water quality, investigators recognized that testing to confirm whether satisfactory conditions had been achieved could be carried out either on the contaminating discharge itself or in the receiving stream. Since the commission regarded stream improvement as its primary objective, it concluded that the improvement of effluents was only a secondary consideration. This reasoning led to the view that quality standards, the dissolved oxygen test in particular, should not be applied directly to effluents but rather to such discharges when mixed with the river water under ordinary conditions.[30]

If the goal was river improvement, implementing a quality standard posed further challenges where waterways were subject to successive discharges of effluent. Under such conditions, rivers might be expected to deteriorate in quality from source to mouth. Accordingly, at least in theory, "the maintenance of river quality at the standard specified by the commission would mean that the quality of discharges [would] have to vary inversely with river quality, with more strict discharge standards being applied to rivers of low quality." The commission balked at this interpretation, not only on account of the inequities that dischargers would experience if variable standards of effluent quality applied at different points along a waterway, but also because this approach would virtually eliminate any incentive for those discharging into comparatively pure rivers to treat their effluent effectively. Instead, the commission proposed one fixed standard, suitable for most locations. Variations – either to greater or lesser requirements – might still be implemented in exceptional situations. Where dilution was very great, the royal commission actually recommended that effluent standards could be relaxed or suspended altogether.[31]

The royal commission's recommended effluent standards ultimately addressed two quality indicators – suspended solids and the oxygen-absorbing capacity of the discharge – both related to the degree of dilution anticipated in local circumstances. Although these, like the proposals of the Rivers Pollution Commission half a century earlier, were not formally adopted in legislation, in practice they became highly influential and offered at least a rough benchmark when pollution cases reached court.[32] Standards elsewhere varied from one jurisdiction to another, sometimes oriented around the use of "best practicable means," sometimes prescribed in greater detail. But following the work of the British royal commission, there was a stronger emphasis on the results to be achieved and a correspondingly diminished tendency to impose particular technological options. Waterways need not be pure to be pure enough.

A critic of this legacy has summarized the consequence of the royal commission's conclusion on effluent standards and its implications. Sharon Beder observes that by endorsing effluent standards to be achieved by whatever treatment process was selected, the royal commission "made the competition between processes on the basis of technical superiority irrelevant." There was no advantage to be gained by pursuing a higher degree of purity than was necessary. "Good enough" superseded "optimal" as a performance standard, with the result that "the skill of the engineer now lay, not in achieving a high quality effluent but rather in achieving an adequate quality of effluent for as little money as possible and letting nature do as much of the work as possible."[33] As earlier observers had expressed the latter point, "When properly used, the forces of natural purification aided by dilution represent an important economic asset, supplementing those of artificial sewage and water treatment."[34]

Plain sedimentation of municipal sewage experienced a triumphant revival, Beder concludes, "because it was good enough, not because it was technically superior" or resulted in a better effluent. Instead, sedimentation was attractive for its simplicity and (when costs of secondary treatment were not incurred) for its relative economy. In settings such as ocean outfalls, where nothing more than primary treatment was contemplated, and wherever short-term expenditures were resisted (even by communities facing the eventual need for two-stage facilities), sedimentation was certainly the cheapest solution.[35]

British and Commonwealth engineers – who had never adopted sewage-based fertilizers as a professional priority – moved promptly in the wake of royal commission approval of artificial methods of secondary treatment to follow the example of their American counterparts in entirely abandoning sewage farming with its extensive land requirements. Trickling filters and activated sludge became standard procedures for secondary sewage treatment.[36] Indeed, as early as 1902, the Local Government Board had abandoned its attachment to land-based disposal and agreed, in special circumstances, to contemplate funding for artificial treatment systems discharging into waterways.[37]

Good Enough for Now

With water supplies known to be vulnerable to contamination from sewage and industrial wastes, and with new technology promising to alleviate river pollution substantially, its rapid adoption might have been anticipated. In Canada, where, according to a 1916 survey, 204 of the 279 sewerage systems discharged sewage without treatment,[38] there must have been some incentive to implement treatment, given that two-thirds of the country's roughly five hundred waterworks relied on lakes and streams, many of which were officially described as "possibly polluted."[39] Sources of pollution in the United States were equally ubiquitous. Municipalities themselves contributed massively to surface-water contamination, since the sewerage infrastructure rarely included waste treatment facilities. In 1892 – at roughly the time of Sedgwick's investigation into the transmission of typhoid – only twenty-seven American municipalities employed any sewage treatment. By 1900, that number had risen to sixty, serving roughly a million people; a mere 4 percent of the population of sewered communities flushed treated water into the waterways. There was certainly pressure for improvement.

Armed with greater confidence in their understanding of certain diseases, public health officials campaigned vigorously for sewage treatment to prevent the contamination of water supply sources. A major Canadian report on water and sewerage for the national Commission of Conservation estimated that fifteen hundred lives could be saved annually if sewage were treated sufficiently to eliminate bacteria. Treating sewage was more fundamental than treating water supply, it insisted, for even with filtration at intake, "there is great danger of overloading the filters if the source of supply is grossly polluted by raw sewage."[40] In the words of another advocate of sewage treatment, "It is better to have an unpolluted water, and it is better to have to filter or purify a slightly contaminated water than an altogether contaminated water."[41]

Sewage treatment had powerful advocates in the United States as well, as state officials from New York, Pennsylvania, and Minnesota urged that untreated sewage not be permitted to enter streams serving as sources of water supply. Their objective had gone beyond preventing "nuisance": it was to combat routes of infection so as to prevent disease, principally typhoid. A 1908 report on stream pollution presented at a conference of state and provincial boards of health consolidated criticism of the practice of using streams for sewage removal, even adding recreational losses to public health concerns as a reason for protecting surface-water quality. The report called for double safeguarding – sewage treatment combined with filtration of water supply. A Stream Pollution Committee expressed its disapproval of any discharge of organic matter into streams used as water supplies and advocated at least partial sewage purification under the supervision of state boards of health. To encourage jurisdictions lacking effective legislation to control sewage, the committee further recommended that "State Boards of Health . . . present at every opportunity to the people generally, the importance of the questions involved."[42]

The same recognition of the need for public support was evident in Canada, where Advising Sanitary Engineer T. Aird Murray lamented that legislation on public health was "of little or no avail" without popular understanding and support. Attempting to rally such support for sewage treatment, Murray rhetorically asked whether Canadians were prepared to endure the stigma of the second-highest typhoid death rate in "the civilized world." Would they remain "content that our beautiful lakes and rivers shall be turned into sewage disposal areas and open sewers?" "Not a bit of it,"[43] the zealous engineer insisted. But Murray had overestimated the civic-mindedness of his compatriots. It seemed often enough that they would rather not find the money, and were resigned – if not entirely willing – to suffer the neglect. Several factors either diverted attention away from sewage treatment or presented obstacles to improvements. Technical, legal, political, and economic factors also constituted formidable and persistent challenges.

Even along prosperous and heavily populated stretches of the eastern seaboard of the United States, repeated efforts to maintain water quality showed very limited results. This was particularly true around New York City, where several major rivers converged. The population of New York's metropolitan area surged from 2.2 million to 5.5 million between 1880 and 1910,[44] while industrial activity, including more and more noxious processes such as metal refining and manufacturing, chemical production, and eventually petroleum refining, intensified. Districts adjacent to the port, in both New York and New Jersey, were simultaneously the source of an extraordinary volume of locally produced contamination and the destination for the bulk of the water-borne waste descending from inland sources.

In a 1902 study of sewage pollution in the New York metropolitan area, Marshall Ora Leighton warned that, "in inland municipalities there may be developed in the course of time a system of sewerage conducting such a volume of city waste that the largest river on the globe can not dilute it sufficiently to disguise its presence."[45] A number of rivers in the area faced severe pollution pressures, although with somewhat varied consequences. As yet, only the lower reaches of New Jersey's largest river, the Raritan, had been seriously damaged by sewage, but industrial pollution had been documented for over a quarter of a century in the Passaic, the most valuable of the state's drainage systems. Indeed, that river was virtually abandoned in the 1870s and 1880s.[46] Extensive commercial development and water withdrawals for Newark, Jersey City, and other growing communities had simultaneously increased demand for waste removal and purification while undermining the river's capacity to accommodate those wastes. In 1902, "the clutch of pollution" had fastened itself along the lower Passaic, where conditions were described as "insufferable."[47]

In 1896, a commission examining sewerage along the Passaic Valley concluded that the lower Passaic constituted a public nuisance, a health menace, and a growing threat to property interests from the Great Falls at Paterson to below Newark. The district population had risen by 43 percent during the 1880s and a further 22 percent between 1890 and 1895, when the daily discharge of sewage reached seventy million gallons.[48] A state sewerage commission established shortly after release of the Passaic findings declared existing arrangements to be wholly inadequate.

Sewage descending from Newark, Orange, Bloomfield, Montclair, Harrison, and neighbouring towns was driven back up the river by incoming tides. The same water, with increasing quantities of sewage being added, would travel up and down with the tidal flow until a heavy rainfall finally gave it sufficient downward volume to cleanse the lower river by removing the sewage accumulation to the bay. Yet such relief was never more than temporary.[49]

An inventory of losses attributable to pollution included the costs of securing replacement water supplies, severe typhoid outbreaks at Belleville and Jersey City, loss of the value of potential water sales to neighbouring counties, and destruction of the ice supply industry and commercial fisheries, as well as adverse impacts on real estate values.[50] In the warmer months of 1899, conditions reminiscent of London's Great Stink compelled residents living up to half a mile from the Passaic shore to close their windows against the stench; homes closer to the river were virtually uninhabitable.[51]

Even under these appalling conditions, willingness to pay for sewage works – the *sine qua non* of economic value – seemed limited. Sewerage assessments were prime targets for legal sniping. Thus, New Jersey legislation empowering the Board of Sewerage commissioners to levy taxes for capital expenditures up to $9 million and maintenance expenses associated with the challenge of cleaning up the Passaic ran afoul of court action.[52] Both the city of Paterson and one of its resident property-owning taxpayers who had been assessed for public services attacked the Passaic restoration scheme. The good news for those seeking to improve conditions in the river was a clear affirmation of the legislature's jurisdiction over such matters. "To relieve a river from pollution and to construct and maintain for this purpose sewers to the seaboard or to other point or output and to carry away through such sewers all that would otherwise pollute such river," Justice Garrison observed in the Court of Appeal, "is clearly within the power of the central legislative body."[53]

Yet this particular legislation was ruled unconstitutional. It unlawfully delegated to sewerage commissioners a power to tax; and the taxation covered an area whose boundaries did not correspond with those of the sewerage district. As the court observed, neither the taxation area nor the sewerage district were political divisions of the state of New Jersey, nor were they invested with any governmental function. The arrangement was therefore contrary to fundamental principles of state law requiring "that the district to be taxed shall be coterminous with a district to which some right of local self-government is given."[54] This noble constitutional principle did not, of course, prevent one community from taxing another indirectly by discharging sewage downstream. By way of eventual resolution, a major outfall for the twenty-two communities of the Passaic Valley was completed in 1924, providing for very extensive diffusion under the waters of New York Bay through 150 separate nozzles spread over three and a half acres.[55]

On the Hudson, south of Albany, teams of workers with horse-drawn sleds in 1901 removed 300 thousand tons of ice from the frozen river. Valued at $8 million, this ice harvest was the principal source of supply for the city of New York. Even a poor year brought $3 million in revenues. "For what fraction of this sum annually," observers interested in the quality of an important source of supply pointedly inquired, "could all the Hudson Basin sewage be purified?" Nevertheless, Leighton dismissed the need for remedial attention to the Hudson: "If we survey the different tributaries and note the points of contamination and the extent thereof we shall very soon realize that in nearly every stream there is such enormous dilution that, taken individually, there is hardly any reason to believe that any material damage has been done." The Hudson's overall volume was so extensive, he believed, that the flow "so largely dilutes the sewage poured into it that not infrequently the effects thereof are not traceable by sanitary analysis." The likelihood of the Hudson deteriorating to the level of the Passaic was "very remote." What's more, its middle section (downstream from Troy, the head of natural navigation and tidewater) was more valuable "in its present contaminated condition than if regulations were enforced which would maintain its purity." The assessment presumed that any attempt to enforce regulations along these lines would interfere with shipping to such a degree that great economic losses must ensue. The analysis brought Leighton within the ranks of those who believed that "river pollution is less expensive than the avoidance of it."[56]

Without so clearly articulating the calculation, New Yorkers had evidently subscribed to such a view from the origin of the Manhattan settlement, where the first recorded sewer construction dated from 1696. Even twentieth-century officials who were deeply disturbed by the long-term consequences of the unfortunate legacy that had been passed down to them were little inclined to be judgmental, for it "was then but natural and probably sound engineering to empty the sewers into the harbor since, even as late as the last century, the waters surrounding New York city appeared to be amply able to assimilate all the sewage discharged into them."[57] With passing generations, as the volume of wastes discharged without consideration into the harbour increased, their insidious detrimental effects had gone largely unnoticed.

Refuse dumping had also enjoyed the passive endorsement of public opinion. Oyster Island was an official dump site as of 1857, although two decades later its use was curtailed and solid refuse diverted to the southeastern side of Staten Island. By this time, New York State legislators had prohibited dumping wastes into the Hudson and East Rivers, Upper New

York Bay, and parts of Raritan Bay. Subject to the approval of the state shore inspector, dumping in Lower New York Bay south of the Narrows was authorized. State interest in such activities, however, had little relation to contamination and rather more to do with revelations by a US Coastal Survey in 1871 concerning refuse blocking certain harbour channels.[58] In 1886 the New York Chamber of Commerce, again with an eye to commerce and communication rather than contamination, pressed successfully for a prohibition in national rivers and harbours legislation against dumping in New York Harbor. Lacking penalty provisions, however, the measure was of limited effect. Shortly thereafter the War Department assumed extensive powers over the port pursuant to the *New York Harbor Act*.[59]

The *Sludge Acid Act* (1886) was New York State's legislative response to the environmental impact of industrial discharges. Shellfish were the principal victims of such discharges; their fate seems to have motivated the statutory initiative.[60] By the end of the century, the extent of the deterioration was undeniable, finally arousing public opinion[61] sufficiently to produce the suggestion that New York City's outfalls be extended from the bulkhead to the pierhead line to encourage dispersion.[62] The New York Bay Pollution Commission, appointed by the state legislature in 1903, completed a sanitary survey of the harbour waters and reported on the Passaic Valley sewer outfall in 1906. If the need for sewage treatment was now apparent, the reports offered no specific plan for abatement. The task of formulating such a response fell to the Metropolitan Sewerage Commission,[63] a newly constituted agency that, at least in principle, offered the possibility of a comprehensive response.

The commission's reports appeared in three volumes between 1910 and 1914. They documented extensive contamination from sewage and industrial sources in the Upper Bay, along the East and Harlem Rivers, and around Newark Bay, particularly where the Passaic and Hackensack Rivers discharged. Sewage sludge covered much of the harbour's bottom, a finding that could hardly have been surprising in view of the 620 million gallons of untreated sewage discharged each day by municipalities in the New York metropolitan area.[64]

With the inquiry under way, a number of New York residents sought to apply the *Refuse Act* of 1899 to forestall or prevent proposed sewer construction. The legal initiative was foiled, however, when the judge advocate general resolutely determined that application of the *Refuse Act* was limited to navigational concerns and that pollution was a matter for state action.[65] With no state action readily forthcoming, sewage volumes increased to the point that by 1914 the raw sewage of six million people was being discharged into New York Harbor, large quantities of it accumulating on the bed. Further studies showed that tidal effects merely diluted rather than replaced harbour waters, since only about 23 percent of the average volume of fresh and salt water present in the harbour at high tide actually passed out through the Narrows during the ebb.[66] Still, New York – the beneficiary of distant, well-protected water supplies – would not make significant progress on the wastewater question for some time. For different reasons, other communities were beginning to experience a similar sense of immunity.

Clean Water and How to Get It

Alongside impressive investigations into sewage disposal options such as the work of the British royal commission, the New York Harbor inquiries, or efforts by Canada's Commission of Conservation, important developments were under way in drinking-water treatment. Emerging approaches to safe water involved protective and remedial water treatment measures at intake, through filtration or other means. Water treatment rapidly became an alternative to safeguarding sources of water supply.

Filtration had originally served to address concerns about colour, odour, and clarity. These objectives had underpinned a pioneering system in Paisley, Scotland, dating from 1804, another in Glasgow three years later, and the initiative of London's Chelsea Water Works in 1827. From 1856, Berlin's water supply from the Spree River had been passed through sand filtration beds before entering the mains.[67] Growing awareness of bacteriological risks had encouraged other European communities relying upon surface-water sources to implement this protective measure. In Hamburg, where tidal flows on the Elbe returned city sewage upstream to the municipal intake, the water supply was not filtered, and the city was ravaged by typhoid in 1892. When Altona, downstream from Hamburg's sewers but with filtration in place, suffered much less severely from the outbreak, the message that filtration was useful in preventing disease resonated widely.[68]

In the United States, as the twentieth century began, water treatment still remained the exception, despite some isolated experiments with sand filtration. Around the time of Sedgwick's research into the typhoid epidemic at Lowell, only six communities filtered their water supplies. The first of these, Richmond, Virginia, had introduced filtration as early as 1832, but, before widespread appreciation of bacterial risks, many other communities had disregarded advice to follow suit. When Albany adopted filtration in 1899, it dramatically lowered death rates from typhoid, in contrast with nearby Troy, which took no preventive steps.[69] As of 1900, when 40 percent of the population used public water supplies, only 6.3 percent had access to filtered water.[70]

The field had developed rapidly, as Rudolph Hering recounted in 1901 to a US Senate Committee inquiring into the question of filtration of water supply at Washington, DC. When called upon in 1883 to advise Philadelphia on a new city water supply, Hering had recommended unfiltered water taken from the Blue Ridge Mountains. Subsequently, in view of advances in filtration that allowed much less satisfactory sources to be purified sufficiently for domestic use, he recommended the Schuylkill and Delaware Rivers as far more feasible alternatives for present and future generations of Philadelphians.[71]

Experimentation with chlorination in North America began in 1893, about the same time that similar tests were more widely under way in parts of Europe.[72] Middelkerke, Belgium, in 1902 and Lincoln, in England, three years later, were among the first communities to chlorinate municipal water supply on a continuous basis. The first continuous chlorination of an urban water supply in the United States was introduced in 1908 at Jersey City's Boonton Reservoir. When Jersey City, claiming that the water it received from the Jersey City Water Company was not "pure and wholesome," demanded that the company construct an intercepting sewer and treatment plant, the company refused, opting instead to take the initiative in testing chlorine gas and bleaching powder. Judicial acceptance of chlorine as an effective disinfectant at Jersey City encouraged rapid adoption of the process elsewhere.[73] Chicago, still resisting sewage treatment, introduced chlorination to its entire water supply in 1916 and was successful by 1919 in achieving the lowest typhoid rate in the United States at one per hundred thousand. This was down from an alarming sixty-seven cases per hundred thousand in the 1890s and fourteen per hundred thousand in 1910.[74] Toronto, having ultimately chosen to retain Lake Ontario as the source of its intake and sewage discharge, introduced chlorination in 1910. The city later pioneered such refinements as pre-chlorination, super-chlorination, and de-chlorination.[75]

Filtration and chlorination were enormously beneficial from a public health perspective. Rates of infection from water-borne diseases fell dramatically. Yet the possibility of treating municipal water en route to the consumer had a profound impact on the treatment of wastewater en route to the environment. If the threat of disease could be directly removed from municipal water supplies through water purification, it might become less important to treat sewage before discharging it into natural waterways. The choice between water treatment and source protection strategies such as sewage disposal, as Martin V. Melosi has expressed it, "highlighted how sanitary services were caught in the transition from the filth to the germ theory of disease."[76]

The success of mechanical and chemical water treatment technology increased the availability of local supplies while it reduced incentives to protect water sources from contamination. "Today," noted one turn-of-the-century observer, "the purification of polluted water for domestic use has reached that state of perfection at which it has become the practice of reputable engineers to take polluted water from a stream at the very doors of the city and purify it, rather than to expend large sums of money in conserving an unpolluted supply miles away in a sparsely settled district."[77] In 1903, the *Engineering Record* advanced the opinion that it was often "more equitable to all concerned for an upper riparian city to discharge its sewage into a stream and a lower riparian city to filter the water of the same stream for a domestic supply, than for the former city to be forced to put in sewage treatment works."[78] Sewage treatment was clearly in contention with water purification as a public health strategy and for municipal support.

The use of chlorination or filtration greatly increased the security of urban populations from water-borne diseases. Simultaneously, however, new chemical and mechanical safeguards for public health reduced incentives to maintain the quality of water supplies. This photograph, taken in the early twentieth century, shows the interior filter of the Toronto Filtration Plant.

Amongst the voices heard on the subject of sewage treatment and water purification, few were more influential than that of Allen Hazen. His well-recognized experience included several years in Massachusetts at the Lawrence Experimental Station in the 1880s and in Chicago as manager of the "perfect sewerage" arrangements installed at the World's Columbian Exposition. He had continued to investigate methods of refining sewage treatment procedures still further. However, in the early 1900s, Hazen experienced something of a conversion. Although continuing to consult on sewage treatment, he firmly criticized attempts to purify sewage as a means of improving public water supplies.

Keeping sewage out of rivers, was, in Hazen's view, neither practical nor necessary. In light of the benefits of equivalent expenditures for other social objectives, he claimed, it was "not even desirable" to protect rivers from sewage. There were numerous situations where local nuisances might justify the expense of sewage treatment, but in the majority of cases where sewage was discharged into rivers, "there does not result any local nuisance which would justify, to prevent it, the expenditure of the money necessary to purify the sewage, or, even if the work could be done for it, of one-tenth of the required sum." Local nuisance, as Hazen saw it, still entailed physical factors such as "the discoloration of the water, the presence of floating substances objectionable in appearance, the deposition of sewage mud on the bed of streams, and the production of offensive odors."[79]

Hazen recognized that there would be situations where sewage purification efforts designed to address nuisances might simultaneously enhance the quality of a water supply, but, he added, "this is the exceptional case, and not the common one." He made this assessment in what became a popular primer, *Clean Water and How to Get It,* which he wrote aboard the SS *Aorangi* on the Pacific Ocean and dedicated to the Brisbane Board of Waterworks, whose consulting commission had allowed him to promote the culture of flushing on another continent. "To set about cleaning up the rivers of the country for the purpose of improving the quality of the public water supplies would involve the purification of sewage from thousands of cities and towns where that was the only reason for the purification, or, in other words, where there was no local nuisance produced by the discharge of crude sewage." To Hazen, the bottom line was clear: "The discharge of crude sewage from the great majority of cities is not locally objectionable in any way to justify the cost of sewage purification."[80]

From the engineering perspective as well, he suggested, it was preferable to purify water supplies rather than to purify sewage. For the benefit of municipal officials and administrators, Hazen argued that this approach was "very much cheaper" because there would be smaller volumes to handle and because the cost of purifying water was much lower than the cost of purifying an equivalent volume of sewage. Among further advantages was the fact that "the methods of water purification are more efficient in stopping germs of disease than are the methods of sewage purification," and he insisted that "because all the water used can be with certainty treated," there was more security in the water treatment model. Sewage purification facilities could rarely treat all the sewage from the areas they served, because of storm overflows and "street wash" that did not always pass through the sewers, and on account of "the thousand minor pollutions that practically cannot be stopped." All in all, it was "both cheaper and more effective to purify the water, and to allow the sewage to be discharged, without treatment, so far as there are not other reasons for keeping it out of the rivers." The ratio of expenditures seemed an extraordinary testament to the economic virtues of water treatment and flushing. "Usually," Hazen anticipated, "one dollar spent in purifying the water would do as much as ten dollars spent in sewage purification." All this meant that waterworks administrators "must, and rightly should, accept a certain amount of sewage pollution in river water, and make the best of it." Purifying water rather than sewage was the only way to master the situation: "Success in supplying good water cannot be otherwise reached."[81]

Others also elaborated the affirmative case for water treatment. George Chandler Whipple, a student of Sedgwick's, set out to persuade his readers that impure water supplies affected municipal pocketbooks as well as the health and comfort of communities. Whipple's analysis, *The Value of Pure Water* (1907), focused on a different bottom line: he calculated that the loss attributable to each typhoid death in a community could be put at roughly $6,000. This represented the average life value of $4,635 per typhoid victim, combined with a modest estimate of treatment expenses for an additional fifteen to eighteen cases of infection associated with each death. Further reflections on the costs of water treatment led Whipple to conclude that water purification was "a sort of life insurance for the people." For each unit decrease in the typhoid death rate per hundred thousand population, Whipple imagined a ten cent per capita saving. Citing Albany as an example, he suggested that, as recent purification efforts had lowered the municipal typhoid death rate by seventy-eight per hundred thousand, the annual per capita saving in terms of life value was $7.80. To capitalize this amount in an era where life insurance was available at $17 per $1,000 "would represent an insurance policy of about $460 per year for each inhabitant, or $2,300 for each head of a family." Contemplating the financial perspective alone, and disregarding the humanitarian dimensions of the question, he concluded, "no city can afford to allow an impure water supply to be publicly distributed."[82]

Growing acceptance of the benefits of water filters specifically designed to remove disease-causing germs served to rationalize taking the burden of the bacterial threat off the sewage side. Thus, argued the authors of a Boston study, "It is unnecessary to place on the sewage-purification works the extra burden of bacterial removal . . . It is scarcely fair to hamper the essential task of sewage disposal by demanding a bacterial purification which can be better attained by subsequent special treatment in water filters." Sewage purification efforts should be devoted only to the problem of nuisances likely to result from organic decomposition.[83] In the absence of such concerns, sewage treatment might be foregone.

By the First World War era, it seemed that water treatment would prevail overwhelmingly against the sewage treatment alternative – as much by default as by design. Public health officials, whose predecessors had once promoted waste removal by means of municipal sewerage to guard against miasmatic perils associated with organic decomposition, now emphasized that bacteriological insights dictated the purification of water supplies. Contemporary engineering wisdom, epitomized by Hazen, garnered support amongst municipal officials with an eye on expenditure levels and debt. The further thought that the primary beneficiaries of municipal sewage treatment would be downstream residents – outside the jurisdiction – frequently sealed the fate of that option for at least a generation. Whether or not surface waters served as water supply sources, the convenience of using them for waste disposal almost universally overrode any dimly perceived benefits of preserving the integrity of natural waterways.

Pittsburgh, for example, at the confluence of the Allegheny, Monongahela, and Ohio Rivers, had drawn its water from the two former waterways since 1828. By 1900, seventy-five communities with a combined population of some 350 thousand were discharging untreated sewage into the two rivers. Indeed, with both intakes within city limits, Pittsburgh was exposed to the discharge of its own sewers. The city endured over five thousand typhoid cases each year in the decade spanning 1900, with immigrants from eastern Europe particularly vulnerable. Having recently left sparsely populated rural regions, they were disinclined to believe warnings about the quality of local water, and lacked the means to safeguard their own supplies even if they had acknowledged the risk.[84] Basing his calculations on estimated death rates and his elaborate formula for assessing water quality costs, Whipple concluded that Pittsburgh was depleting its "vital assets" or human resources by some $3,850 thousand each year through continued reliance on Allegheny River water.[85]

Eventually, in 1907, to protect the water supply of its half million residents, Pittsburgh introduced a slow-sand filtration plant, thereby reducing its typhoid death rates dramatically from approximately a hundred per hundred thousand during the final decades of the nineteenth century to twenty-two by 1910. But when called upon by state health officials to develop a comprehensive sewage collection and disposal plan in the interests of downstream communities, Pittsburgh declined to act. Its decision to continue discharging the untreated contents of a combined sewerage system dating from the 1880s was supported by engineering opinion. Hazen, with Whipple, now his new associate, compared the costs of new treatment facilities, including conversion from combined storm- and wastewater sewerage to a separated system, with modifications to existing arrangements.[86] For example, upstream storage reservoirs could supplement flows at periods of low volume in the river, and thereby prevent or moderate nuisance problems. Basing their stance on cost comparisons, Hazen and Whipple continued to champion dilution. Moreover, the projected costs of treatment would exceed state law governing allowable levels of municipal indebtedness. The consultants could find no precedent for a large city purifying its sewage in consideration of its neighbours' downstream water supply, and were of the view that downstream water treatment would in any event be necessary. As if the point needed emphasis, Hazen and Whipple noted that Pittsburgh would not even benefit from treating its own sewage. All in all, they summed up their argument, "no radical change in the method of sewerage or of sewage disposal as now practiced by the city of Pittsburgh is necessary or desirable."[87]

Thus, as historian Robert Gottlieb so aptly expresses it, the triumph of water treatment over sewage treatment had the deleterious effect of decoupling the contamination of rivers, lakes, and streams from the issue of drinking-water quality.[88] With water treatment targeting the risk of bacteriologically transmitted disease at the intake end of the water system, environmental protection seemed much less necessary. A new generation of municipal officials and their public health and engineering advisors were willing to tolerate surface-water contamination from both sewage and industrial waste on the assumption that with filtration or chlorination these wastes had little significance for public health.

And sewage treatment presented its own array of problems and complications.

Separating Water from the Waterways

The "Big Odor" Case, Excrementitiousness, and the Price of Sludge

After years of controversy over deteriorating waterfront conditions dating from well before James Mansergh's critique, Toronto finally approved an interceptor sewer system in 1908 to carry sewage for treatment in settling tanks at Ashbridge's Bay.[89] Construction of the sewer system connecting to a sewage treatment plant on Morley Avenue soon got under way. Completion of the undertaking in 1913 was generally hailed as a major advance, but from the outset residents of the city's east end severely criticized the venture. It was soon evident that the response to one form of nuisance pollution might very well give rise to another. Complaints from residents and deputations to city officials produced expressions of sympathy and triggered municipal investigations into the plant's operation, but they failed to resolve the situation. Neighbourhood residents adversely affected by the latest municipal improvements embarked on legal action.

Samuel E. Fieldhouse, a food, confectionery, and ice cream merchant, had the misfortune to carry on business from premises just opposite the

The introduction of a large-scale interceptor sewerage system in Toronto was soon followed by the construction of sewage treatment facilities. The new treatment facilities were almost immediately the subject of citizen protests and litigation initiated by Samuel Fieldhouse, whose confectionary establishment was greatly inconvenienced by sewage operations.

lakeshore location of the Morley Avenue plant. Fieldhouse was among those who had petitioned civic officials to remedy the situation, but when minor modifications failed to produce satisfactory improvement, he took civil action against the city in November 1915. He alleged that the plant caused a nuisance and that the city had been negligent in its construction and operation.[90] He specifically charged that the nuisance and associated water pollution endangered public health, had destroyed his business, and had rendered it "unbearable" to live on his premises. Fieldhouse claimed damages, together with an abatement order against the city and an injunction to bring the nuisance to an end.[91] Despite Toronto's formal denial of legal responsibility, the Fieldhouse claim was amply supported by municipal documentation. Senior municipal officials acknowledged the existence of the nuisance, and the municipal board of control had gone so far as to investigate possible claims against the experts from New York and Birmingham who had advised on the Morley Avenue plant. That specific inquiry produced a sobering response from Toronto's commissioner of works: "The advice of the experts, relative to sludge disposal was not followed, and the condition they foresaw if sludge were deposited contiguous to the premises, has eventuated."[92]

Compounding the awkwardness of Toronto's position was the fact that whatever approval provincial health officials might once have expressed for the city's plans had evaporated. After effectively conducting a raid on the plant, one provincial health inspector unhesitatingly asserted that the complaints were "well founded, as the pollution of the atmosphere by this plant cannot help but be a nuisance and menace to the health of the nearby residents who are compelled to breathe it." For good measure, he added, "Undoubtedly some different method of treating and disposing of the sludge is required and should be insisted upon without unnecessary delay."[93]

Approximately $6 million, according to municipal estimates, would be required to eliminate the problem. There was little enthusiasm to proceed with such costly remedial action, especially in wartime, and the city was of the opinion that there was actually no adverse impact on the drinking-water supply. A convenient "do nothing" argument also emerged in the proposition that even if the $6 million were to be spent, "new discoveries or experiments in treatment might soon render the whole plant out of date."[94]

The "Big Odor" case, as the *Toronto Daily Star* labelled the proceedings, became a cause célèbre when it reached the court of Chief Justice Mulock in December 1917. One hundred and fifty witnesses were assembled, and "blueprints without end festooned the judicial desk." Mulock, increasingly given to crustiness, took an active part in the trial, reportedly at one point offering the city his services as a sanitary engineer on the principle that anyone could do a better job.[95] Mulock identified two sources of offensive odours: concentrated sewage or sludge that was allowed to settle on a nineteen-acre disposal site, and effluent drained off through a defective and inadequate outfall pipe and a storm overflow line to Ashbridge's Bay.

In response to the city's assertion that the sewage treatment facility had been authorized, the chief justice underlined an important distinction between statutory authority to establish a sewage plant, and authority to create a nuisance by operating it. And, he added, to make the point clear, "inability to operate it without causing a nuisance does not . . . furnish an excuse for . . . creating a nuisance." Mulock confidently concluded that Toronto's new sewage plant was a cause of nuisance and that that nuisance could be traced largely to official negligence, demonstrated clearly enough in the municipality's long-standing failure to repair the deficient outfall pipe.[96] The trial result was precisely what Fieldhouse had sought and what the city had feared: Fieldhouse won $2,000 in damages, an injunction prohibiting the city from operating the plant so as to cause a public nuisance, and an abatement order requiring remedial action by 1 May 1918.

Toronto appealed Mulock's judgment, and while a decision on the appeal was pending, civic leaders attempted – unsuccessfully – to negotiate a settlement. Although an official attributed the difficulty to "the uncompromising attitude of the plaintiff,"[97] we should not overlook the city's own difficulties with compromise. As late as 30 April 1921, a court official reviewing Fieldhouse's entitlement to damages from the date of the trial to the date of abatement fixed the amount at $3,820, noting that the long-awaited day had not yet arrived.

Toronto's initial reluctance to comply with the original finding of the court had inspired the city to contemplate desperate measures. The municipal council – before an unfavourable appeal decision – had determined to petition the provincial government for legislation to authorize the operation of the sewage treatment facilities on a retroactive basis. A few months later, civic officials had second thoughts, and the application was reconsidered.[98]

Toronto's inclination to seek legislative relief for its sewage arrangements was the course of action that had been urged on Birmingham more than half a century earlier, and it was hardly the last time that some municipality or other would entertain legislative salvation in the face of adverse judicial results. What was Birmingham up to now, at the dawn of the twentieth century, several decades after Charles Adderley used the magic sword of the common law to slay the dragon of pollution?

The community of Tamworth, situated on the River Tame about twelve miles downstream from the Birmingham Sewage Farm and associated works at Saltley, initiated proceedings against the city in 1899. In contrast to Charles Adderley's earlier common law action, the Tamworth complaint focused directly on statutory standards. The trial was postponed several times, initially to allow the city to implement modifications recommended by an engineer who appeared to have the confidence of all parties. The effort proved futile, finally forcing Birmingham, now a city of nearly three-quarters of a million, to examine new sewage arrangements, including septic tanks and a bacteria percolation system. By the time the dispute came to trial in 1907, Birmingham was scrambling to complete new facilities. The sewage farm now covered twenty-eight hundred acres, containing roughing tanks, septic tanks, an irrigation network, lagoons, and sixteen acres of new bacteria beds. It was a £250 thousand effort. Commendable, yes. Good enough?

In 1861, in the context of authorizing local councils to establish sewage outfalls beyond their own district borders, Parliament had added limitations. This new authority, Parliament proclaimed, should not be construed "to give power to any local board to construct or use any outfall, drain, or sewer for the purpose of conveying sewage or filthy water into any natural watercourse or stream until such sewage or filthy water be freed from all excrementitious or other foul or noxious matter, such as would affect or deteriorate the purity and quality of the water in such stream or watercourse."[99] There was little elaboration of the statutory prescription for some time, although Sir George Jessel thought the message to municipal officials was "pretty plain" in 1874: "You shall not send your sewage water into a natural stream until you have made it wholesome water – until you have got rid of all the noxious matter in it."[100] The legal system does not thrive on "pretty plain" meanings, particularly in situations where the context within which that meaning must operate is undergoing transformation. Drawing directly upon the 1861 local government legislation, the *Public Health Act, 1875* reaffirmed the earlier prohibition, extending its application to any canal, pond, or lake.[101]

To appreciate the problem lurking in Tamworth's confrontation with Birmingham – one of the more significant conundrums in the philosophy of sewers – it is helpful to consider local geography. The River Tame, rising just northwest of Birmingham, swings south toward the city. In the nineteenth century, it flowed through populous manufacturing districts before reaching the boundary of the Saltley sewage farm, itself at the eastern edge of Birmingham. Shortly after passing the sewage works, the Tame was joined by its tributary the Rea, a river that flowed through Birmingham, including its manufacturing sections. From there, the Tame proceeded north to Tamworth. Key to the litigation was the fact that, upstream from the point at which Birmingham effluent entered them, both the Tame and the Rea were already polluted with sewage and manufacturing waste.[102]

By the date of trial for Tamworth's complaint, Birmingham's elaborate waste treatment facilities were both notorious and renowned. Engineers and others concerned with water and waste frequented Birmingham on professional pilgrimages, in search of revelation, though often settling for much less. The possibility that Birmingham's flagship arrangements violated national legislation, the *Public Health Act,* was accordingly a matter of general importance. But what did the *Public Health Act* mean when it said that municipalities were not authorized to convey filthy water or sewage into any natural stream or watercourse "until such sewage or filthy water [was] freed from all excrementitious or other foul or noxious matter such as would affect or deteriorate the purity and quality of the water in such stream or watercourse"? This was undoubtedly a call for treatment, but it begged the question: what level of treatment – and what result – would satisfy the legislative expectation?

To the long-suffering instigators of the complaint, the statutory prohibition was directed against "sending into the river that which would deteriorate the purity and quality of water in a natural stream or watercourse, that is to say, the water in its natural unpolluted condition."[103] Yet to the defendants, the section's very wording presupposed that it was permissible for sewage or filthy water to go into the stream. Municipal authorities had only to free it from enough sewage to prevent it from affecting the purity and quality of river water in a manner amounting to deterioration. "The test of whether the effluent affects or deteriorates must be determined by the character of the water in the stream itself as you find it."[104]

The first judge to hear the competing arguments, Justice Kekewich, gave what he considered a reasonable interpretation: the legislation "necessarily implies that excrementitious or other foul or noxious matter may be conveyed, for it is not said that none such shall be conveyed into any natural stream or watercourse, but only that there shall not be conveyed into it such as would affect or deteriorate the purity and quality of the water." Kekewich thought it semantically sound to assume that Parliament intended to prohibit deterioration, that is, the "making worse" of the purity and quality of the stream or watercourse. On this basis, he determined that "no one can be said to have offended against the prohibition if as a matter of fact there has been no such deterioration."[105]

A further opportunity to consider the water quality standard set by the statute arose in the Court of Appeal. Cozens-Hardy, master of the rolls, put the question of interpretation quite simply: "Now, is this an absolute prohibition against conveying any filthy water into any natural stream, or is it only a prohibition against conveying filthy or sewage water in such a manner as to prejudicially affect or deteriorate?" He approvingly recited the trial judge's view before concluding, "It seems to me hardly reasonable to say that a man can be guilty of a statutory offence under this section if he has improved the water by that which he casts into it." What better indication of the state of early-twentieth-century waterways than the impression conveyed by this suggestion that the condition of a stream might be *improved* by the introduction of more wastewater? By way of elaboration, Cozens-Hardy denied that a statutory offence would occur "if the defendants . . . put into the river some water which does contain a certain amount of filth, although it is so much better than the rest of the water which was in before that upon the whole it improves the character of the water." In other words, he asserted that this would not "deteriorate the purity of the water; on the contrary, it benefits it, and we must read the word 'affect' in the sense of prejudicially affect or deteriorate the purity and quality of the water."[106]

Fletcher Moulton L.J. also considered the standard that Birmingham and other municipalities were expected to reach by sewage treatment: "It must be sufficient that it is so changed by subsequent chemical change that it becomes harmless from the point of view of the public and does no damage to them." The test adopted by the legislators, he believed, was an excellent one: "You may not discharge into a stream that which makes it worse." This rule entailed important consequences: "If by carefully applying purifying processes to the upper waters of the stream these waters become more and more pure, then those that discharge into the lower waters of that stream must improve their methods so that they do not interfere with that improvement." In particular, the condition of a stream was not to be worsened "by discharging water into it that has received excrementitious matter . . . What you must not do is to discharge into the stream water which will make it worse."[107] In an era when almost all sewage was water-borne and "must therefore go down natural channels to the sea," any legislative restriction on the discharge of sewage was of great importance. Courts, Fletcher Moulton cautioned, should avoid "pushing beyond their legitimate meaning the legislative provisions as to purity" or they would run the risk of making the task of sanitary authorities "impossible to perform."[108]

The proclamation not only endorsed the discharge of effluent, but pointed to a duty of the courts to ensure that expectations for water quality would not get too high. It would also appear that communities were now being offered a financial incentive (in the form of reduced requirements for waste treatment) to discourage improvements in stream quality above them. Indeed, it was increasingly common to read that the prohibition on discharging filthy water applied only to waterways that were not already foul. Farwell L.J. also crisply dismissed the plaintiffs' argument as one that "assumes the absolute purity of all streams, watercourses, and canals, ponds, or lakes, an assumption to which I cannot assent."[109] And so it was that in judicial hands the statutory regime, like the common law before it, accommodated the perceived practicalities of the day.

Of course, now that the meaning of the legal test had been established, the standard still had to be applied to Birmingham. To assist it with this task, the Court of Appeal dispatched Sir William Ramsay, a distinguished Nobel laureate in chemistry at University College, London, to investigate current conditions, since Birmingham's program of reconstruction had continued during the course of the judicial proceedings and was now headed for a price tag of £.5 million. Ramsay was specifically instructed to report "whether the sewage effluents ... were ... entirely or to any extent freed from all excrementitious or other foul or noxious matter such as would affect or deteriorate the purity of the water in such river."[110]

Ramsay's report on excrementitiousness was to the point. He assured his audience that if excrementitious matter meant actual excrement, whether feces or urine, there was none to be found. If, however, Parliament understood excrementitious matter to include partly or wholly oxidized feces or urine, Ramsay acknowledged the presence of some such matter. By way of encouragement, he added that this definition applied, though to a lesser degree, "to the water drunk by the greater part of the population of London, which is drawn from the Thames and the Lea, and these rivers are polluted by large volumes of sewage effluent from the towns on their banks." He was persuaded that the combination of sewage purification before effluent entered the rivers and treatment of water supply by means of storage and sand filtration was safe: "No danger is to be anticipated from drinking it." In keeping with well-recognized legal practice, Ramsay, the chemist, then formulated his own question for response: "Are the waters of the Tame and the Rea before they receive the effluent from the Birmingham Sewage Works purer or less pure than after the effluent has been added?" In his view, Birmingham's effluent did not render the river more foul. The effluent had been sufficiently treated that it did not in his judgment "affect or deteriorate the 'purity' of the water in such river" either at the point of discharge into the stream or in what came to known as a mixing zone within a distance of six hundred yards downstream.[111]

After all this attention, the Tame saw its first fish in decades shortly before the First World War. However, in the course of that great conflict, with its unprecedented industrial demands, this waterway – like so many others – became a casualty of war. By 1918 it was known again as an open sewer.[112] Resource limitations and wartime priorities help to account for Britain's failure to maintain remedial momentum, but in the United States, where these considerations were initially less relevant, the introduction of sewage treatment faced another unexpected set of constraints.

In their enthusiasm to introduce activated-sludge treatment, several American municipalities had overlooked the niceties of royalty payments, and it became necessary to test Walter Jones' entitlement to patent protection. Chicago, after finally beginning to address sewage treatment, suffered yet another legal setback. Activated Sludge Ltd., holders of US patent rights for the innovations developed by Jones and Attwood Ltd., sued Chicago for unauthorized use of the technology. Chicago might have settled the claim for $90,000. Instead, after several years of resistance in the courts, the city found itself burdened with roughly $1 million in fines.[113]

Milwaukee, however, bore the full legal brunt of the effort by Activated Sludge Ltd. to protect its intellectual property when the validity of Jones' original patents on sewage treatment by aeration dating from 1913 to 1914 was recognized in 1934. Jones' entitlement survived challenges claiming that he and his Manchester collaborators had simply appropriated the inventions of researchers at the Lawrence Experimental Station in Massachusetts. His rights even held up against a very similar US patent obtained in 1915 by Leslie Frank of the United States Public Health Office, who had dedicated his invention to the people of the United States.

Activated Sludge Ltd. not only succeeded in confirming Walter Jones' patent rights, but it also demonstrated that Milwaukee had infringed those patents, and secured a court order enjoining the Wisconsin city from operating its sewage treatment facility. An appeal court overturned that aspect of the judgment, imposing financial compensation alone. As the court observed, a permanent injunction would close the treatment plant, leaving the community no alternative but to run its raw sewage into Lake Michigan, threatening the health and the lives of half a million people. In such circumstances, "we think no risk should be taken."[114] As of 1938, 150 of 203 activated-sludge plants were paying royalties of about twenty-five cents per capita to Activated Sludge. Yet others, such as San Marcos, Texas, shut down treatment facilities or reconsidered plans to install activated-sludge technology. Some of the latter adopted alternatives such as trickling filter systems; others chose to defer action until applicable patents had expired. Washington, DC, chose the latter course of action.[115]

Innovations and improvements in sewage treatment technology did not, for a variety of reasons, necessarily lead to vigorous implementation. Chlorination took some of the pressure off the public-health side. Foot-dragging in the face of expense could always be expected. Wartime disruption and competition for labour and resources certainly played a role, and there were often, indeed, practical obstacles to address. More generally underlying the resistance to investment in preventing contamination of the waterways were emerging or re-emerging assumptions about the acceptability of pollution.

10

Streams Are Nature's Sewers

WATER POLLUTION PERSISTED, despite the ability to eliminate contamination from municipal water supplies, arguably even worsening as increasing volumes of industrial effluent mixed with growing discharges of municipal sewage, treatable, but less in need of treatment from the perspective of public health. Purity and wholesomeness were less often perceived as natural characteristics of water: they were the engineered result of the transformation of untrustworthy raw material. Although complaints about the deterioration of lakes and waterways continued to arise, the modern urban resident became largely indifferent to their natural condition. Would the legal system and policy makers acquiesce to contamination, recognizing as lawful the de facto state of affairs that had emerged? What justification of such an outcome would be forthcoming to make it acceptable?

How a Stream Becomes a Sewer

When Missouri challenged Chicago's appropriation of Great Lakes waters to flush municipal and industrial wastes toward the Mississippi River in the early 1900s, Justice Oliver Wendell Holmes – to his credit – recognized the magnitude of the underlying question: was the destiny of the great rivers "to be the sewers of the cities along their banks or to be protected against everything which threatens their purity"? Reluctant to decide the fate of the waterways prematurely, "at one blow by an irrevocable fiat,"

Holmes embraced the common law tradition of furnishing minimalist answers to monumental questions. Perhaps, as a New Englander accustomed to flushing as second nature, he had underestimated the current state of deterioration and overlooked the almost inevitable implications of the incremental alternative. By the time Illinois incorporated the Lake Michigan diversion via the re-engineered Chicago River into its sewage removal network, many waterways had already been severely degraded for generations. Others faced serious threats from the deeply entrenched belief that an essential function of rivers was to carry off the residues of human activity.

The consequences were widely recognized. A Pennsylvania governor protested the assumption by property owners during the 1890s that they were entitled to dump anything they wished into water bodies fronting their properties.[1] The *Lawrence Evening Tribune* in 1891 lamented that the Merrimack had "lost its virtue because of the fact that to all intents and purposes it [had] become a gigantic sewer."[2] To suggest that a once-natural waterway had become nothing more than a sewer – in the Merrimack case, a gigantic one – was simply an observation in the late nineteenth century. Such remarks were often merely expressions of resignation, perhaps extending to discouragement and regret. Whatever the case, later generations might well ask how their predecessors could observe the process of deterioration without feeling the loss. How could they have found it distressing and not take action against the cause?

For much of the first half of the nineteenth century, Scottish courts, applying their own distinctive legal principles, endorsed the proposition that it was the function of a "public river" to carry off waste. This notion had insinuated itself into the jurisprudence by means of a bold and long-standing proposition that had been accepted in 1804 when the skinners of Inverness successfully defended their practice of washing hides in the River Ness. The foundational statement, dating from 1661, set out that though an individual would be constrained by law from undertaking an activity detrimental to the property of a neighbour, "yet might he well send away any stagnant water, corruption or filth by a public river." This was so because in the case of such waterways, "one prime use is to purge the earth of all corruption and to carry it into the sea." Accordingly, "the corruption, not only of men and beasts, but of the earth, as of minerals, coal-pits, lime, and all others," might, in the absence of law or custom to the contrary, be "freely turned" into them.[3]

Two decades after the skinners successfully defended their right to foul the Ness, the public rivers doctrine allowed the Earl of Moray to drain the sewage of three hundred tenants into the Water of Leith against the objections of Stockbridge residents, whose primary and domestic uses would ordinarily have taken priority. Not until the 1850s, in the aftermath of a vigorous challenge to sewage irrigation at Edinburgh, was the principle that public rivers were dedicated to pollution checked. By then, however, the unfortunate assumption had taken hold in some regions to the detriment of many rivers in the country. Although courts eventually repudiated the open-ended public river doctrine, pollution might also find justification as an ordinary or reasonable use of waterways, among other claims.[4]

As Chief Baron Pollock explained in *Wood v. Waud* (1849), certain ordinary uses of waterways were sufficiently accepted as part of life that they might accumulate lawfully regardless of their environmental impact: "If the stream were only used by the riparian proprietor and his family, by drinking it, or for the supply of domestic purposes, no action would lie for the ordinary use of it." It was equally the case that "if a field be covered by houses, the only use by the inhabitants might sensibly diminish the stream, yet no action would, we apprehend, lie any more than if the air was rendered less pure and healthy by the increase of inhabitants in the neighbourhood and by the smoke issuing from the chimneys of an increased number of houses."[5] So it might be that ordinary domestic uses, multiplied many times, could lawfully contribute to the progressive deterioration of a waterway.

The Massachusetts Supreme Court reached such a conclusion, perhaps even lowering expectations as to the level of purity that downstream residents might be entitled to expect. Although the downstream riparian enjoyed a natural right to water "in its pure state, fit to be used for the various purposes to which he may have occasion to apply it," that right would have to yield to the equal rights of those upstream. Their lawful use of a stream for milling, irrigation, watering cattle, and so on would "tend to render the water more or less impure." The cultivation and fertilizing of lands bordering the stream or its sources, as well as the presence of farmhouses and outbuildings, would "unavoidably cause impurities to be carried into the stream." Population growth would only compound the problem, for "as the lands are subdivided and their occupation and use become multifarious," the pace of deterioration would intensify. The water might very well be rendered unsuitable for many uses to which it had previously been put, "but so far as that condition results only from reasonable use of the stream in accordance with the common right, the lower riparian proprietor has no remedy." As the population of towns or villages increased along the riverbank, "the stream naturally and necessarily suffers still greater deterioration." Roads, streets, gutters, and sluices discharging surface water that carried "abundant sources of impurity" would create a situation "against which the law affords no redress by action."⁶ Thus did the common law – highly deferential to custom, practice, and relationships among individuals – accommodate gradual deterioration, and effectively abandon any pretense of maintaining natural conditions.

What is more, the distinction between ordinary and extraordinary uses might vary according to the characteristics of a particular locality. Along certain streams, therefore, permissible "ordinary" uses of the waterway might extend to urban and industrial wastewater removal. As one British judge even remarked, "It may be that the question what is an extraordinary use, depends upon the development of the trade in the neighbourhood, and on the use to which the water of rivers is put in the adjoining district . . . The diffusion of trade may make a great change as to what constitutes an extraordinary use of running water."⁷ This unusual possibility was accepted hesitantly, reflecting its exceptional nature.

More common were situations in which, simply by the passage of time and through neglect on the part of those who might have objected, a polluter acquired a right to continue to contaminate a waterway. Thus, litigants trying to preserve water quality were sometimes shocked to learn that the battle had been lost long before they reached court. Like his father before him, the Earl of Harrington, the occupant of Elvaston Castle on the Derwent a few miles below Derby in the English midlands, objected to the discharge of raw municipal sewage into the watercourse on his ancestral properties. The adverse effects were amply described in evidence, with a report by the county medical officer, Dr. Barnise, offering something of the flavour of the case. Responding to Harrington's complaint, Barnise examined the castle's domestic water supply, which was taken from a lake fed by the Derwent. The water pumped to the castle "positively stank," he clinically noted. Barnise reinforced his assessment with reference to "hundreds of dead fish (chiefly tench)" giving off an offensive smell in the vicinity of the pumping station. A black stinking mud covered the sides of the lake and the feeder stream; the water was "bubbling with putrefaction." Barnise diligently pursued his investigation upstream to the Derwent itself, where he saw "pieces of excreta floating by." His conclusion, given the proximity of the supply lake to the castle, and the presence of a case of cholera in Derby, was that it was unsafe for Harrington to occupy his own ancestral premises.⁸

After years of fruitless protest and complaint, Lord Harrington convinced the county council to take action. In 1898 his persistence culminated in an order under the rivers pollution legislation requiring Derby to cease polluting the Derwent. Derby successfully obtained extensions of the deadline for compliance while the ornamental lake at Elvaston Castle deteriorated further. Harrington eventually took action personally to seek an injunction against the sanitary authority.

Justice Buckley recounted the sorry tale of Derby's sewerage, a story corresponding closely to the Massachusetts court's account, which would have been widely recognized around the English countryside. "As the population of the borough has increased; as new connections have been made to previously existing sewers; as new sewers have from time to time been laid; and as privies have from time to time been converted into water-closets, the volume of the sewage and the pollution of the river have continually increased."[9] The steadily worsening situation that had finally found its way to court could be traced back to at least 1872. Unfortunately for Harrington, this fact placed his claim to clean water in direct conflict with the counter-assertion that Derby residents had implicitly established their right to pollute.

Householders who had been discharging their sewage for more than twenty years had acquired a right – described as *prescriptive* – to continue to do so. It was true, Justice Buckley conceded, that a prescriptive right did not confer on the corporation or sanitary authority of Derby the right to pollute, but it would be impossible for the court to issue an injunction that had the effect of interfering with the prescriptive rights.

A second class of pollution rights-holders was also entrenched – by statute, if not on the basis of prescription: "As soon as the corporation have made new sewers or have permitted or directed connections into them or conversions of privies into water-closets with consequent discharge of further sewage, there have arisen . . . rights in individuals under the statute to continue to discharge through the sewers which in fact have been constructed."[10] The rivers pollution order, which Harrington had obtained from the county council but whose implementation schedule he seemed powerless to influence, also worked against him. Buckley cited the existence of that order as a further reason to deny an injunction: if Parliament had provided an effective alternative remedy, the court should not intervene. Although the court awarded substantial damages for the undeniable impacts experienced at Elvaston Castle, the statutory scheme, presumably designed to enhance water quality, had actually undermined individual efforts to protect it. As we shall see, that dynamic and the inherent possibility that legislative standards might be lower than the levels of environmental protection property owners could secure for themselves, has become one of the most contested features of modern water law.

As legal commentators on both sides of the Atlantic were careful to emphasize, prescriptive rights were not without limitation. On behalf of his colleagues, Lord Watson put the matter plainly: "A proprietor who has a prescriptive right to pollute cannot . . . use even his common law rights in such a way as to add to the pollution." Or, as he elaborated for the benefit of an American audience, the use of a stream as a sewer may to a degree give rise to a presumptive right for such use, "but it gives no greater right than past usage."[11] Thus, a city that had used a stream to discharge a certain quantity of sewage enjoyed no right to discharge a greater quantity. Changed circumstances along a river might result in further limitations on established discharges. For example, a decrease in stream flow resulting from upstream diversion, or simply lack of rain, would increase the burden of pollution from the same discharge of sewage, possibly producing a nuisance where none had existed before.[12] But if there was a right to sue to preserve the existing level of water quality, incentives to do so were rather limited. Rarely would it be considered worthwhile to preserve the less-contaminated condition of a waterway whose utility had already been heavily compromised by a prescriptive claim.

An important exception to prescriptive rights existed where the nuisance, rather than being merely private as was Harrington's complaint about Elvaston Castle, took on a public character. In considering sewage pollution of the water supply of York, Pennsylvania, the state Supreme Court pronounced in 1899, "No prescription or usage can justify the pollution of a stream by the discharge of sewage in such a manner as to be injurious to the public health . . . to deposit [sewage] in a natural water course, in close proximity to a source of supply from which the water is used for domestic purposes . . . is a public nuisance." One scholarly assessment, however, is that courts typically lacked the administrative capacity to decide cases where complex scientific and medical information was at issue. Their contribution, accordingly, was largely confined to situations involving damage to property or the creation of nuisances; by contrast, it was "relatively unimportant in cases involving hazards to health through waterborne disease."[13] Opportunities to remove contamination by filtration and chlorination further reduced the prospect of judicial intervention.

Another dimension of the re-engineered landscape provided a loophole that also proved troublesome to water quality. The fact that continuing discharges of municipal and industrial waste to *sewers* were more likely to be permissible than discharges to *streams* did not escape the attention of polluters. Those sued in relation to pollution, or charged with water pollution offences, sometimes found the line between streams and sewers particularly significant. They were especially anxious to be on the "right" (that is, "wrong") side of it in order to secure immunity in disposing of waste and refuse. For, whenever a stream crossed the notional divide that separated natural waterways from sewers, any inhibitions against further contamination were lost. And as more streams were conceptually transformed into channels for waste removal, it seems to have become easier to view others as destined for sewerage also, as some examples from the north of England illustrate.

When officials of the Matlock Board of Health, for instance, installed sanitary drain pipes in a watercourse running across a property in Derbyshire to a small brook, the owner, Wheatcroft, brought a trespass action. For many years this watercourse, or ditch, had received sewage from nearby houses, most of which eventually made its way to the brook. In defence of its actions, the Matlock board insisted that it had merely effected improvements to a sewer in the exercise of a responsibility and power conferred upon it by the *Public Health Act, 1875*. When this particular spat reached him for resolution, Justice Denman had no hesitation in finding that Wheatcroft's watercourse was indeed a sewer pursuant to the legislation, falling conveniently, it appears, within the court's understanding of the phrase "sewers and drains of every description."[14]

The questionable dividing line between natural waterways and sewers also occasioned controversy in the Tyneside town of South Shields. Here, Miss Falconar, the proprietor of a brewery on the Dean Burn, sought to defend what little of the natural character remained of the Dean, a modest brook formed from the confluence of Harton Burn and Colliery Burn. The Harton and Colliery Burns had each performed a number of years' service in carrying off domestic waste from houses and cottages in the vicinity. Yet Miss Falconar believed herself entitled to the undiminished and uncontaminated flow of the Dean. She took exception when South Shields workmen channelled the Harton and the Colliery into brick sewers, thereby diverting their flow to provide drainage for a number of new dwellings before being reunited with the Dean. Municipal authorities even threatened to cover the latter as well, and to incorporate it within the sewage network.

Without actually citing the Matlock incident, urban sanitary officials in South Shields had adopted the Matlock argument: that is, they contended that the Dean had long since taken on the character of a sewer vested in their hands pursuant to the *Public Health Act*. At trial, however, the chancellor of the County Palatine of Durham noted that the drainage of slops and refuse into the brooks had been modest until very recent construction had increased the sewage flow. He granted an injunction to save the Dean.

At the appeal, Lord Halsbury (who, as general editor, lent his name to a landmark synthesis of English law) concluded that the Dean was in fact a sewer of long standing. Rather than merely being entitled to take measures to eliminate it from the inventory of local risks to public health, the sanitary authority, as Halsbury saw it, was actually *obliged* to do so. Lord Justice Lindley, although his approval of the 1876 rivers pollution legislation was on record, agreed, describing a process whose tragic inevitability could have been captured only by time-lapse photography: "In days gone by the stream was a purely agricultural stream, but by degrees it had changed its character completely and it had become a sewer in the ordinary sense of that word – viz., a channel for the reception and carrying away of sewage. It was now a dirty, filthy sewer."[15] The stream, of course, had not changed its character: it was merely the passive recipient – increasingly passive – of household waste and excrement. It had been actively corrupted, as had so many once undefiled watercourses, through the acquiescence of the common law.

Similar contests over the character of small waterways arose in the context of various statutes that made it an offence to discharge pollutants into streams. Streams, the intended beneficiaries of the legislative protection, might be defined to include "any river, stream, watercourse or inland water (whether natural or artificial) . . . including rivers, streams, canals, lakes and watercourses." However, the definition would typically be couched in language that allowed for exemptions: for example, "any sewer vested in a local authority" or "watercourses . . . mainly used as sewers."[16] The lamentable significance of these categories and distinctions shortly became all too apparent.

The legal transformation of streams into sewers had occurred with sufficient frequency that by 1900 judges might confidently remark, "As regards the law, there is . . . no doubt that a natural stream or watercourse may cease to be a stream or watercourse, and may become a sewer into which certain persons may acquire certain rights to drainage."[17] It had even been established that "a tidal stream may be so used as to have become converted into a sewer,"[18] and cases of this character continued to arise.

Early in the twentieth century, Scotland's Lanark County Council challenged practices at Airdrie and Coatbridge, towns that routinely discharged their wholly untreated sewage into small streams, or burns, eventually making their way to the Clyde.[19] Putting the case as plainly as they could, the municipalities' legal counsel told the House of Lords that the burns, allegedly polluted, "are not streams, but sewers, and there is no law against the pollution of sewers."[20] Lord Loreburn conceded the singularity and ingenuity of the proposition before dismissing it as among "the most hopeless arguments" he had heard for a considerable time.[21] Loreburn was entirely satisfied that the local officials were committing an offence against river pollution under the 1876 legislation and had not brought themselves within any of its exceptions. He took up the assertion put forward by Airdrie and Coatbridge – "Permit us to prove that these burns are sewers, and if we can prove that they are sewers, surely it cannot be an offence to pour sewage matter into the sewers" – and set out to demonstrate the intolerable implications of its logic. They were asking the court to accept their claim to have committed "in an aggravated degree" the very offence with which they were charged as an excuse for committing it. "The object of the act is to prevent streams being turned into sewers, and this is what they propose to do and have done."[22]

Although the *Airdrie* judges objected to the transformation of streams into sewers, the sentiments they expressed did not dissuade others from attempting to convince courts that a natural waterway serving as a sewer should formally be regarded as one. George Legge & Son, for example, incurred great expense to repair a culvert installed on the company's premises almost fifty years earlier by a previous business operation. Seeking to offload both costs and responsibility, Legge identified the local sanitary authority as a possible source of salvation. For many years, at least twenty houses had been discharging sewage to surface-water drains connecting with ditches leading into the culvert. More than twice as many new dwellings had secured unobstructed access to the same system shortly before complications necessitated the expensive reconstruction. Might what was once a natural watercourse flowing through Legge's culvert have been used for waste disposal long enough that it could by now be deemed a sewer? Such a conclusion would place responsibility for the cost of repairs on the borough of Wenlock.

To avoid prolonged immersion in the detailed niceties of the stream of waste implicated in Legge's claim, the trial judge posed a more general question: was the change of status from a natural stream to a sewer within the meaning of the *Public Health Act, 1875* possible in law? Without referring to *Airdrie*, which might have offered some guidance, the trial judge concluded that such a transformation was certainly possible. But this was not Miss Falconar's nightmare, where years of minor but lawful sewage discharges preceding statutory prohibitions had so altered the character of the natural waterway as to endow it with sewer status. In contrast, the sewage flushed through Legge's culvert was there in contravention of the *Rivers Pollution Prevention Act*.

The general question of law – could a stream become a sewer? – subsequently reached the House of Lords, where the proposition was unanimously rejected.[23] Their Lordships saw no basis to imagine that repeated incidents of unlawful pollution, if sufficiently frequent, could become legal. Lord Macmillan reflected that it would be remarkable to imagine that one could legalize stream pollution by contravening the legislation designed to prevent that pollution. "In the case of a sewer vested in the local authority, the owners of adjoining property are entitled to discharge their sewage into it, so that, on this argument, if a natural stream in consequence of illegal pollution becomes in fact sufficiently foul to merit the legal appellation of a sewer, and so to vest in the local authority, the further pollution of it ceases to be an offence, and becomes a right."[24] He rejected the attempt to prove "that what was in law a protected stream [had] become in law an unprotected sewer, simply by reason of infringements of the law designed for its protection."[25]

The West Riding Rivers Board took seriously its responsibility for waterways in the Yorkshire district. It diligently resisted the almost commonplace claim that any waterway into which some enterprise discharged effluent was no longer a stream enjoying legislative protection, but had instead become a sewer into which it was entirely lawful to discharge liquid wastes.[26] In proceedings that culminated in a decision of the House of Lords, the West Riding Rivers Board put to rest the fashionable defence argument that the historic discharge of industrial liquids had created a common law prescriptive right to continue the pollution.[27]

This decision in a case involving the Butterworth Company was written by Lord Macnaughton, who had reviewed riparian rights at the end of the century. His formulation of the doctrine, juxtaposing "natural flow" and "reasonable use" language, has been an enduring one.[28] In the Butterworth episode, Macnaughton observed that the company had made no effort to render its effluent harmless to the waterway, nor did it propose to do so. The company's excuse, as Macnaughton understood it, was simply that, after discharging its poisonous matter into a sewer belonging to the district sanitary authority, it had no further responsibility: "If the sanitary authority pour it into a stream they, not we, are guilty of an offence against the statute." To Macnaughton the company's argument was no more than "an idle contention," for it was self-evident to him that "if a person sets in motion poisonous liquid in a course and direction which must take it into a stream, the person who sets the liquid going causes it to flow or fall into the stream, whether it passes through a conduit vested in somebody else or not."[29] He was equally unwilling to accept Butterworth's alternative assertion that the company, through long-standing usage of the sewer, had thereby acquired a prescriptive right to continue the practice. Certainly no prescriptive right to foul a stream could arise in the face of legislation prohibiting such action. And the company had, in fact, avoided such a claim. But Macnaughton was equally of the view that one could not extend a right to pass certain wastewater through a sewer into a right to discharge offensive and unauthorized substances.

The number of streams legally transformed into sewers by virtue of neglect on the part of those who might have acted in their defence will never be known. Other small waterways were deliberately buried, covered over, or structurally altered in confirmation of their sewer status.[30] These practical considerations contributed to a strengthening perception that, notwithstanding the overturning of Scotland's public rivers doctrine, the function of many waterways – and certainly of coastal waters – was to receive the ever-increasing volume of wastewater discharged into them. Such a use of water was also on occasion presented as beneficial.

THE BENEFITS OF CONTAMINATION

Writing for the United States Geological Survey, Edwin B. Goodell, in his 1905 *Review of Laws Forbidding the Pollution of Inland Waters in the United States,* singled out a late-nineteenth-century dispute between the Sanderson family and the Pennsylvania Coal Company. The final judgment in that case, he wrote, was then the sole exception (at least in riparian states) to the general rule entitling victims of water pollution to damages or an injunction against the polluter. The Pennsylvania coal producer, whose wastewater outpourings had rendered the downstream water supply useless for household purposes and even for watering farm animals and gardens, succeeded after a lengthy series of appeals in eliminating an injunction imposed upon it by the trial judge. In the final round of litigation, in 1886, Justice Clark, in whose home district of Indiana County coal interests were prominent, cast the deciding vote to support the majority opinion: the contamination was simply the consequence of a natural product, coal, being discharged in its "natural state." The outcome has been attributed both to the changing composition of the Pennsylvania Supreme Court and to the defendant's "deep pockets," which allowed it to prolong the litigation in its quest for a more satisfactory result.[31]

In the ongoing effort to reconcile industrial activity with protection of waterways, other features of the decision, notably the attempt to weigh the Sanderson family's losses against the benefits of contamination, reappear intermittently. In reaching their ultimate conclusion, the majority dismissed the Sandersons' injury as a trifle, "a mere personal inconvenience." The judges found that, when compared with the economic harm that a proliferation of similar claims would inflict on the coal industry, the complainant's suffering was a singularly inadequate basis for awarding damages.[32] One formulation of the outcome presented the dilemma in the starkest terms: riparian owners, who had established their homes well before the advent of the coal mine, would be required to give way to the interests of the community in developing mineral resources when the operation of the coal mine produced "acidulated waters" that rendered the stream entirely useless for their domestic purposes.[33] This line of reasoning, in which the interest of injured property holders is weighed against some collective or community interest, was precisely what English judges had cautioned against in the celebrated Birmingham sewage decision of 1858.

The "balancing doctrine," as Christine Rosen explains in reviewing the practice in several US states, allowed the court "to weigh the costs of imposing an injunction on a polluter against the benefits of abating the pollution. When the costs exceeded the benefits, it prohibited the imposition of the injunction."[34] But if there is more than one way to skin a cat (a proposition I am prepared to take on faith), it is also true that judges were able to balance costs and benefits in a variety of ways. In Rosen's view, the judges in early balancing cases approached costs and benefits in a rather simplistic manner. In comparison with modern cost-benefit analysis, the approach was essentially descriptive, involving little attempt to price either costs or benefits, or to discern indirect and secondary consequences of pollution abatement.[35]

Where benefits of an injunction from the perspective of the victim who had brought suit were assessed against costs that polluters would bear if forced to abate, the approach might be described as *private balancing*, a process that takes no account of costs and benefits to society generally. An alternative perspective, *social-cost balancing*, involved a comparison of benefits of abatement to the victim or victims with anticipated costs that an injunction and abatement order might impose on society overall. Those costs were often thought to be industry-wide or broader still, on the assumption that injunctions might have the domino effect of curtailing industrial activity in general. If this were to happen, the costs would be enormous: injunctions might stifle economic development entirely, thus depriving all urban residents of the benefits of industrial activity. Viewed in this light, "pollution was the price one paid for the amenities of city life and economic and social progress in general."[36] The minor inconveniences suffered by individuals in exchange for the benefits of industrial investment and employment thus added up to a collective acceptance of fairly significant contamination.[37]

Whether judges concluded that the so-called domino effect would entail catastrophic results depended largely upon their assumptions about pollution abatement. Judges commonly believed that, short of closing a polluting business, abatement was impossible. On occasion, where complainants insisted on a full plant shutdown, partial injunctions were not even considered; but plaintiffs were quite often seeking only an abatement of the nuisance to which they were being exposed. Yet, if judges did not believe that pollution-abatement technologies could be installed while business continued to operate, injunctive relief would be denied, particularly when the ominous threat of the domino effect had entered the mind of a judge who had no inclination to take account of the overall impacts of pollution on society.[38] Consequences of this outlook were particularly noticeable in Pennsylvania, where judges – anticipating that compulsory abatement would almost invariably lead to the closure of a shop or factory – were quick to attribute unacceptably high costs to the injunction remedy.[39] To deprive people of their businesses in such a way appeared to the judges unjust and morally wrong.[40] Within such a framework, when property rights in defence of the environment contended with property rights in defence of industry and perceived social well-being, the latter would prevail. With variations, judges elsewhere also employed the balancing doctrine to favour polluters. In 1900, for example, a New York court actually presented such balancing as a rule: "A court of equity always considers the equities presented in the particular case in which it is asked to interfere," it explained, adding that injunctive relief would be denied "where an injunction would cause serious injury to an individual or the community at large, and a relatively slight benefit to the party asking its interposition."[41]

Although general economic benefits might not always prevail, there were usually opportunities to generate a little sympathy. Claims about balancing interests thus continued to appear alongside other arguments in defence of pollution. An interesting case involved New Brunswick's Nepisiguit River. The concentrating processes of the Canadian Iron Corporation involved a good deal of leakage into Austin Brook, a tributary of the Nepisiguit located upstream from salmon ponds and spawning beds. In a contest that in some respects pitted a very traditional perception of natural flow against what appeared to at least some commentators as reasonable use of the waterway, a fishing club pursued an injunction claim against the mining corporation. The mining company, with the provincial attorney general's support, offered four arguments in defence of its operations, all generally indicative of a fundamental disregard for the quality of New Brunswick's waters. The defence argued, quite without effect in the end, that the company enjoyed a right to discharge effluent from its concentration operation by virtue of its mineral lease. If the people of New Brunswick wanted a mine, this proposition clearly implied, they must inevitably and automatically accept the consequence – pollution; they had indeed sanctioned its creation, therefore the licence to do business was a licence to pollute.

Having failed to establish any lawful authority to damage the river, the Canadian Iron Corporation went on to deny that the river had even suffered damage, for in its view the evidence showed that nothing it had introduced to the water was injurious to fish. This too was (and remains) standard operating procedure, for if anyone is to be held responsible for harm to the environment, it is necessary to establish that actual harm has occurred or at least that the threat of harm existed. Who has to show harm – and how convincingly they have to show it – remain vital elements in disputes of this kind. Even if harm has occurred, defendants have been inclined to suggest that it wasn't all that serious. The Canadian Iron Corporation didn't miss its cue: "If the spawning beds were injured, the only injury would be that the eggs deposited there were destroyed." This is not exactly what we call the long-term view, but the argument that concern over the well-being of fish didn't automatically extend to the well-being of fish eggs enjoyed extraordinary resiliency.

Expert witnesses for the defence had offered the standard account of flushing and dilution. Some suggested that any discoloured water from the mills was eventually sufficiently mixed in the stream that "it was not perceptible and that the water was clear." Other experts insisted that samples they examined "were pure and would not affect the fishing." Rounding out the defence quartet was a final dismissive plea reminiscent of the fate of the Thames decades earlier. Invoking the priorities of the age, the mine asserted, "It would be an unfortunate thing if the law is found such that a large industry has to give way to fishing for sport." For their part, the anglers insisted upon "pure, clear and unpolluted" water.[42]

Within a few years, American judges had repudiated balancing. The new perspective rejected the notion that injunctions might – or even should – be withheld simply because the costs of abatement for the polluter – and those whose livelihoods depended on the polluter – outweighed the benefits to the victim. In the American Progressive era, the earlier view had become a source of moral indignation and outrage: it was no longer acceptable to imagine, as the balancing doctrine had allowed, that merely by virtue of a business's comparatively large economic value to investors, employees, and consumers it could secure an entitlement to pollute subject only to a compensatory payment to injured plaintiffs.[43]

Christine Rosen attributes the courts' altered outlook to judges' increasing sensitivity to several major developments. First, there was a growing recognition of the heavy social and private costs of pollution, particularly in relation to public health. Moreover, technology to control and abate pollution was more readily available, and the costs of such initiatives, though possibly substantial, were not prohibitive. Finally, public attitudes regarding the acceptability of pollution had shifted significantly around the turn of the century.[44] It can hardly have been merely coincidental that in precisely this period the environmental burden of human waste had effectively been shifted from land to water.[45] Cumulatively, these general trends encouraged judges to re-assess the influence of the balancing doctrine in favouring virtually unbridled manufacturing or resource development.[46]

Balancing in general had been discarded in a number of jurisdictions, and the extreme application in *Sanderson v. Pennsylvania Coal* had found no favour elsewhere even though coal, not being subject to putrefaction, was once considered to inflict "no more injury than that caused by the same quantity of sand."[47] When the Sanderson case was advanced as an example before the House of Lords in England, it was thoroughly discredited in principle. One judge, however, taking his cue from the Birmingham court, remarked that the value of the mining interests might have provided sufficient grounds to appeal to the legislature for relief from the claims of downstream riparians, confining them to financial compensation as a remedy for injury suffered.[48] As the twentieth century moved forward, this model was increasingly favoured, supported by forcefully articulated claims about the suitability of waterways for waste disposal by municipalities and industry alike.

Streams Are Nature's Sewers

The earlier environmental impacts of the lumber trade and industrial activities such as tanneries, woollen mills, and paper making were dwarfed by subsequent developments. By the 1880s, value added by American manufacturing and processing had already exceeded the value of agricultural production,[49] and population increased by over 350 percent between 1850 and 1920.⁵⁰ Similar expansion had occurred in Britain, with population nearly trebling between 1850 and the First World War. New types of industrial activity, particularly chemical production, compounded the problems of water use. In addition, rain washings from tar-sprayed roads carried tar acids such as phenols, cresols, naphtha, and waste oil from automobiles into streams.⁵¹

The conviction that water was a natural repository for wastes gained surprising currency during the interwar period, both on economic grounds and more generally. "Some streams," argued E.B. Besselièvre, a sanitary engineer with the Dorr engineering firm of New York, "have been so long used as carriers of industrial wastes, that it would not be economically practicable to require the elimination of all wastes, or to attempt to use the streams for potable water supply." In other cases, he suggested, mildly polluted streams experiencing a more moderate flow of industrial waste might be reclaimed even to the point of becoming potential sources of water supply.⁵²

In advising economy-minded municipal governments, engineers like Besselièvre often emphasized that where "the cost of purifying sewage before turning it into a river, or the expenses entailed in securing some other place of final disposal, far exceed the cost of providing a pure-water supply from highland or other sources," it would be "manifestly better," in the absence of other rights to be considered, "to use a stream as a sewage course rather than a source of water supply." The governing principle in these situations, it was argued, should be the "general public economy."⁵³ So, it was not only by inadvertence that streams turned into sewers. The decision was frequently deliberate, calculated on the basis of values not unrelated to Scotland's once fashionable public river doctrine, and eventually the outcome of official policy.

Besselièvre particularly commended Pennsylvania to delegates at a conference of the American Institute of Chemical Engineers. The state, he said, had "taken the lead in recognizing the basic rights given by long continued usage, and [had] taken into consideration the economic feature of compelling complete reclamation of a stream, as against the enormous cost to the industries of accomplishing this."⁵⁴ Thus, though balancing may have lost favour in the courts, its influence was becoming apparent in legislative initiatives affecting waterways.

Pennsylvania's governor from 1923 to 1927 and again between 1931 and 1935 was Gifford Pinchot. He had come to office as a professional forester widely known for his contributions to resource conservation at the national level in the United States. Pinchot was careful to distinguish conservation from hoarding resources or holding back production. To him it meant, rather, a "wise use" of that which was to be conserved: "The conservation of natural resources which have been given to all the people by the Almighty, means their use in a way that will benefit the greatest number of people for the longest possible time."[55] When the Pennsylvania Administrative Code of 1923 created a Sanitary Water Board, its guiding principles seemed to offer a promising mechanism to advance the new governor's utilitarian vision of conservation in the context of water.

W.L. Stevenson, chief engineer for Pennsylvania's department of health, explained that the new board's efforts to achieve maximum benefits for the greatest number of people followed a scale of priorities.[56] Ranked in descending order, those uses were as follows: waterworks serving the public for domestic and municipal purposes; the conveyance of sewage and industrial wastes after suitable treatment for purposes of sanitation; manufacturing and industrial consumption and power development; navigation; and recreation. The inclusion of recreational uses (a category now broader than fishing), which had not enjoyed much favour in an earlier era, is noteworthy, yet still more so is the remarkable ascent of flushing into second spot. On this scale, at least, waste removal now ranked well above navigation and even power development.

Pennsylvania legislators noted the range in quality of state waters, from the pristine purity of small streams flowing through virgin forests to grossly polluted waterways draining industrialized valleys where intense municipal development had occurred. They accepted these differences in conditions as key indicators for "determining the required degree of treatment of sewage and industrial wastes" in conjunction with the "natural powers of streams to inoffensively assimilate and dispose of polluting matters by dilution."[57] The Sanitary Water Board, given authority over all intrastate waters, designated streams into three classes. Waters that were so grossly polluted as to be beyond consideration as sources of potable water supply could be abandoned to their fate as carriers for industrial wastes. Streams that might eventually be looked to as sources of human water supply should be kept free of certain toxic wastes and would require partial treatment of other wastes discharged into them. Waterways already in use for water supply purposes would receive only those wastes that had undergone complete treatment.[58] In the absence of forceful advocates for other values such as fisheries or recreational use, this restricted calculus, largely confined to water supply and waste removal, somehow managed to avoid critical scrutiny.

Only in exceptional circumstances would the Pennsylvania Sanitary Water Board order municipalities to install intercepting sewers and sewage treatment facilities. Instead, the modus operandi was to ascertain the present and likely future use of the stream at issue and then to determine the minimum degree of treatment required to maintain the stream in a suitable condition. On the industrial side, the Pennsylvania approach was collaborative, for practical as well as philosophical reasons. State officials had rapidly discovered in a pilot project, for instance, that neither they nor their scientific counterparts in industry had reliable scientific data on the actual capacity of streams to assimilate tannery waste. Nor was either group aware of practicable ways of treating tannery waste so that the effluent could be safely discharged to the streams. Their goal was to locate a margin within which rivers could be degraded – but not too degraded. Yet, in offering a rationale for cooperative effort on the part of industry and the state, Stevenson frankly acknowledged that all this activity was taking place without any real appreciation of the harm that it might entail.[59]

Stevenson later claimed success for Pennsylvania's Sanitary Water Board at the level of cooperation between state authorities and industry. The board can be credited with inaugurating the practice of developing formal stream-pollution agreements with industrial sectors during the 1920s. During the decade, such agreements or other cooperative measures were implemented with tanneries, pulp and paper mills, coal mines, petroleum refiners, and the operators of "byproduct" coke ovens, whose phenolic wastes contributed offensive tastes and odours to several municipal water systems.[60]

To Stevenson, and clearly to a good many of his peers at a 1925 gathering of the American Institute of Chemical Engineers, deterioration of the landscape and waterways had been an inevitable consequence of urban and industrial development, a consequence largely acceptable according to a policy calculus that balanced environmental losses against the overall benefits of progress. "The loss of the pristine purity of the streams draining developed areas is the bill which man must pay for the comforts and advantages of modern urban life."[61] This expression of the attitude underlying formal policy bore a remarkable resemblance to arguments about balancing and the benefits of contamination that had been so confidently asserted from time to time in judicial proceedings.

The question of whether water quality standards could be applied throughout the United States was most complex. Indeed, when completing its work in 1939, the Federal Water Pollution Advisory Committee dismissed national water quality standards as impractical. Forthrightly asserting that

"Stream systems are Nature's sewers," these experts championed a comforting new alternative to the nineteenth-century vision of self-purifying waterways. The committee rejected national standards for water quality on three counts. First, variations in natural water quality had to be acknowledged; second, stream flows and thus their ability to accommodate waste fluctuated. And third, there were significant differences in human requirements within drainage basins and from one basin to another. These factors, cumulatively considered, satisfied the inquiry that an attempt to apply fixed or uniform standards of water quality on a national basis would not advance the abatement effort. Instead, the committee considered it more practicable to set standards for selected portions of drainage basins. Standards in each case would be set to "express the best balance between the stream's use for receiving and assimilating sewage and other waste and its use for other purposes, aesthetic and economic."[62]

British policy makers and advisors equally addressed serious interwar pollution of the nation's waterways in a series of almost continuous inquiries operating from shortly after the completion of the work of the Royal Commission on Sewage Disposal in 1915. Widely noted deterioration, partially attributable to wartime disruption, would require substantial investment in new works and facilities. Trade wastes from industry and manufacturing had become increasingly troublesome despite research into treatment technologies. The defence of employing the "best practicable and reasonably available means" of preventing pollution remained in place, as did the requirement for ministerial authorization of legal proceedings against manufacturers. In considering whether to permit legal action, the minister of health was required to determine not only that reasonably practicable means of rendering the pollution harmless were available in the actual circumstances of the case, but also that "no material injury will be inflicted by such proceedings on the interests of such industry." The most substantial initiative of the era was new legislation, the *Drainage of Trade Premises Act, 1937*, to facilitate the discharge of trade effluent into public sewers for eventual treatment by local authorities.[63]

In Canada as well, interwar indications of official initiatives to accommodate pollution and willingness to allocate waterways to waste disposal could be found. Anthony V. Delaporte, a young Canadian sanitary official from Ontario attending the same conference as his Pennsylvania counterparts, expressed the opinion that his province, still largely agricultural in character, had comparatively few pollution problems. Where complaints about stream pollution did arise, he said, it was departmental policy "to work out a solution for the manufacturer, either in the shop, or in our own laboratory before making any recommendation concerning the matter. In all instances we try to impose the minimum of treatment."[64] Pulp and paper manufacturing, Delaporte acknowledged, was one industry in which no satisfactory treatment had been implemented. Examples abounded of rivers whose banks were coated with commercial grades of pulp in valuable quantities. He could only assume either that industry managers had judged that the value of the pulp would not pay for its recovery, or that they were unfamiliar with recovery techniques.

Delaporte, along with those inclined to remark that streams had changed their character, appears almost to blame waterways themselves for the downstream impact of industrial waste. Most complaints, he noted, arose in the summer months, when minimum runoff coincided with high temperatures. These conditions might result in a nuisance if wastes requiring a large amount of oxygen were discharged into a small stream. "It is not always the waste itself which creates the nuisance," Delaporte insisted, "but the new conditions upset the stream balance, and favour the growth of certain organisms, which in turn cause a very definite nuisance." He offered by way of example the case of a large distillery discharging waste into a river: "This waste itself was inoffensive, but during low periods the fungus growth in the river caused a serious nuisance in a city several miles away."[65]

Delaporte explained that his Sanitary Division was recommending that streams be allocated for industrial uses, adding, "I have just reported on one stream as suitable for one manufacturer." No industries had yet been protected, although Ontario was clearly permitting many northern mills to discharge without treatment into the Hudson Bay watershed on the assumption (more accurate with respect to river volumes than to the number of Aboriginal residents) that there was "considerable dilution and little population."[66]

One questioner was anxious to determine whether the practice of allocating a stream to manufacturing in order to influence industrial location provided "an indefinite or perpetual license to pollute the stream." Putting the question another way, he inquired, "Is this allocation going to be an insurance against increased expense in future years?"

"Yes!" Delaporte replied, and he set forth his department's policy of minimum interference with industrial usage and maximum accommodation to industrial pollution. Existing legislation, he observed, offered protection for industry against legal actions "in no little degree." Downstream riparian owners were ostensibly shielded from "gross negligence" in regard to the disposal of industrial wastes, but it was departmental policy "to refrain from participation in any common law action unless some municipalities' sources of domestic water is involved, and there is no other solution." As in Pennsylvania, government sought cooperation with the manufacturing sector, and recognized that "arbitrary standards applied indiscriminately do not lend themselves to a permanent solution advantageous to the numerous interests involved." Waterways had multiple and changing uses: "In some situations the volume and quality of water and type of introduced waste is such as to make the re-use by lower owners advantageous, if the lower owners be protected by a minimum degree of treatment of a waste discharged higher up." Manufacturers might, in such circumstances, be required to take measures not originally contemplated, but they would never "become the prey of pettifogging litigation."[67]

Litigation, pettifogging or otherwise, had played an essential role in conflicts over water quality that emerged alongside the nineteenth-century sewerage revolution and the ever-increasing volume of industrial discharges that followed in its wake. The record in terms of water quality was undeniably mixed – occasional triumphs and intermittent setbacks emerging from the application of basic doctrine and technical rules in the context of particular waterways and their patterns of use. The extension of sewerage infrastructure multiplied opportunities for legal conflicts over water quality demonstrating from time to time that the public culture of flushing was gaining recognition and acceptance as its domestic equivalent had previously done. Some courts eventually succumbed to the repeated temptation to elevate waste removal on the scale of water uses.

In Virginia, persistent friction between oyster producers and municipalities whose untreated wastes flowed relentlessly downstream to tidal waters erupted into passionate conflict in the early decades of the twentieth century.[68] Virginia public health officials officially cited (without quite endorsing) the opinion of a firm-minded commentator who referred to coastal contamination as "a crime against nature," all the more indefensible for being committed by "highly cultured people."[69] When an oysterman named Watson and other holders of private oyster leases along the bottom of Hampton Creek learned in 1914 that the sale of oysters from their area had been banned because pollution made them unfit to eat, they sued the municipality of Hampton. The claim, successful in lower courts, asserted that the city's sewage discharge was an unlawful trespass against leaseholders' interests.

Yet, as the matter advanced to the appeal level, it became clear that Watson and his fellow oystermen had no cause for celebration. To the Virginia Supreme Court of Appeals, the city was engaged in the public function of disposing of its wastes. The sewerage function was authorized by statute and not subject to a statutory treatment obligation. The municipality's interest outranked private rights respecting oysters, even where those rights also derived from the state: "Sewerage systems for all thickly settled communities have become an imperative necessity," the court proclaimed, adding that the need for such arrangements now constituted "a public right, which is superior to the leasing by the state of a few acres of oyster land." Oracular pronouncements from the bench bolstered the legal analysis: "The sea is the natural outlet for all the impurities flowing from the land, and the public health demands that our large and rapidly growing seacoast cities should not be obstructed in their use of this outlet, except in the public interest." Here the law embraced seemingly unquestionable assumptions about the natural order and its intended contribution to the convenience of civilization: "One great natural office of the sea and of all running waters is to carry off and dissipate, by their perpetual motion and currents, the impurities and offscourings of the land."[70]

Undeterred by this confident judicial vision – dishearteningly reminiscent of the ancient Scottish public rivers doctrine – Frank W. Darling, another litigious oysterman, took on Newport News in a similarly one-sided contest against the alleged "imperative necessity" of contamination by flushing. Darling was also firmly instructed that the city's right to flush was prior and superior to any legislation on oysters. In establishing the private oyster lease program, legislators must have assumed the planting of oysters would not be continuous in "an area so certain ultimately to be polluted." Riparians have always drained refuse into the sea, "which is the sewer provided therefore by nature." When it reached the Supreme Court of the United States, the case provided Oliver Wendell Holmes an opportunity to revisit the question he had avoided in the original Chicago diversion litigation: were rivers destined to be the sewers of municipalities along their banks, or would they be protected from threats to their environmental purity? In *Darling v. Newport News,* Holmes observed that anyone who agreed to lease land under water, whether for oysterbeds or otherwise, would be expected to assume the risk of pollution of the water. Who could imagine that "for a dollar an acre . . . or whatever sum . . . paid," Darling had acquired a property interest immune to that risk? Surely no one would expect that by issuing the lease the state committed itself "against using its legislative power to sanction one of the very most important public uses of water already partly polluted, and in the vicinity of half a dozen cities and towns to which that water obviously furnished the natural place of discharge."[71]

If a little corruption might once have been tolerated by the law, municipal flushing had become to some courts an imperative necessity giving rise to a public right. This was hardly a situation to be called in question, since rivers had been provided to carry off the outpourings of the land. Some, however, imagining other reasons for water and waterways, did indeed question the willful degradation of rivers, lakes, streams, and oceans in the public interest.

11
Riparian Resurrection

DESPITE THE CONFIDENT ASSERTION that streams are nature's sewers, voices were raised to protest the continued deterioration of twentieth-century waterways. On occasion, such protests took the form of direct legal challenges against those responsible, sometimes on the basis of traditional riparian principles and sometimes in reliance on legislative arrangements. Now and again, those challenges succeeded, sometimes dramatically. The successes initially provoked a new round of response: in a few instances, legal remedies against pollution were actually removed. In other cases, effluent dischargers benefited from clearer legislative authorization or from excessively vague or lenient standards. Debate eventually returned to a new point of departure: if waterways were to be spared the fate of sewers, what was to be done, how, by whom, and why?

COUNTER-CURRENTS

Twentieth-century societies that had managed an immense war effort, established elaborate transportation networks for trains, automobiles, and the airplane – and could even put up the odd skyscraper – might have been expected to do a better job on water quality. With promising technology identified, what more could be involved than recognizing the problem and allocating appropriate resources? Yet, as George Chandler Whipple, a knowledgeable American observer of the 1920s, complained, "There is now greater indifference to stream pollution, a greater laxity in enforcing laws, than was the case before the World War."[1] A British counterpart, H.G. Maurice, recalling his disappointment at the failure of a "simple and practical" scheme for cleaning up the estuary of the Tyne in the interwar years, ruefully remarked in 1944 that the cleanup had never really got past the planning stage because "it would have cost half as much as one day of the last war – about the equivalent of eight hours of the present one."[2] Those who were critical of the persistent flow of contamination offered an array of perspectives, lamenting lost amenities, highlighting continued public health concerns, and condemning lack of investment in remedial infrastructure or a failure to enforce protective measures.

Pennsylvania's response to water quality conditions in the 1920s may have been portrayed as willingness to face up to payments on the outstanding "bill" owed for the well-being of "modern urban life" with some measures of pollution control and even restoration. Ironically, mounting pressure for water protection in the state was attributed to automobile use. Car travel allowed growing numbers of urban residents to venture outdoors for swimming, hunting, and fishing. The consequence in one official's view was "an insistent demand on the part of the public for absolutely clean streams." Where people had once objected only to gross stream pollution, they tended increasingly to express outrage at "all visible evidences of any pollution of streams where they desire to find recreation."[3] A contemporary Ontario engineer outlined his province's approach to an issue that was affectionately known in chemical-engineering circles as the "Isaac Walton situation," although the sportsmen's league of that name was still in its infancy. Rather than using a system of river classification like Pennsylvania's, Ontario operated a loose zoning arrangement. In the northern reaches of the province where fishing abounded and tourism was promoted, "there is now no pollution at all, practically, and none would be tolerated." More severe deterioration was experienced in the south, where development was concentrated, but the province knew how to cater to the anglers: "Our touring agencies and the railways arrange so that the Isaac Waltons get up to this [unpolluted] section and not on the polluted streams. When an Isaac Walton gets on a polluted stream, it is a mischance."[4]

Founded in 1922 by a group of Chicago sportsmen, the Izaak Walton League of America (IWLA) invoked the memory of the seventeenth century's most renowned angler in a sophisticated campaign to safeguard sport fishing against pollution and other threats. Within five years, at the invitation of President Coolidge, the league had concluded a national water quality survey that attributed three-quarters of the water pollution problems of the United States to domestic sewage. The IWLA's activities soon included local remedial initiatives, lobbying for legislative reforms, and extensive research.

The British Field Sports Society was similarly engaged in assessments of mid-twentieth-century pollution and its detrimental impacts on fisheries. When it investigated the state of streams at the instigation of alarmed anglers, distressing conditions came to light. Once-sparkling streams had deteriorated dramatically: "There are . . . beautiful, clear trout streams, from which all life has mysteriously disappeared; brooks poisoned with acid or black with coal-dust; rivers whose surfaces are covered with soapsuds for miles below a single paper mill; waterways which are veritable ulcers in the midst of 'England's green and pleasant land.'"[5] Such laments were by no means isolated, as the effects of increased demands on water, untreated municipal sewage, and industrial wastes became apparent.

One contributor to *Blackwood's Magazine*, identified only as GKM, the son of a country doctor, mourned the dwindling of a small trout stream that had flourished in the southern England of his youth during the 1890s. Seasonal rituals along the river had been central to existence in the countryside, whether an annual flooding to water pasture lands, the exercise of historic netting rights, or boyish efforts to hook a trout. But, returning to his fabled landscape after a career overseas, GKM noted the effects of a bore hole, sunk deep into the great chalk range, a feature of the landscape which, "like a huge sponge, absorbed the winter rains and discharged them all summer to the springs of the river." As industrial and individual consumption grew, GKM's river "began to shrink." There were no fish in the reach, the author reported in 1947, noting the gradual diminishment of the waterway over more than a quarter of a century. The river's plight was scarcely noticed by the younger generation: "Soon there will be none left that knew it in its health and vigour." GKM's grim forecast was Orwellian indeed: "So a generation will arise that has never seen a river, and children will be taken to see one as they are taken to see the sea."[6]

Aldo Leopold, a Wisconsin-based consulting forester, former president of the Ecological Society of America, and a conservation adviser to the newly formed United Nations before his death in 1948, echoed the nostalgia of British observers. "Perhaps our grandsons, having never seen a wild river, will never miss the chance to set a canoe in singing waters."[7] Leopold was among the most influential forerunners of a subsequent generation of environmentalists who cautioned that, in the midst of unprecedented human accomplishments, "we had forgotten the earth . . . in the sense that we were failing to regard it as the source of our life."[8] Drawing inspiration from the teachings of Albert Schweitzer, Leopold elaborated an ethic rooted in an ecological conscience that redefined the human species from "conqueror of the land-community to plain member and citizen of it." It was vital, Leopold stressed, to envisage waters as having more functions than simply "to turn turbines, float barges, and carry off sewage."[9] Waters, like soil, were the foundation of cycles of life and energy; when they were polluted, drained, dammed, or diverted, their essential contributions were undermined.

Complaints about the impact of pollution on fish were not uncommon, even if the response must often have been disappointing. Judge Wells, writing to game and fisheries officials in 1912 as an angler, it seems, rather than in his official capacity, protested the presence of machine oil in the Welland River. He assumed that the oil was being discharged by industries into drains leading to public sewers, which in turn flowed into the river. Inquiries made to two possible offenders, the Plymouth Cordage Company and Canadian Steel Foundries Limited, resulted only in denials; officials failed to determine the source or route the pollution might be travelling. Eventually, the attorney general was advised that to secure reliable evidence it would be necessary "to have some person specially employed, but such person would have to be carefully chosen."[10] Toothless recommendations of this kind – embodying their own excuse for inaction – offer complacent satisfaction to sender and recipient alike.

Although not without champions, fish were hardly the subject of universal sympathy. When, for example, legal action was contemplated against a tannery that was polluting a creek in the community of Oshawa, Ontario, a local doctor voiced his objection to this company's being singled out. If this tannery were subject to prosecution, he said, then "there is an action against any firm." Property owners had not complained, and the town of Oshawa itself was discharging sewage into the same creek (although it had spent some $20,000 to construct treatment facilities). "We are all trying to solve this problem," the doctor continued, before taking aim at the type of individual who seemed to have provoked the threat of prosecution. "This complainant who, like many others, likes to sit on the stream with a pot of Beer on one side and a pot of Bait on the other, and where there has been no good fishing for 40 years, would like to have the state go to an expensive suit, to gratify his imaginary 'fish preserve.'"[11]

This physician's image of the legal universe embodied familiar sentiments. You can't charge anyone because we are all guilty, he appears to say, and if we all do it, there is nothing wrong. To him, mere anglers had no significant status in possible proceedings, certainly not in comparison with property owners, the more legitimate complainants in such matters.

Yet the doctor's protest equally speaks to the acquiescence of riparians, and to the sad fact that once rivers had been denuded of fish, there was little inclination to envision restoration. Oshawa's creek was no longer a natural waterway, but it was normal enough.

Toronto's Don River was perhaps equally hazardous to fish, but results obtained in 1925 from monitoring dissolved oxygen levels were downplayed by the supervising chemist, who interpreted them only as an indication that the stream was carrying the maximum pollution it could carry "without nuisance." Earlier experiments had demonstrated that fish would not survive at levels below 3.5 cc of oxygen per litre, and researchers were becoming aware that waters with borderline oxygen levels during the day might deteriorate to a dangerous condition at night when photosynthesis ceased.[12]

Although the levels of oxygen necessary for survival of wildlife were exceeded in the upper portions of the Don, a very worrisome situation prevailed in the lower reaches. An increase in the volume of oxidizing solids upstream – or simply hot weather – could produce "an active nuisance" in the lower stretches.[13] Rather than triggering a cry of alarm over a situation in urgent need of attention, the monitoring results merely produced warnings about the possibility of an "active nuisance." Concern seemed to focus on the scientific technicalities of identifying the river's capacity for resilience or tolerable levels of waste loading. The level of official interest was roughly that of an interrogator supervising the torture of prisoners: apply maximum doses of pain falling just short of the point of death, which would terminate the suffering while frustrating the objective of the torturing. The Don, among hundreds of other North American waterways, was the victim of the cumulative impact of decades of abuse and had been left to its fate.

In the absence of significant fish kills, limited attention was directed toward a broad range of environmental consequences, while remote or isolated industrial sites largely escaped scrutiny.[14] Nonetheless, a few influential critics singled out industrial wastes for more intense scrutiny. George Warren Fuller, who had begun his highly influential career at the Lawrence Experimental Station furthering Allen Hazen's studies in water purification and sewage treatment, was among those to press the point that industrial wastes needed "far more effective attention than has been the case hitherto."[15] The effect of industrial wastes on water supply took on the dimensions of a major concern in 1923, particularly after the American Waterworks Association revealed that at least 248 water supplies across the United States and Canada were affected.[16] Dr. John Emerson Monger, then state director of health for Ohio, explained some of the newly recognized impacts of industrial discharges on water quality: when chemical wastes destroyed the plankton that helped the natural decomposition of organic materials, the pollutants formerly removed by the plankton could be carried great distances downstream and harm the water supply of an "unoffending" community.[17]

A 1925 Ohio initiative attracted favourable comment within the public health constituency.[18] Said to be "the most advanced piece of public health legislation bearing on the subject of stream pollution" in the United States, it conferred on state health officials the same authority over industrial wastes as they exercised over municipal sewage. Ohio's Public Health Council could make regulations according to the varying needs of state watersheds, thus permitting sources of water supplies to be zoned and protected. New industries whose wastes might pollute streams, as well as established industries proposing material changes in the character or volume of their wastes, were expected to file treatment plans for health department approval. An optimistic summary of the anticipated result was that the statute would limit Ohio pollution "to its present dimensions" and enable the orderly consideration of the circumstances of each individual watershed. Most encouragingly, stated Dr. Monger, "Unpolluted water supplies are thus protected against pollution by new industries." Adequate waste treatment arrangements were now a precondition in Ohio for a new industry seeking to discharge "any waste at all" into streams that served as a source of a water supply. Nor could an existing industry alter the character or quantity of its wastes without providing for adequate treatment. With public health officials empowered to oversee each instance of proposed industrial discharge, Monger continued, "The very worst that can happen is that the problem will get no worse than it now is."[19]

Though perhaps not as elaborate as Ohio's initiative, attempts to check industrial pollution through controls established under the auspices of health officials or departments were effectively the norm.[20] At the level of implementation, however, the public health response to industrial pollution frequently foundered. Industry, of course, was inclined to resist reform on economic grounds. And, in comparison with the disease-related threats of sewage, which were acknowledged and familiar to the public, the significance of industrial contamination remained largely unknown. Indeed, some of the ingredients of industrial wastewater were even considered to be beneficial for their role in "sterilizing" bacteria. Moreover, industrial discharges posed complex treatment challenges often only beginning to attract the attention of researchers.[21]

Solutions to even quite conventional forms of industrial pollution proved somewhat elusive. A typical challenge confronted one public health inspector when a vice president of Canadian Milk Products sought precise directions on how to respond to a sixty-day abatement order. "Would you," he asked, "please tell us how to abate the nuisance... If there is anything that we can do that will be successful, we shall be only too pleased to carry out any wishes you may express in this regard." He promised "instant consideration" for any forthcoming suggestions, assuring the inspector of his company's good intentions: "We would like to have any waste that we put into flowing water just as clean as we possibly can." His request for specific direction was accompanied, however, by a confident assertion that no effective treatment procedures were actually available. Able chemists and engineers – as far as the company could determine – had failed to conceive a process that rendered the waste "absolutely without odour." The correspondence concluded with an exculpatory observation to the effect that "a little" inconvenience to area residents was unavoidable. In the heat of summer, any milk plant would naturally give off some slight odour. Not even "utmost vigilance," which Canadian Milk Products most assuredly devoted to the situation, explained its vice president, could ensure better results, for on occasion, he confessed, "our men are possibly careless and odours do arise from the creek into which our waste empties."[22]

To the milk producer, the presumption of pollution was surely the conventional position and beyond reproach. Occasional water pollution (if it even counted) was universal in an important industry whose social benefits outweighed whatever incidental harm might arise. A best-efforts attempt to control the situation was as much as could be expected, for the company could hardly be held responsible for the incidental carelessness of its employees. There was therefore nothing morally wrong in the conduct complained of, and certainly nothing criminal in nature. Any official wanting more should identify effective remedial measures, and fairness seemed to dictate that if other similar operations had been unsuccessful in meeting a higher standard, there was no basis for imposing such obligations. Actual enforcement was not really expected.

Indeed, an overview of water pollution laws and state enforcement practices across the United States in the late 1930s revealed a multitude of deficiencies. Some jurisdictions had made no provision for coordinating the efforts of various agencies, or had simply failed to confer any enforcement authority; in others, exemptions undermined the effectiveness of legislation. It was not uncommon to find that financial limitations resulting from statutory or constitutional shortcomings prevented municipalities from undertaking appropriate waste treatment measures, or that no provision had been made to foster comprehensive solutions throughout metropolitan areas. Enforcement was also frequently impeded by weaknesses in the definition of pollution. Certain statutes were "too vague to permit accurate judicial interpretation" of concepts such as "tolerable pollution"; others were "too rigid to permit reasonable enforcement."[23]

Anglers, public health officials, and a handful of pioneering environmentalists had formulated criticisms of municipal sewage and industrial waste. They also offered some remedial guidance directed at institutional arrangements, the economic attractions of maintaining water quality, and the importance of underlying values. But these observations had not coalesced into a comprehensive critique and remained of limited effect. When specific complaints made their way to the courts, however, they gained much-needed attention for broader remedial programs.

Riparian Resurrection

Not long after the massive extension of the sewerage systems in European and North American cities, a leading Canadian judge declared drains and sewers to be a necessity, indeed a right. Such facilities were generally constructed, Justice Thibodeau Rinfret explained, pursuant to statutory powers and resulted in some respects from the exercise of common law rights enjoyed by local ratepayers, as represented through municipal governments. Though exercised for the benefit of all inhabitants, such collective rights were not without limitations. Indeed, they were restricted by correlative obligations and could not be exercised (at least not without appropriate compensation) in a manner prejudicial to others. The judge specifically emphasized that statutory powers should not be construed as authorization to create a private nuisance – "unless the statute expressly so states."[24]

This abstract exposition on society's right to flush was occasioned by conflict between Edmonton and two city residents, Malcolm Forbes Groat and Walter S. Groat, descendants of the Alberta capital's first settler, who lived on the north bank of the Saskatchewan River. Their property was bounded by a stream that flowed through Groat Ravine Park, an extensive tract of countryside preserved within the burgeoning western city, and into the Saskatchewan River. But a refuse dump, storm sewers, and a poorly maintained sewage-pumping station had significantly degraded the stream, which had once – like so many others – been "pure and healthy."[25]

Supreme Court Justice Rinfret's formulation of the principle governing drainage of riparian lands seemed unremarkable. The riparian proprietor, he observed in 1928, enjoys a common law right to drain his land into a natural stream. But such a right cannot be exercised so as to injure those downstream. With a dramatic flourish, Rinfret then produced a veritable rallying cry for the common law cause of clean rivers: "Pollution is always unlawful," he wrote, "and, in itself, constitutes a nuisance." Unlike the more or less contemporary remarks of US Supreme Court Justice Oliver Wendell Holmes and others about the inevitability of pollution, those of Rinfret (though he did accept municipal sewerage as a necessity) showed an apparent willingness to impose severe constraints on the culture of flushing.

What exactly might constitute pollution was more problematic than Rinfret's rhetorical landmark acknowledged. Would the judge's stark proclamation mark the end of flushing (at least of untreated effluents), such that waterways were to be preserved in a natural state, or was some level of waste discharge, even accompanied by a degree of deterioration, to be tolerated in the name of competing principle? If the latter were to be accepted, how would that level be determined? Would this be accomplished in relation to reasonable practices, social convention, or economic convenience, for example, or with regard to such environmental consequences as the actual condition of the waterways and their inhabitants?[26] Notwithstanding the rhetoric, a firm point of reference for an absence of pollution was not easily discerned.

Justice Rinfret was prepared to show a little sympathy to municipalities dealing with sewage, but not too much: "While the courts will naturally be slow to grant an injunction against a public body carrying out an important public work, they cannot lose sight of the fact that in this case there is an existing nuisance."[27] In the end, he gave Edmonton two years to find a more suitable means of dealing with the situation.

Rinfret's colleague Justice Lamont addressed another possible form of excuse. As was then customary, Edmonton's municipal powers, conferred by the city's charter, authorized sewers and drains. However, ruled Lamont, "Apart from statutory authority . . . the city cannot by flushing its streets collect these impurities and by means of a storm sewer pour them into a stream the waters of which the plaintiffs have a right to take for domestic or other purposes."[28] Such formal statutory powers, without more, did not confer a right to pollute with impunity. Except for one dissenting voice, that of Justice Smith, Canada's Supreme Court agreed that there was no justification for Edmonton to contaminate the Saskatchewan River.

If Edmonton lost – an assessment beyond dispute to subsequent readers of the judgment – who won? To put this enduring question on a somewhat higher plane, one might ask whether private litigants such as the Groats are effective instruments of environmental protection. Are they perhaps, as some have argued, the most effective instrument? Or is private litigation by nature tinged with self-interest and predominantly inspired by that impulse? In this case, not only did the court give Edmonton two years to resolve the situation, thereby allowing the water quality problem to persist and the river to further deteriorate, but some of the abatement options suggested by Justice Rinfret would simply have moved the contamination downstream. And, as the dissenting judge pointedly noted, the Groats' stewardship of their own land was hardly conducive to maintaining for downstream riparians the "pure and healthy" condition whose loss to themselves they had lamented. Justice Smith thought it was of some relevance to mention the Groats' "barnyard, piles of manure and toilet on the bank" of the stream, as well as the fact that their dozen cattle and forty horses enjoyed access to it.[29]

In Britain as well, riparian owners and others interested in preserving their beloved rivers explored the options. John Eastwood, Bow Street magistrate and angler (not necessarily in that order) concluded from a personal analysis of more than a dozen pieces of legislation ostensibly offering protection to the rivers of the United Kingdom that "none of them was any good." Convinced that the common law – and more particularly actions by anglers and riparian landowners to enforce their rights – would satisfactorily protect waterways from municipal and industrial pollution, Eastwood set out to overcome the obstacle that had plagued such actions for decades: the daunting cost of the judicial process. He was supported in his ambition by a belief that abatement technology had improved dramatically and that "our sense of values" was completely altered as a result. It was one thing "if a vital industry [could] get rid of its effluent only by poisoning a river," but quite another where means to render the effluent harmless were available. "Is the industry entitled to destroy the pleasure of millions merely for the sake of cheaper production?" Clearly not, he reasoned, for rights and duties were transformed by the possibility of avoiding environmental harm. Though industrial enterprises had responsibilities toward their shareholders, as local authorities had to their ratepayers, both were subject to a "wider duty . . . to the general public."[30]

Several thousand longhand letters later, Eastwood and a few early collaborators in the Anglers' Cooperative Association (ACA)[31] were in a position to finance anglers' litigation through a Guarantee Fund amounting initially to £10,000.[32] With the backing of the ACA, and in the company of the Earl of Harrington, whose family had enjoyed several generations of exposure to pollution along the Derwent and Trent, the Pride of Derby and Derbyshire Angling Association took on three formidable opponents:

the Corporation of Derby, British Celanese Ltd., and the British Electricity Authority. These three polluters contaminated portions of the Derwent and Trent Rivers in a multiplicity of ways, although each defendant protested that its own activity was not sufficient to create a nuisance. However, their cumulative effect was clear. Inadequate municipal treatment facilities discharged sewage into the stream where it joined organic matter in suspension resulting from the British Celanese operation. The electrical authority – after withdrawing about 120 million gallons daily to cool its condensers – added to the burden on the river by discharging enormous volumes of water at substantially raised temperatures.

Lord Justice Alfred Denning, one of the twentieth century's more forthright and colourful common law judges, took the opportunity of the angling association's suit to remind his audience of the court's usefulness: "The power of the courts to issue an injunction for nuisance has proved itself to be the best method so far devised of securing the cleanliness of our rivers."[33] In reviewing the evolution of municipal government liability for sewage nuisances, he noted that local authorities could not be held responsible when the inadequacy of a sewage and drainage system resulted from population growth and construction over which they had no control: "They obviously do not create it [the nuisance], nor do they continue it merely by doing nothing to enlarge or improve the system." Such reasoning had allowed local authorities to avoid liability for overflowing storm sewers for several decades. The situation was significantly different, Denning continued, when local governments undertook building themselves or authorized others: "They know (or ought to know) that the increase in building will cause the existing sewers to overflow, yet they allow it to go on without enlarging the capacity of the sewage system. By so doing, they themselves are helping to fill the system beyond its capacity, and are guilty of nuisance."[34] Substantially increased local government authority over building permits – and indeed local government involvement in housing construction – entailed increased responsibility for the impact of resulting sewage, he said. And this was true notwithstanding any difficulties local governments might face in persuading their residents to comply with operational requirements at the domestic level. Newspapers welcomed the result and forecast, too optimistically, a halt to pollution, a "disgraceful evil" that had turned many British rivers into "open sewers in which through large parts of their courses scarcely any form of life [was] possible."[35]

Successful legal proceedings such as the *Pride of Derby* litigation and an expanded membership base substantially enlarged the anglers' funding pool. Between 1948 and 1963, the ACA intervened in several hundred water pollution incidents, claimed credit for a significant number of settlements, and was involved in obtaining forty injunctions. In its confrontations with polluters, success was more forthcoming against industrial dischargers than against local governments.[36]

Anglers and fishing interests in North America also more actively complained about water conditions affecting their pastime or livelihood. One notable example was the case of the Kalamazoo Vegetable and Parchment Company (KVP) of Michigan. After KVP reopened a kraft-style paper mill in 1946, the condition of Ontario's Spanish River deteriorated rapidly. Despite an agreement between Ontario and KVP that the company would not deposit more refuse than necessary in the river, KVP was soon discharging at least three and a half tons of chemical-laden fibrous materials each day.[37]

Fishermen and camp owners along the lower Spanish protested that foul odours emanated from the water and that ice from the river was too tainted for domestic purposes. Even after being boiled, the water was unfit to drink. Nor would cattle drink river water in sufficient quantities to maintain their normal milk supply. The complainants judged the water unfit to bathe in; they observed that fish were either killed or driven elsewhere; and they bemoaned the destruction of wild-rice beds that had formerly attracted ducks. The facts, as ascertained by Chief Justice James Chalmers McRuer of the Ontario High Court, were in substantial agreement with the allegations. McRuer not only awarded damages in compensation for the demonstrated losses, but accepted as well the claim for an injunction to prevent KVP from further interfering with the rights of its downstream neighbours. In so doing he specifically rejected the suggestion (conventional in industry circles) that KVP's economic significance was relevant to the determination of the dispute. "If I were to consider and give effect to an argument based on the defendant's economic position in the community, or its financial interests," McRuer reflected, "I would in effect be giving to it a veritable power of expropriation of the common law rights of the riparian owners, without compensation."[38]

To give KVP a chance to make alternative arrangements to dispose of its effluent, McRuer suspended the injunction for six months. During this interval, statutory amendments were introduced to authorize courts to refuse to issue an injunction against a polluting mill when the advantages of the mill – "direct and consequential" – to its community outweighed the "private injury."[39] In setting private injury against industrial advantage, the particular form of balancing encouraged by this legislation disregarded the significance of harm to general, community, or public interests that are so customarily ignored in such situations.[40]

In the midst of the legal conflict over KVP's pollution of the Spanish River, Premier Leslie Frost, a former minister of mines whose own constituency contained a number of resource-based communities, rose in the legislature on 24 February 1950 to underline the seriousness of stream pollution. "We do not," he insisted, "hold lightly the rights of individuals to protect their interests in the courts of this land. That is, in itself, a very important matter." The premier equally acknowledged the "high importance" of employment while recognizing that with industrialization and population increase "a certain amount of pollution" would be unavoidable. "At the same time," he added, "we are determined to hold that to the least possible limit." This exercise in verbal gymnastics was balance-beam work of Olympic calibre. Observers might be forgiven for failing to appreciate how the premier proposed to resolve an apparent impasse between the water quality rights of riparians and the tendency of industry and population growth to produce "a certain amount of pollution."[41]

When the KVP matter reached the top of the Canadian judicial pyramid, the Supreme Court of Canada firmly declined to be influenced by parliamentary afterthought. Justice Patrick Kerwin determined that the legislative effort had been in vain, for the jurisdiction of the Supreme Court could not be extended by a provincial legislature, but only by the federal government, which had created the court. Kerwin found no other basis on which to exercise discretion to restrict the remedy to damages alone. As if the matter might have escaped KVP's notice in this instance, and perhaps to banish any possibility of self-doubt, he reaffirmed that "the rights of riparian owners have always been zealously guarded by the Courts."[42] There would soon be other opportunities to assess this claim.

In the mid-twentieth century, the Ontario village of Richmond Hill proudly completed a sewage disposal facility along a branch of the Don River. Relative modernity proved to be no guarantee of effectiveness. Sewage effluent and stormwater transformed a clear and sparkling stream in which children swam, cattle drank, and fish and watercress abounded into something much less agreeable. Witnesses documented the smell of sewage and the presence of toilet paper, condoms, and other waste matter along the banks – and the disappearance of aquatic life.[43] Partly on the grounds that a creek running through the property carried open effluent from the Richmond Hill disposal works, the Ontario Municipal Board (the body responsible for overseeing such matters) refused an application by a downstream resident of the Don to develop her property. Annie Stephens, the owner, had hoped to subdivide her rural acreage to accommodate suburban housing. She invoked the judicial process in search of damages and an injunction to prevent continued operation of the offending facility.[44] In

1955, the court ruled in her favour, awarding damages to compensate for injury or loss, and imposing an injunction – whose effective date was deferred – against the municipal sewage operation.

Echoing claims made decades earlier by the city of Toronto in the classic *Fieldhouse* litigation, Richmond Hill argued that municipal sewage operations were essentially immune from liability for causing pollution on the basis of the statutory authority under which they had been established. The trial court handily dismissed this claim. In the court's view, immunity could extend no further than the inevitable consequences of statutorily authorized undertakings. Since the municipality had manifestly failed to conduct its sewage disposal operations with a facility adequate to the task, the resulting pollution was hardly inevitable. Other measures – a larger plant or a sewage farm, for example – might have been employed. The cost of flushing had just gone up. Moreover, the trial court made clear that such costs were not its concern. The judiciary would confine itself to the rights of the parties, leaving to legislators the utilitarian quest for the greatest good for the greatest number – subject, of course, to the rule of law.[45]

When Richmond Hill officials chose to appeal, in hopes of persuading a higher court that their sewage arrangements should not be subject to legal attack, the village suffered a further rebuke.[46] Acting in contravention of provincial public health legislation, it had failed to notify its downstream neighbour, Markham Township, of its plan to construct the sewage facility. Therefore, Justice Laidlaw concluded, the Richmond Hill facility had been built without lawful authorization. But, invoking a hypothetical situation in which the village *had* followed correct procedures, Laidlaw went further to reflect upon the relationship between the provincial health department's power to approve such sewage facilities and the broad, historic prohibition against contaminating waterways that Ontario – and so many other jurisdictions – had enacted in the midst of late-nineteenth- and early-twentieth-century typhoid outbreaks. What exactly might the 1906 legislators have had in mind in proclaiming that "No garbage, excreta, manure, vegetable or animal matter or filth shall be discharged into or be deposited in any of the lakes, rivers, streams or other waters in Ontario, or on the shores or banks thereof"?[47] Half a century after it came into effect alongside equivalent proclamations enacted elsewhere, this particular legislative time bomb exploded around the Ontario Department of Health, the agency responsible for its administration.

Officials of the Department of Health presumably believed that the 1906 prohibition no longer needed strict enforcement. It had been enacted, after all, they recalled, during the interval between the discovery of typhoid's bacteriological transmission through the waterways and the general recognition of chlorination as a means of safeguarding municipal water supply more directly. This was of no concern to the court. In the words of the judge, the Department of Health "cannot in any case disregard the express prohibition . . . It has no express authority to authorize the doing of something in direct violation of that section." Only a very clearly expressed power "could make lawful what the Legislature declared in express terms to be unlawful."[48]

Clear legislative signals were certainly attractive. In their absence, twentieth-century litigation to safeguard waterways from sewage and industrial pollution faced the uncertainties of the common law. Riparians protesting against effluent discharge found it increasingly difficult to establish that the practice was not a reasonable, and thus lawful, use of rivers and streams. In the words of a New Jersey appeals court, "it cannot be said that the discharge of treated sewage effluent into a running stream is *per se* an unreasonable riparian use in today's civilization."[49] In the United States, the proliferation of permits to treat and discharge sewage also seems to have deterred common law actions.[50] It is somewhat ironic that twentieth-century civilization's widespread acceptance of industrial wastewater was tripped up in the United States by historic legislation, much as a long-forgotten piece of Ontario's public health regime served as a wake-up call in Canada.

In the 1960s, in the midst of intense controversy over the Storm King power proposal on the Hudson watershed, folk singer Pete Seeger lyrically exhorted enforcement officials to "Bring Back Old 1899," a reference to the 1899 *Refuse Act* on rivers and harbours. The old statute's provisions concerning refuse and obstruction of navigable waters had an unexpected role to play. Seeger chanted,

> McKinley, that staunch old Republican,
> He signed it right on the line.
> Then for seventy years they ignored it,
> The law called "Eighteen Ninety Nine."

Somewhat more pointedly, Seeger remarked that "Until people start to love their river, it's going to be a sewer."[51] A rediscovery of the *Refuse Act* of 1899 and, in particular, that it offered financial reward in the form of half of any fine assessed, payable at the discretion of the court to "the person or persons giving information which shall lead to conviction,"[52] stimulated a surge in public interest. Anglers' associations, among others, sought to take advantage of this option until courts concluded that the sharing of fines was available only in the context of a successful criminal proceeding instituted against the polluter by public officials.[53]

After seventy years, the *Refuse Act* emerged as a cause célèbre in environmental circles. The scope of its prohibition against discharging "any refuse matter of any kind or description whatever" was vigorously contested. Industrial polluters faced the brunt of these prosecutions, for, although the act expressly exempted refuse matter flowing from streets and sewers "in a liquid state," its provision against discharging refuse matter had established a "primitive absolute." By permit, dischargers might be exempted, but as late as 1970 a mere 266 out of thousands of industrial dischargers operating across the United States actually had formal authorizations.[54] In twenty-two states no permits had ever been issued.[55]

Some industrial polluters argued that as dischargers of liquid matter they were eligible for the sewage exemption; others protested that, as components of national rivers and harbours legislation, the controversial provisions were really designed to protect navigation and could not apply to pollution that did not obstruct waterways. Both objections were overcome during the 1960s when US Supreme Court decisions restricted the liquid-discharge exemption to actual sewage and broadened the concepts of obstruction and refuse.[56]

Leading the court's expansive interpretation of the 1899 legislation was Justice William O. Douglas, once a champion of major development initiatives but in his later years a man deeply committed to environmental values and a thoughtful student of Henry David Thoreau, John Muir, and Aldo Leopold. Leopold's advocacy of a "land ethic" profoundly influenced the final years of the judge's career, underpinning Douglas's protest against the destruction of the natural world: "We allow engineers and scientists to convert nature into dollars and into goodies. A river is a thing to be exploited, not treasured. A lake is better as a repository of sewage than as a fishery or canoeway. We are replacing the natural environment with a synthetic one."[57] An exasperated minority on the court sternly rejected Douglas's sentiments: according to this emerging line of interpretation, Justice Harlan exclaimed, "dropping anything but pure water into a river would appear to be a federal misdemeanour."[58]

None of the four hundred water polluters charged in the United States during the 1960s was accused of dropping pure water into rivers.[59] Nor was pure water a typical complaint in the litany of cases pursued in Britain by the ACA and its allies. In Canada, following Annie Stephens's success against the comparatively new sewage treatment works at Richmond Hill, government officials identified dozens of similar facilities that would also be vulnerable to injunctions. As legal challenges against water pollution became more frequent or threatened to become so, governments were once again compelled to select a strategy to clean up the waterways or to forestall complaints.

The Undertow

The most extreme response to critics of water quality was to eliminate or severely constrain their ability to take legal action against polluters. Indeed, legislatures and administrative authorities in Europe and North America took a variety of measures to lessen the impact of formal regulatory controls and common law claims on municipal and industrial polluters. The Hertfordshire County Council had managed in 1937 to secure a private act of the British parliament insulating the Colne Valley Sewage Board from legal action. No riparian owner or other party injured by the sewage works, the legislation explicitly declared, would henceforth be entitled to any right of action for an injunction or damages against the sewage board. John Eastwood's Anglers' Cooperative Association had been bested by this early measure, but successfully protested later government initiatives designed to extend to other polluters similar immunity against common law claims.[60]

Comparable legislative restrictions on citizens' attempts to preserve water quality were more successful in Ontario, and extended beyond statutory protection for the operations of the KVP pulp mill.[61] Even though the *Toronto Globe and Mail* insisted that only "aroused citizens" were capable of controlling municipal pollution,[62] the fact that as many as sixty municipal sewage systems were vulnerable to injunction claims drew the attention of the provincial government to the overall statutory framework regulating water quality. In legislation providing specifically for the dissolution of the injunction against Richmond Hill, the province infused new life into the series of municipal defences that the courts had swept aside. Amendments to the *Public Health Act* established that sewerage projects carried out in accordance with terms and conditions imposed under the act would be "deemed to be under construction, constructed, maintained or operated by statutory authority."[63] The right to seek compensation or damages was preserved, but jurisdiction over such claims was transferred to the Ontario Municipal Board.[64] Furthermore, sewerage projects operating in conformity with their approvals and applicable orders were exempted from the broadly crafted historic prohibition of 1906, which Justice Laidlaw had revitalized.

If municipal authorities were relieved, not all commentators welcomed the changes. One newspaper decried the probable elimination of the injunction remedy in such situations unless some private person undertook "to petition the courts for a declaration that such a law violates the Common Law rights of British subjects."[65] Property holders in the United States enjoyed a greater degree of protection by virtue of constitutional safeguards, although they remained vulnerable to the public interest in sewage pollution.[66]

Circumscribing legal remedies against pollution was not the only available route: it was also possible, of course, to envisage technical innovation and massive expenditures on infrastructure as a means to safeguard water quality. Such a response certainly had institutional support from those poised to undertake the implementation of remedial wastewater works. A series of preparatory conferences in Chicago, Cincinnati, and New York, and a strong impulse from George Warren Fuller for a forum to discuss sewage and industrial wastes, led to the creation of the Federation of Sewage Works Associations in 1928. Within a decade, twenty-five associations representing over twenty-four hundred individual members from across the United States had been joined by international affiliates, among which were the Institute of Sewage Purification and the Institute of Sanitary Engineers in England, and the Canadian Institute of Sewage and Sanitation.[67] A formidable task lay ahead.

Inventories of sewage and wastewater systems across the United States during the 1930s had documented the scope of the work required, as well as the unprecedented efforts needed to rectify the deficiencies. At the start of the decade, when 60 million out of 122 million Americans had sewerage, barely a quarter of the urban population was provided with sewage treatment.[68] As of 1933, federal grants and loans made available through Depression relief programs and public works initiatives functioned in combination with reinvigorated efforts on the part of health officials to increase the number of sewage treatment plants. It was claimed in 1939 that more had been accomplished in relation to municipal pollution abatement in the previous six years than in the preceding quarter-century.[69] Some 1,200 communities constructed such abatement facilities between 1935 and 1939, a year in which 848 sewage disposal systems (including enlargements and modernizations) were either completed or begun. One estimate suggested that in 1939 alone, the number of people served by sewage treatment grew from forty-one million to slightly under fifty-three million, a remarkable increase of 29 percent. The gain over a three-year period, according to the *Engineering News-Record*, was an astonishing 52 percent increase in the number of urban residents provided with sewage treatment.[70]

This commendable achievement must nevertheless be put in perspective. Despite the extraordinary progress of the late 1930s, sewage treatment was still available only to slightly more than half the urban population of the US as the Second World War was getting under way.[71] The US Public Health Service estimated at that time that 95 percent of the urban population was served by public sewer systems. However, 42 percent of the sewered population remained without treatment of any kind. Among the 58 percent with some form of treatment, a handful of states – Maryland, Minnesota, Texas, Illinois, and Wisconsin – offered treatment to as many as 90 percent of the sewered population. A "Census of Sewerage Systems Completed by Public Health Service" on the basis of 1938 data reported that just a little over half of treated sewage benefited from some form of secondary treatment such as activated sludge. The remaining treatment facilities offered primary treatment only.[72] In 1946 the domestic sewage load on inland waterways alone was estimated as the equivalent of raw sewage from a population of forty-seven million people.[73] The immediate postwar era was also the period when, for the first time, the industrial pollution load on American waterways exceeded municipal sewage discharges.[74]

The Izaak Walton League of America, often in alliance with the Sierra Club and the Audubon Society, tried valiantly to direct public investment toward the protection of water quality and away from popular though destructive large-scale river development schemes.[75] As well as questioning the big projects, the IWLA put forward alternatives to satisfy the craving for postwar economic renewal. Citing a 1944 report from the federal security agency, Kenneth Reid, the executive director of the IWLA, highlighted alternative investment opportunities in infrastructure. New sanitation facilities with estimated costs in the range of $3.5 billion "would largely pay for themselves in savings of health and economic loss." Such facilities were needed "as quickly as supplies of labor and materials and other practical considerations, permit their construction."[76] The postwar backlog was corroborated by other sources. By some estimates, as many as a hundred million Americans were in need of improved water and wastewater services.[77] As state sanitary engineers convened in Washington in May 1951, the US Public Health Service reported that over a ten-year period, 6,600 new municipal treatment plants or additions to existing facilities would be needed in order to reduce water pollution and conserve water resources. Construction costs were estimated at around $2.5 billion, with almost that much again required for updating and expansion over the decade.[78] Whether expenditures of this order were costs or opportunities was open to debate.

From unappreciable beginnings and through "a gradual and insidious process," the IWLA's Reid exhorted, a "crisis of first magnitude" had emerged, menacing public health and striking at "the very foundations of our entire economic and social structure." Some rivers may have endured the wastes of a modest number of small industries without serious lasting effects, but as larger facilities contributed to the pollution of certain streams, nature's assimilative capacity was severely overtaxed; a river became "a vile, dead thing – a liability and a scourge to the people living in the watershed, instead of a natural feature of beauty and usefulness."[79]

Canadians, too, were alerted – if less eloquently – to the accumulating consequences of underinvestment in wastewater infrastructure. For roughly a quarter-century preceding the mid-fifties, badly needed sewage treatment expenditures were deferred by depression and war. Gratifying improvements in public health attributable to chlorination of water supplies were offset by the deterioration of waterways. As the Second World War drew to a close, 30 percent of Ontario's municipal sewage was being discharged without treatment, a statistic that compared favourably with the national record. Data derived from Canada's 1956 census suggested that barely half of the country's sixteen and a half million urban residents were served by sewerage. Of the sewered population, roughly two million had primary treatment; three million had secondary treatment facilities in place.[80] New construction was almost universally required to address the complete absence of treatment or to replace outdated facilities that were overburdened by the expansion of population and industry.

In postwar Britain, reconstruction requirements were equally daunting, whether bomb-damaged facilities were being replaced or overtaxed infrastructure upgraded. Luton's discharge to the River Lea in Hertfordshire, for example, had grown from a million gallons of sewage daily in the First World War era to seven times that quantity in 1946.[81] The need for extensive new investment was obvious, not simply from the perspective of human health and convenience, but in light of alarming environmental losses.

Whatever its inspiration, much of the massive investment in new municipal and industrial wastewater facilities following the Second World War was carried out in conjunction with an elaborate system of permits, licences, or authorizations. These approvals varied somewhat in form and style as between jurisdictions, but their intended functions were much the same. They were ostensibly designed to impose controls on polluters in the interests of water quality and yet by definition they authorized flushing in the public interest. In Britain, where the term "consent" applied, an approvals system was introduced in the context of an emerging public debate[82] and alongside the work of the Central Advisory Water Committee and its subcommittee on preventing river pollution, whose recommendations for a more comprehensive approach to water ultimately bore fruit.[83]

The *Rivers (Prevention of Pollution) Act* (1951), replaced its forerunner, the original rivers pollution statute dating back three-quarters of a century. The declared object of the initiative was maintenance or restoration of the wholesomeness of rivers and other inland and coastal waters. Among several promising innovations, a new scheme of discharge permits or "consents" and the reconstitution of river management boards attracted attention. David Kinnersley, an authority on British water management, acknowledges the formation of the river-basin agencies as a positive step, but notes several weaknesses related to the willingness of waste dischargers to conform to the discipline of the consent system. First, the consent-based model was prospective in design. New or future discharges to inland waters were subject to the consent regime immediately, but existing discharges to such waterways and to tidal and coastal waters were exempt. Entrenched rights to pollute were thereby confirmed.[84] In addition, in deference to commercially sensitive information, the consent process was highly confidential. In Kinnersley's view, the long-term effect, when those affected by pollution eventually demanded information about polluters, was to discredit the regulators.[85]

The persistent problem of standards was addressed by empowering individual river boards to prescribe the characteristics of the poisonous, noxious, or polluting effluents whose discharge they prohibited. This flexibility was justified, contemporary observers claimed, on the assumption that uniform standards could not sensibly be applied. Uniform standards were seen as threatening to either "permit much more serious pollution of the clean rivers or place an impossible immediate burden on industry and local authorities in industrial areas." It was hoped, on the other hand, that standards suited to local circumstances would make it possible "to safeguard rivers that are now clean and to bring about a gradual improvement in those that are not."[86] The power to set standards for poisonous, noxious, or polluting discharges was later withdrawn, however, before it was ever exercised.[87] This not only left such standards indeterminate but also deprived the public of the only avenue ever contemplated by the legislation for a prosecution without the prior approval of the minister.[88].

Standards were criticized as lenient, oriented more toward the capacity of dated and overtaxed sewage works than to the quality of regional rivers, or even, on occasion, modified to accommodate deficient treatment works. As the Anglers' Cooperative Association commented disdainfully in Britain, "You don't make the effluents comply with the standards. You make the standards comply with the effluents."[89] Similar indignation was voiced in the United States, where one assistant attorney general remarked of his state's environmental laws, "Enforcement [has become] a process of whittling down the obligations of the polluter to the point where he can meet them."[90] In the absence of clearly defined legislative standards for water quality, the task of determining whether the pollution under consideration was authorized, or actually harmed health and welfare, or did so in an unreasonable manner, fell to courts and administrative agencies.[91]

The Ohio Water Pollution Control Board, for instance, adopted regulations to establish a set of water quality standards for seven categories of water usage including public water supply (highest), industrial water supply (lowest), agriculture, and cold-water fisheries.[92] For each category, scientific tests and certain minimum conditions were set out. Arnold Reitze, an Ohio law professor, dissected the scheme's limitations. Inherent in the acceptance of multiple categories embodying various levels of water quality, he discerned, was the assumption that water did not need to be pure. Ohio's seven descending categories implied that the quality of the lowest was very low indeed. If these arrangements were merely temporary expedients en route to systematic elevation of standards, the scheme might have been acceptable, but as it was, in Reitze's view, Ohio had enshrined a zoning system that legitimized continued low-quality water. Furthermore, the regulatory arrangements paradoxically ensured that future water uses would continue the pollution of the state's waterways by industry. Reitze also criticized the technical limitations of the scientific tests for each category: because they were not wide-ranging enough, their application could create only vague and inconsistent results. For example, the Public Water Supply criterion set no limitations on toxic substances, and the Recreation criterion had only a limitation on bacteria. The Agricultural Use and Stock Watering category had no specific limitation regarding any of the nine tests enumerated, but merely repeated the general "minimum conditions applicable to all waters at all places at all times."[93]

Damning illustrations highlighted ways in which water could satisfy designated criteria but nevertheless be disturbingly unfit. It appeared, for example, that the criteria for public water supply could be met by water that was polluted with mine acid runoff, or that water deemed suitable for one of the aquatic life categories could have a bacteria content so high as to be dangerous to humans. Standards designated for recreational use were such that water-skiing would be permissible in an effluent of hot pickling acid.[94]

Shortcomings abounded in other regulatory arrangements, where inconsistencies, a lack of rigour, vagueness, and uncertainty rendered it difficult to maintain – let alone improve – water quality. Severe limitations were widely noted in the formulation and enforcement of permits and approvals for waste discharges. A task force examining the situation in British Columbia's Fraser River found that pollution-control permits were often drawn up in ambiguous or vague terms that lacked the specific information upon which effective enforcement would depend. Investigators observed that when regional biologists and technical staff in either the provincial waste management service or Canada's federal fisheries office had identified present and future problems, such as leachate, their recommendations had been disregarded. The result was frequently "either illegal pollutions or severe environmental damage from permitted sites." Vaguely worded permits undermined enforcement action, as did deficiencies in procedures for reporting non-compliance.[95]

Throughout, there was a persistent current of concern about just what pollution was. One observer claimed that a "relative" view of water pollution had become dominant: "It assumes a free use of water for waste disposal up to a point of 'unreasonableness,' however legally defined." Moreover, the burden of establishing harm to marine resources or proving a deleterious impact on other water uses had become the responsibility of enforcement officials; it was now the obligation of governments to undertake research on water quality impacts. The relativist perspective also maintained that enforcement must be a local concern because the unique characteristics of the receiving water, the economics of the discharging plant, and the prevailing political tolerance level were crucial to any decisions to compel treatment or process change. Relativism resisted high standards: "The rhetoric of the 'relativists' often views effluent standards as 'treatment for treatment's sake' and condemns the proponents of these standards as alarmists. The position is based, in short, upon a series of assumptions incompatible with the 'no discharge' provisions of the *Refuse Act*."[96] This critique effectively summarized the disheartening implications of business as usual in what Hubert Humphrey, speaking in Pennsylvania at Gannon College in October 1966, referred to as an "effluent society,"[97] a society that had become unthinkingly accustomed to the aquatic translocation of its residuals. Even thinkers had accustomed themselves to an urban condition that was ultimately dependent on acceptance of nature as a wasteland.

Aldous Huxley, strolling with fellow author Thomas Mann along a beach south of Los Angeles just before the outbreak of the Second World War, was struck by the sight of myriad small whitish objects reminiscent of dead caterpillars. On closer inspection, the caterpillars revealed themselves to be condoms, "ten million emblems and mementos of Modern Love," an "orgiastic profusion" that had poured out of Los Angeles' nearby raw sewage outfall.[98]

Fifteen years later, Huxley noted marked improvements along the California coast, which were directly attributable to the Hyperion Activated Sludge Plant. Here, he wrote, by means of compressors, aerobic circulation, and digestion, something "hideous and pestilential" was transformed into "sweetness and light." The light took the form of methane, which fuelled nine 1,600-horsepower engines; the sweetness was represented by an odourless solid, selling in dried and pelletized form for ten dollars a ton to local farmers as fertilizer. Significant postwar expenditures on sewerage and wastewater treatment did occasionally produce noteworthy accomplishments. While he welcomed the transformation, however, Huxley cautioned, "The art of living together without turning the city into a dunghill has [to be] repeatedly discovered." Our environmental victories "are in no sense definitive or secure."[99]

12

Governing Water

Aldous Huxley's understanding that environmental victory – living together without turning cities into dunghills – is always precarious was widely apparent to contemporary mid-twentieth-century observers. Reporting on dirty rivers in 1959, the *Economist* welcomed all-party support in the British parliament for a private member's bill proposing to extend oversight of water quality beyond non-tidal waters to encompass tidal estuaries.[1] This was, of course, but the latest response to flushing's inherent tendency to safeguard one environment at the expense of another. To be sure, complex amendments lay ahead to address such intriguing uncertainties as the legal curiosity, "At what point [from the perspective of sewage disposal arrangements] does an estuary cease to be an estuary?" However, the *Economist* was pleased to remind its readers that social services more or less originated with sewage, and "it is high time that it was restored to something like its former primacy."[2]

The magazine's exhortation rested on the belief that "The discharge of untreated sewage anywhere, by any authority, ought to be as intolerable to public opinion as the robust old habit of crying 'Gardy loo!' and emptying the close-stool out of the window."[3] Although the gap has closed substantially on the sewage front, the impact of the impulse to flush and forget, leaving the ultimate challenge of sanitation and security to the natural water systems that form the basis of human existence, has been remarkable. The negative effects of flushing wastes to waterways have continued to surface repeatedly, whether we are considering domestic detergents, industrial metals, or agricultural runoff.

The environmental consequences of synthetic detergents initially became evident at sewage treatment facilities, where they were found to resist biological degradation; indeed, they were harmful to certain bacteria that were essential for purification processes designed to encourage biological action. They created persistent and unsightly foams that interfered with the transfer of oxygen from the atmosphere at the water's surface. In addition, by emulsifying solids that would otherwise have been precipitated, they undermined the efficiency of settling tanks.[4] In time, the presence of phosphates in most heavy-duty formulations produced more widespread damage. Between 1947, when heavy-duty detergent formulations came to market, and 1970, annual sodium tripolyphosphate production rose from a hundred thousand tons to over a hundred million tons. Passing through most treatment facilities and flowing without impediment through the sewerage conduits of communities still lacking treatment plants, phosphates nourished algal growth, stimulating it to excess in a process known as eutrophication. They thus undermined the quality of the aquatic environment, particularly in vulnerable waters such as shallow Lake Erie. Here, as the problem came under scrutiny, 137 thousand pounds of phosphorus were being added daily, some 72 percent of which came from municipal wastes, two-thirds of that amount attributable to detergents.[5]

It had been understood for many years that certain constituents of the pulp and paper process could be toxic to fish and other aquatic life, should they be found in "abnormally high concentrations." Yet around mid-century, prevailing opinion still favoured the formulaic response, "Most of the mills whose effluents possess this undesirable quality are located on streams where large quantities of water are available for dilution and this prevents the toxic constituents from reaching concentration harmful to the aquatic environment." A dilution factor of twenty to one, not far removed from the rough and ready nineteenth-century assumptions on which Illinois had based the Chicago diversion plan, was presumed sufficient to prevent harmful levels of concentration.[6] However, the concept of toxicity underwent a transition, for, as the *New York Times* editorial page explained in the summer of 1970, even though mercury was a known poison, its toxicity had not previously been linked to waterways: "Until recently neither Government officials nor scientists gave much thought to the possible harmful effects of mercury-containing wastes dumped into sewer systems by industrial plants. There was evidently a widespread assumption that mercury was insoluble and would lie forever quietly and inertly at the bottom of any body of water it reached."[7]

When a graduate student in zoology at the University of Western

Ontario reported finding mercury in pickerel in March 1970, Ontario closed commercial and sport fishing in the St. Clair River and Lake St. Clair. Canadian officials immediately alerted their American counterparts. By this point, twentieth-century use of mercury in the United States was estimated at 163 million pounds, with contemporary annual consumption in the range of 4 to 6 million pounds. Widely used in battery cells, bleach production, pharmaceuticals, the electrical industry, and paints, mercury was also presumed to be essential to pulp and paper manufacturing. Chloralkali plants alone discharged fourteen hundred pounds daily to the waterways, with roughly 10 percent of that amount traceable directly to ten facilities, the most serious offenders being found in Puget Sound in Washington State, the Niagara River in New York State, and the Androscoggin River in Maine.[8]

The companies responsible were promptly charged under the venerable *Refuse Act* of 1899, but the prosecutions were soon withdrawn in exchange for commitments to immediately reduce or eliminate mercury discharges.[9] The implementation of new production techniques led rapidly to substantial decreases across the United States.[10] Canadian mercury users, including chlor-alkali plants, were equally successful in reducing mercury losses dramatically as soon as the problem had been identified.[11] By that time, however, an uncertain legacy of contamination was already embedded in river sediments.

The use of animal wastes to enrich the soil has been commended to generations of farmers for its beneficial effects, for its contribution to the disposal of manure, and for its essential conformity to the natural cycle. Yet postwar changes in farm animal production practices accelerated through the 1960s to substantially alter the situation. As larger numbers of animals were confined on feedlot operations of industrial proportions, often with limited adjacent cropland, immense problems of waste disposal emerged. One steer of average weight (450 kilograms) would typically excrete 43 kilograms of nitrogen each year. A feedlot containing 32,000 such animals would produce 1,400 metric tonnes of nitrogen over the same period, roughly equivalent to the nitrogenous wastes of 260,000 people. By 1970, more than forty such operations existed in the United States, up from five in 1962.[12] Potential impacts from runoff wastes, in addition to the pollution of streams and surface waters, included infiltration and percolation to groundwater, the contamination of wells, and the destruction of fish habitat.[13]

Unrelenting evidence of deterioration gradually encouraged more sustained political reflection. There was little agreement, however, on the appropriate late-twentieth-century response to the challenge of governing water and safeguarding its quality. Which level of government, employing what instruments or mechanisms, was best suited to assume responsibility for what level of water quality, to what end or purpose? Would water quality be recognized as a core objective, or treated as an incidental consideration subordinate to other priorities?

Local, regional, national – and, ultimately, multilateral and international – solutions would all be invoked to address the long-delayed realization that there is no such place as away. At each level, however, it has been difficult to assert the central importance of water quality against an array of competing demands and assumptions, including the allure of flushing.

Do unto Others What You Can Get Away With

In the 1950s Ontario premier Leslie Frost acknowledged, like many of his counterparts, that pollution was "causing increasing embarrassment in [Ontario's] relationships with neighbouring provinces and states." Political leaders largely continued to believe, however, that the distribution of water and the elimination of pollution were essentially a local problem. "Both are very clearly duties and responsibilities of municipal government."[14] Yet pollution was far from prominent on the local public works agenda. Legislators, after observing that 83 of the 197 Ontario communities with public sewage systems still lacked treatment facilities of any kind, complained about the "widespread lack of appreciation by many people of their responsibility in providing adequate and safe disposal of all sewage and wastes."[15] Even with the encouragement of provincial officials, several communities staunchly resisted long-overdue expenditures.[16]

Although Toronto, where Lake Ontario water quality continued to deteriorate, had essentially disregarded waste treatment provisions, it had never been prosecuted. Suburban municipalities routinely dumped sewage into the Don and Humber Rivers with impunity; York Township had never been pressed to upgrade a treatment facility that had been designed for about half the population using it. Instances of such disregard for the requirements "could be multiplied," reported the *Toronto Globe and Mail*,[17] so that at the national level in Canada, immense sums were now essential for construction and maintenance of adequate sewage treatment facilities. A few years later, having observed the futility of citizen complaints, the newspaper insisted that municipalities should be required to have "as much care for public health beyond their borders as they are required to exercise on behalf of their own residents."[18] Municipalities were often willing to proceed with remedial works only if pressed to do so by some external authority. "Many members of Councils have stated they will do nothing until they are compelled to act. They are not averse to these programmes, but they feel someone else should assume responsibility especially since they claim the work is chiefly for the benefit of those living downstream. It is felt that a rigid policy requiring action would be productive of much progress."[19] The virtues of such external encouragement were subsequently recognized and appreciated in many related settings.

Investigations of the conditions of the waterways of England and Wales also revealed many instances of severe contamination for which local authorities shared significant responsibility. Here, too, however, deference to local authority and reluctance to enforce the law firmly persisted. Of thirty-one requests to prosecute under England's rivers pollution prevention legislation received by the minister of health between 1930 and 1939, sixteen were resolved without resort to the courts when the subjects of complaint agreed to remedial measures. The minister consented to fourteen prosecutions while declining in one due to wartime circumstances.[20] Local interests continued to weigh in against enforcement through the mid-century years, especially when remediation would be costly. "There has been a reluctance to encourage prosecutions of water quality offenders and to recommend policies which would result in an increase of costs to the local Authorities."[21] Yet the persistence of gross pollution throughout the industrial heartlands of Lancashire, Yorkshire, Tyneside, and the Midlands was also attributable to the fact that industrial discharges, falling between administrative stools, had often escaped official notice.[22] To one scandalized observer, industry's disregard of provisions designed to control pollution was "so consistent as to suggest unconcealed contempt."[23]

An advisory committee on water pollution operated between 1935 and 1939 in the United States. While ritually acknowledging local responsibilities for water protection, the committee carefully identified two factors unique to pollution abatement that demonstrated the need for "special stimulation" to overcome residual resistance. First, abatement does not ordinarily benefit the creators of pollution, who would generally be expected to assume the cost of treatment. Cities typically discharge waste to the detriment of downstream water users rather than to their own detriment. Accordingly, pollution abatement differs substantially from civic undertakings such as street programs, water supply, public buildings, and other facilities in that it "fosters a natural tendency to defer action . . . so long as the interests of the polluters are not damaged or the downstream users do not protest formally or vehemently." Second, the tangible benefits of pollution abatement and prevention expenditures are not easily identified. "It is difficult for a taxpayer to understand that he or his fellow citizen benefits from a decrease in the count of coliform organisms at a waterworks intake in a relatively clear-appearing stream." Pollution abatement merely prevents damages, in contrast with other public activities that produce tangible goods and services: "The citizen sees little that is tangible in the prevention of an epidemic, or the maintenance of a normal fish population, or reduction of heavy water-treatment costs."[24] Pennsylvania governor James H. Duff advised fellow governors that the nation had failed to recognize "that the safe disposition of body and industrial wastes is as necessary and proper an expense of municipal government as the maintenance of police and fire departments, and often the failure to provide for it is much more dangerous to the community welfare."[25] By this point, the spread of contamination to underground waters could already be observed.

A.E. Berry, a Canadian long-time observer of the dilemma, made similar remarks on the low level of public support for necessary urban expenditures: "If sewers are available to carry the sewage from the premises, the householder is little concerned about pollution of a stream with which he has little contact." The separation of waste disposal practices from their environmental consequences obscured the importance of investment in waste treatment or pollution prevention. Thus, Berry explained, "the ratepayer is asked to finance many municipal projects which are more tangible to him in their effects on his mode of life. Thus he is inclined to put off and to delay financing these outfall sewers and sewage treatment plants."[26] As Berry and others noted, the citizen and taxpayer related more readily to linear markings on the surface of the planet than to its organic integrity.

As intermediate levels of government initiated measures of their own, there were certainly indications of increased expenditure and more systematic activity. The results fell short on many fronts, however. Facilities remained inadequate in hundreds of Ohio communities, for example, even after more than $1 billion had been allocated to sewerage and wastewater in the years leading up to the mid-1960s. The Cuyahoga River, already infamous for its surface fires, poured into Lake Erie bacterial concentrations from five to twelve hundred times the maximum level then deemed safe for swimming. At only three beaches along the Erie shore could bathers confidently assume freedom from the risk of infection.[27]

In Wisconsin, spending on pollution control rose dramatically during the same period, and by 1966 all but one of the sixty-three communities that in 1949 discharged raw wastes to Wisconsin watercourses had installed at least some form of treatment – although often reluctantly. In one case,

Fond du Lac was caustically chastised in the state courts for its lackadaisical attitude: "It should not be hurried or nudged along in abating this nuisance because it has existed so long now there is no need to hurry and the city will eventually take care of the problem. Essentially, the attitude is that the city because it is a city is not subject to the law."[28] Even where progress had been made, state personnel were inadequate to maintain effective surveillance, and officials responsible for resource development insisted that further investments would be essential if they were even to maintain the status quo in the face of stronger waste streams that were anticipated: "Much greater treatment efficiencies will be required to clean up the state's streams – to effect substantial curtailment of the present strengths of wastes being discharged – and this will require major expenditures for new, more efficient types of waste treatment facilities."[29]

In New York State, as a consequence of municipal resistance to the local tax implications of pollution control measures, it appeared that "no noticeable improvement in the purity of New York streams" had occurred between the creation of the state's Water Pollution Control Board in 1949 and the passage of pure waters legislation in 1965. The new measures made various forms of financial assistance available to municipalities and industries to promote wastewater treatment.[30]

Strong traditions of local government in some jurisdictions, a lack of constitutional clarity in others, and a persistent inclination to separate the existence of legal authority from any obligation to exercise it, left waters vulnerable to abuse. Thus, when the late twentieth century experienced a renewal and intensification of discussion as to the most suitable location of governmental responsibility for water quality – among other aspects of environmental protection calling out for attention – many questions needed to be addressed.

Theorists revelled in the intricacies of properly allocating environmental powers amongst competing local, regional, and national claimants.[31] Shouldn't those who created pollution assume its costs and consequences? And wasn't it important that communities preserve the capacity to identify their own preferences and priorities, even if they favoured a little more employment and a little more pollution rather than the reverse? All else being equal, some asked, should preference not be given to the level of government most closely associated with the pollution problem in order to avoid the cumbersomeness and inefficiency typically entailed by resort to centralized institutions? In this view, the ideal world would give individuals liberty to devote or deny the financial resources necessary to treat and purify their own personal and household wastes prior to dumping, discharging, or depositing them upon the discrete surface acreage allocated them for sustenance and survival, in the knowledge that the full consequences would be theirs to enjoy – or not.

Upstream, Downstream, Back to Basins

Watersheds offer, at least in principle, the inspiration of mutual or common interest in a shared environment as an incentive to formulate agreed water quality solutions, as well as the possibility that practical measures, once implemented, will be able to fulfill their objectives without external interference or obstruction. Leslie C. Frank, a delegate to a US public health conference in the 1920s, spoke for generations of observers before and since when he expressed the view that "the most logical form of jurisdiction would seem to be one delineated by watershed rather than by geographical lines." In advocating such an approach for the United States, Frank confidently assessed the performance of comparable institutions in Germany and England as "relatively effective" in addressing pollution, certainly in comparison with the experience of his own country.[32]

Conservancy boards with responsibility for the Thames and Lee watersheds dated from the mid-nineteenth century, and the general concept of river-basin administration had been extended as of the 1890s to a handful of other waterways including the Mersey, Irwell, Ribble, and Dee, where authority under the historic rivers pollution legislation of 1876 was exercised by joint committees. Leslie Frank was doubtless familiar with these examples, and also perhaps with proposals for a more comprehensive scheme of catchment boards that were to emerge in 1930 to oversee drainage and flood control works. These bodies, in turn, gave way under the *River Boards Act* of 1948 to a new set of thirty-two agencies operating in river basins throughout England and Wales. On a watershed basis, the river boards supervised drainage, fisheries, and water quality, including authority for enforcement action under the 1876 statute and the successor legislation that finally replaced it in 1951. The composition of these multiple-purpose river-basin boards was somewhat unwieldy: an average of twenty-eight representatives of locally elected councils were brought together with appointees from fisheries and agricultural interests. Nor did the actual record of enforcement and water quality improvement signal that an institutional solution had been found for the continuing vulnerability of waterways to municipal and industrial pollution.[33]

The long-serving *Rivers Pollution Prevention Act* had failed, according to

H.G. Maurice, recently retired from the UK Fisheries Department, essentially because there was no one whose "primary business" was to administer its enforcement. He underlined the significance of safeguarding water directly rather than assimilating that fundamental objective within other policy fields. "The conservation of water supplies is a matter affecting fisheries," he explained, "but the conservation of fisheries is not a matter affecting the conservation of water. Take care of the waters and the existing Fishery Boards should have no difficulty in taking care of the fish."[34] Marine biologist C.M. Yonge was another whose reflections on water quality were linked to the state of fisheries, to salmon in particular. He too attempted to focus on the broader issue. The regrettable destruction of salmon would not in itself constitute a national disaster "were it not a symptom of mass pollution which is a calamity of the first magnitude demanding urgent action."[35]

The significant presence of local governments – the principal operators of sewage works – on the river boards created the risk that enforcement actions would be watered down in the interests of harmony.[36] Critical comment suggesting that river boards were reluctant to proceed against polluters under the 1951 legislation because of the influence of local authorities who were susceptible to conflicting interests within their ranks prompted an indignant response. The River Board Association emphasized that permission of the minister of housing and local government was a precondition of enforcement actions and tried to distance itself from that administrative limitation. "Contrary to the wishes of this association this restriction was imposed by Parliament as a temporary measure because, under existing economic conditions, it may not be possible to authorize the capital expenditures necessary to remove the cause of pollution." The association argued that local authorities "are not . . . 'loath to incur the expense of improving their effluents,' but in most cases are prevented from doing so by inability to obtain the permits necessary to effect the improvements to their sewage disposal and sewerage works."[37] Outside observers have generally been more skeptical, however, of an image of keen and willing local officials being obstructed by senior governments in their desire to take preventive or remedial action on behalf of downstream neighbours.[38]

In England and Wales, the consent system governing effluent discharges survived a significant administrative restructuring in the early 1970s. Ten regional water authorities subsumed 29 river authorities, 157 water supply agencies, and nearly 1,400 sewerage and sewage disposal agencies, thereby taking responsibility for the integrated delivery of a wide range of water supply and quality services that had previously been dispersed and poorly coordinated.[39] It was evidently hoped that this reconfiguration would redress the relative lack of attention devoted to water quality in the postwar period, when water supply and water quality questions were addressed separately. During this period, water quality had deteriorated badly in many areas. A river pollution survey conducted for the department of environment in 1970 reported gross pollution along roughly twelve hundred miles of rivers in the industrial heartland. Where some effort had been directed to cleaning up rivers, a number of estuaries had evidently paid the price as coastal waters deteriorated.[40]

The restructuring showed early signs of success. The new Severn-Trent regional water authority, taking over from more than two hundred separate predecessors, set out to establish a water quality baseline as a benchmark for its resolve to make improvements.[41] The initiative faced a severe obstacle in that under the *Rivers (Prevention of Pollution) Act*, as revised in 1961, information concerning consents, consent conditions, and the results of effluent sampling could not be disclosed without permission of the discharger. As the successor body to local authorities, Severn-Trent conferred its own retroactive permission on disclosures, and successfully persuaded roughly sixty of one hundred private companies and public enterprises to agree. The published findings constituted the basis for vigorous denunciation of the prior regime. The former river authorities had permitted many businesses to operate without consent conditions of any kind; numerous municipal sewage works had repeatedly failed to adhere to consent conditions, yet none had been prosecuted. A mere 36 percent of the sewage discharged in the Severn-Trent region had earned a "satisfactory" assessment, with the consequence that significant stretches of the Tame, Trent, Avon, and Stour, among other waterways, were entirely devoid of fish. It was at least satisfying to have such information, long concealed from the public, out in the open, first as a consequence of the unique Severn-Trent initiative and subsequently on the basis of legislative authority.[42]

An opportunity now existed to observe the operation of the new all-purpose authorities combining water supply and quality responsibilities. Severn-Trent's conduct emerged as "a good deed in a naughty world."[43] In David Kinnersley's view, the evidence from England and Wales undermined the suggestion "that authorities which were both environmental regulators and polluters would give high priority to reducing their pollution and to complying with the standards set for their own discharges."[44]

The initial creation and reconfiguration of agencies with responsibilities for aspects of the administration of watersheds or river basins was substantially easier to accomplish in Britain than in the United States and

Canada, where constitutional power is divided and waterways often span the boundaries of established jurisdictions. Those anxious to pursue antipollution and restoration initiatives at the watershed level in the United States thus tended to imagine interstate cooperation as a potential alternative to ineffective internal regimes, on the one hand, and to the intervention of the federal government in water quality management, on the other. The motivation for preserving state powers over waterways was clear enough: "The states are the local people interested in the industries of the state, and by co-operating with the State authorities we should have no trouble." A corresponding aversion to centralization and to the risk that streams would come under the control of the War Department due to its authority over navigable waters was also occasionally in evidence, sometimes even described as "a very dangerous thing."[45] In this context, inter-jurisdictional waters such as the Delaware, New York Harbor, and the Ohio attracted attention from those who imagined that the quality of shared waterways could be improved by interstate cooperation, and that such cooperation would simultaneously forestall federal intervention.[46]

As early as 1922, the state health departments of New Jersey and Pennsylvania had agreed to work cooperatively on sewage treatment along the Delaware River. However, in 1925 and again in 1927, specific legislative proposals for interstate collaboration failed. In an attempt to implement "equitable division" of Delaware waters as ordered by the United States Supreme Court in 1932, New York was brought into the discussion, again unsuccessful, of a wider compact. Two hundred and fifty conference delegates confronting the still "vile condition" of the Delaware in 1936 were vigorously exhorted to renew the search for a solution at the regional level. Henry W. Toll, executive director of the Council of State Governments, expressed astonishment that no regional group of states had ever cooperated effectively in any mutual undertaking.[47]

The Interstate Commission on the Delaware River Basin (Incodel), which ultimately brought together representatives of four states (Pennsylvania, New York, New Jersey, and Delaware), was at work in 1937 on a Reciprocal Agreement for the Correction and Control of Pollution of the Waters of the Interstate Delaware River. The agreement's preamble acknowledged that pollution constituted "a grave menace to the health, welfare, and recreational facilities of the people living in the Delaware River Basin, and occasion[ed] great economic loss." Nevertheless, participating states agreed that variations in geographical conditions and in uses made a uniform standard of sewage and waste treatment inappropriate. They established four zones, classified for regulatory purposes essentially on the basis of land and water uses.[48] The commission congratulated itself that the agreement precluded additional pollution insofar as new municipal sewerage systems were expected to produce an effluent that at least met minimum requirements, and new industry would be forced to meet waste treatment standards.[49]

As had often occurred, the reluctance of a major municipality to cooperate fully threatened the regime's overall effectiveness. Philadelphia, central to the Delaware's water pollution problem, had been in default for some time under an agreement with the Pennsylvania State Sanitary Water Board. Having promised to spend $3 million annually to implement a comprehensive plan for sewage collection and disposal that dated from 1914, the city had not followed through. As of 1939, therefore, almost 80 percent of the city's domestic sewage was still discharged into the Delaware without treatment.[50] Although the voluntary and cooperative means used by Incodel to coordinate federal, state, and local efforts had admirers,[51] two more decades would be needed to overcome financing problems and political obstacles impeding a serious Delaware River cleanup.[52]

The condition of New York Harbor was similarly subject to assessment by any number of commissions, agencies, conferences, and inspections between 1914 and 1931. By 1931, with the formation of a Tri-State Treaty Commission involving New York, New Jersey, and Connecticut, "the first productive step appeared to be taken." This body proposed an interstate commission to control pollution in New York Harbor and adjacent waters, and specifically recommended interstate compacts for cooperative action. The Tri-State Compact, adopted in January 1936 with the signatures of New York and New Jersey (joined in 1941 by Connecticut) and approved by Congress, established the Interstate Sanitation Commission and the Interstate Sanitation District.[53]

The Interstate Sanitation Commission claimed great progress in abating conditions that had contributed to the ongoing deterioration of New York Harbor. Despite wartime interruptions, advances made over a decade or so were triumphantly touted as being so substantial that, barring a further national emergency, "75 per cent of all the sewage of this great area [would] be adequately treated by 1953."[54] Although even that limited performance target would soon appear to be well short of satisfactory, the Interstate Commission in many respects pioneered an operational framework that encouraged similar initiatives.

From the perspective of flow, the Ohio River is the Mississippi's largest tributary, although that flow is subject to extreme fluctuations. Its watershed, a 200-thousand-square-mile basin encompassing parts of fourteen states, presented severe challenges.[55] Following a pollution survey by the US Public Health Service, Surgeon General Hugh S. Cumming took the initiative to convene discussions among state health officials from Ohio, Pennsylvania, and West Virginia that culminated in the Ohio River Interstate Stream Conservation Agreement.[56] The US Public Health Service in collaboration with the Army Corps of Engineers conducted a further survey between 1938 and 1942. They identified 634 public surface-water supplies along the Ohio, 294 of which were subject to pollution from sewage flows estimated at 940 million gallons per day. Two-thirds of this was still untreated despite a flurry of federally supported disposal works built between 1935 and 1940. In addition, the discharge of industrial waste was estimated as equivalent to the sewage of ten million people – topped off by a substantial annual acid load of a good two million tons of mining effluent.[57]

More than a few missteps occurred along the arduous pathway to an arrangement that eventually incorporated eleven of the states bordering the Ohio.[58] State ratification of an interstate pollution contract authorized by Congress in 1936 ran afoul of provisions introduced by several states to ensure that their approval was conditional upon the agreement of others.[59] When a Pennsylvania Senate committee defeated the compact – after the House of Representatives had given the measure its unanimous approval – the goal appeared even more distant. But the enormity of the situation could not be ignored forever. Negotiations finally led to the formation in 1939 of the Ohio River Valley Water Sanitation Commission, which consisted of three commissioners from each state and three from the federal government.[60]

Pending formal acceptance of legislative compacts, interstate commissions elsewhere also encouraged administrative measures for cooperative control of pollution. North and South Dakota joined Minnesota in forming a tri-state commission on water shortages and pollution along the Red River. With federal approval, Connecticut, Massachusetts, and Rhode Island adopted the New England Interstate Water Pollution Control Compact in 1947, creating a pollution control commission with authority to set reasonable physical, chemical, and bacteriological standards for water quality relative to various classifications of water use.[61]

A generous evaluation of collaborative interstate initiatives at the river-basin level would assess their contribution to water quality as mixed. The limitations are hardly surprising in an era where priorities clearly centred on the development of watersheds rather than on their protection. Gilbert F. White, a University of Chicago geographer who advised the Mississippi

Valley Committee, the National Resources Committee, and the National Resources Planning Board before his appointment to Hoover's President's Water Resources Policy Commission, presented the implications of completely engineered development of rivers. Stream flow would be fundamentally altered through storage facilities and diversions "so that the water is available when and where needed, rather than as dictated by natural fluctuations over days, seasons and years." It was the objective of development, moreover, to maximize economic returns. Thus, an "ideally regulated stream would fluctuate in its main channels only to meet fluctuating human demands, the natural variations having been evened out."[62] Others were equally adamant that maximum development could best be accomplished within the river-basin framework. One advocate dismissed absolutely any residual influence of the natural flow theory that had once enabled downstream riparians to moderate the ambitions of upstream developers: "The inutility of this doctrine is so offensive to the modern mind, in the light of the modern emphasis on maximum development, that it is not thought to present any serious obstacle."[63] The underlying perception that water was simply a resource, essentially a raw material contributing to industrial development rather than to the support of life was at the heart of continuing difficulties with water quality. Indeed, prevailing enthusiasm for storage dams, as White and others conveyed it, had virtually severed consideration of the flow or quantity of water from its quality, either for consumption or as habitat. And if streams were now officially nature's sewers, few were willing to confront the prospect that humanity's sewers were widely replacing nature's rivers.

Interstate initiatives to address pollution at the river-basin level had strategic attractions for the participants – they recognized the ultimate limitations of autonomous state action yet could forestall federal pollution controls on interstate waterways. Their advocates found them "a more desirable and democratic method of dealing with the pollution problems of the basin than Washington control or a valley authority usurping the powers of the states."[64] Still, the comparative merits of interstate arrangements and a more pronounced federal role came under increasing scrutiny.

The limitations of interstate arrangements for watershed protection were vociferously expressed by some of the "Izaak Waltons," who, as part of a small but determined chorus of conservationists, showed a sustained and sophisticated concern for the fate of American rivers. In 1939 the Izaak Walton League of America (IWLA) charged that the theoretical promise of the compact model for watershed-based safeguards was seldom, if ever, realized in practice. Supposed safeguards served only as "a legal means of putting off the day of reckoning in pollution control," and in that respect were perhaps without equal. Compact negotiations amongst the Ohio valley states, initiated in 1924 and not concluded until 1938, illustrated the problem. Noting that ratification at the state level and by Congress was still in the wings, the Izaak Walton assessment speculated, "You and I may have long white whiskers and be drinking water imported from Canada by the time that happens."[65]

The IWLA also found substantive deficiencies in the compact model itself. Orders could be put into effect only with the approval of a majority of commissioners from a majority of signatory states; and within a particular state, no order directed toward a municipality, corporation, or other entity could be effected until approved by a majority of commissioners from that state. The IWLA concluded that any state wishing to do so could clean up its own pollution without an interstate compact, and that where states were unwilling to act, such compacts would contribute nothing to the cause of clean streams.[66]

As the league and others emphasized, pollution problems were increasingly national in scope and had "defied the efforts of state and local enforcement agencies [for] many years." Writing in 1945, Kenneth A. Reid, IWLA executive director, remarked that Congress – despite extended effort – could claim no specific accomplishment. To overcome the jurisdictional impasse, the IWLA called for national legislation "to control pollution on a uniform basis throughout the nation." In an attempt to appease the constituency still committed to state and regional initiatives, the IWLA proposal included a provision that the state governments and any future interstate compact would be given "full opportunity to take the needed action for correction of pollution before the federal authority is invoked."[67] They had, of course, already enjoyed – and disregarded – an extended opportunity to do just that.

Widening Perspectives

Observing that public demand for more effective control of river pollution from municipal sewage and industrial wastes was growing, *State Government*, a periodical whose title effectively signalled its intended audience, reviewed ongoing debate over an appropriate US federal role.[68] On the one hand, wildlife and conservation groups typically advocated enforceable regulatory authority for federal agencies. The alternative approach, more generally favoured by health officials, civic representatives, and professional and technical associations, emphasized federal-state cooperation. Federal involvement, these groups suggested, should be restricted to education, research, and financial support. A compromise of sorts was proposed by Kentucky senator Alben William Barkley, a prominent backer of President Roosevelt's New Deal whose career culminated in the vice presidency under Harry Truman (1949-53).

Barkley's initiative called for the creation of a Division of Water Pollution Control within the US Public Health Service. A House amendment offered a concession to the federal control perspective in declaring, "No new sources of pollution, either by sewage or industrial waste, shall be permitted to be discharged into the navigable waters of the United States and streams tributary thereto until and unless approved by the Division." New sources of water pollution discharged without approval were declared to be "against the public policy of the United States." All United States attorneys had authority to launch proceedings to prevent or abate such nuisances and were required to do so when requested by any one of a number of federal, state, or interstate health and pollution control agencies, or any incorporated municipality.[69] Several proposals along such lines came close to enactment in the late 1930s.[70]

The seventy-ninth Congress (1945-46) saw a flurry of legislative activity on pollution control. One of the sponsors, Cleveland Bailey of West Virginia, was even moved to speculate, "Who knows? This may be our 'Lucky Year.'" Bailey put forward a measure requiring that abandoned coal-mining operations be sealed to prevent sulphur waste from flowing into neighbouring streams; he also proposed to support industrial expenditures on pollution control equipment through the economic incentive of tax relief. Another initiative advocated federal funding of a revolving loan fund to support sewage treatment. The Mundt Bill, widely favoured by conservationists, offered even more extensive federal financing; however, in contrast with schemes that would have left enforcement in the hands of officials at the state and local levels, the Mundt initiative authorized federal action to abate pollution where state and local authorities failed to take timely measures.[71]

State Government voiced predictable editorial concern about initiatives at the national level. As well as questioning the constitutional authority of the US federal government even to establish a water pollution abatement program, the journal addressed the risk that prohibitions might actually be enforced. Were enforcement to occur, industrial growth could be hindered. However, invoking the record of historic laxity, the commentator reassured its state-oriented readers that, "If past experience is any criterion, the Public Health Service is not likely to be unreasonable in the enforcement of this section of the bill."[72]

Although 1946 was not the "Lucky Year" for which Congressman Bailey had hoped, 1948 proved to be more promising. Following roughly a hundred unsuccessful proposals over the course of half a century,[73] water pollution legislation finally emerged at the national level in the United States. The *Water Pollution Control Act,* as approved on 30 June 1948, solemnly declared that the policy of the Congress of the United States was "to recognize, preserve and protect the primary responsibilities and rights of the States in controlling water pollution."[74] To that end, the legislation endorsed interstate compacts and other cooperative activity intended to prevent or abate water pollution, and offered technical and financial assistance for programs to treat sewage and liquid industrial waste.

Congress agreed to provide up to $22.5 million annually over a five-year period for distribution by way of loans to municipal, state, and interstate agencies who undertook to construct or improve treatment works. Facilities for water pollution research and training were established in Cincinnati, although actual approval of research funding regularly fell below levels anticipated in the legislation.[75] Interstate water pollution endangering health was designated a public nuisance and made subject to accompanying, though carefully circumscribed, abatement provisions. The surgeon general, who had access to the advice of the multi-member national Water Pollution Control Advisory Board, assumed responsibility for the initiative. In 1952 the program was extended for three years.

What was actually accomplished in 1948 remains a matter of debate. Was it the United States' "first comprehensive Federal legislation addressing the problem of water pollution,"[76] or a highly tentative initiative through which "the federal role was set out on a limited basis"?[77] Certainly, thanks to vigorous senatorial opposition, the measure lacked elements of substance and enforceability for which advocates of strong pollution control had campaigned.[78] When the administration sought a further routine extension in 1955, John Blatnik of Minnesota, who chaired the House subcommittee handling the legislation, proposed a more vigorous anti-pollution bill. The Blatnik initiative set out procedures for federal measures to be taken against instances of pollution to which states had failed to respond. Research efforts and funding – especially for smaller communities under 125 thousand in population – were simultaneously increased. The extent of federal grants, reintroduced to support localities in establishing sewage treatment works, became a matter of some controversy; agreement was eventually reached on a figure of $500 million over a ten-year period.

In 1955-56, thirty-one water studies were under way at the state level across the United States, and several significant national investigations and reviews had recently been completed.[79] As one analyst concluded in 1957, "There have been more major investigations of various phases of water resources than of any other domestic function." National water resources administration in the United States was indeed "perplexingly complex," as the agenda of one remarkable day on the Washington calendar amply illustrates.[80] On 4 February 1957, the House of Representatives dealt more or less simultaneously with both national drought relief legislation and disastrous flooding in eastern Kentucky. Other matters arising included bridge construction across the Potomac, appropriations for works along the Niagara River and the continuing construction of Illinois' Calumet-Sag channel, congressional approval of a Great Lakes Basin Compact, a fish hatchery, and amendments to the *Small Reclamation Projects Act* of 1956. The Senate agenda for the same day encompassed a hydro-electric power proposal advanced by the New York State Power Authority, the allocation of federal revenues from the Columbia River project to irrigation and reclamation, and drought relief for Missouri.

The emphasis of that day – and of the era – was overwhelmingly on promoting the availability and productivity of water resources. Irrigation, navigation, power, and flood control were thus singled out for attention as specific areas of federal concern. Conservation and water quality were of decidedly limited interest alongside other pressing measures that dominated the agenda. Legislators did take a moment, though, to approve a resolution to commemorate the fiftieth anniversary of a state governors' conference on conservation convened by Theodore Roosevelt in 1908; and a contemporary offshoot, a proposed commission on the conservation, development, and use of renewable natural resources, including water, made a brief appearance in debate. The Senate was unanimous in its willingness to permit an editorial on pollution of the upper Potomac to be added as an appendix to the *Congressional Record*.[81] In such an atmosphere, the success of any anti-pollution measure was indeed an accomplishment.

President Eisenhower, despite his reservations about increased federal grants, signed the amended *Water Pollution Control Act* in 1956, but by 1959, then disapproving of federal involvement in pollution abatement, he had begun to urge that the entire program be eliminated. By that time, many municipalities had responded actively to the incentives offered: planning and construction on the sewage front had proliferated, and administrative orders had encouraged certain more recalcitrant communities to abandon dumping in favour of treatment facilities.[82] Democrats responded to

Eisenhower's call for abolition of the program with legislation to further increase grants from $50 million to $90 million a year. Continuing to insist that responsibility for pollution control rested with the states, municipalities, and private industry, the president vetoed the measure in 1960.[83]

From the perspective of enforcement, US federal water pollution legislation produced disheartening limited results. No actions were ever brought to court under the 1948 act, and only one arose in the first decade following the 1956 amendments.[84] Among the impediments were the cumbersome preconditions governing notification, hearing procedures, and, after 1956, the requirement for a conference of water pollution control agencies from states involved in any given example of actionable pollution.[85] Other limitations on the use of the new federal measures stemmed from the need to demonstrate a danger to the "health or welfare of persons" outside the state where the pollution originated[86] and to the fact that courts were called upon to consider the practicability as well as the physical and economic feasibility of abatement.[87]

Adjustments to the definition of interstate waters – gerrymeandering, if you will – also greatly restricted the applicability of the statute. The Public Works Committee of the House of Representatives concluded in 1961 that only four thousand out of an estimated twenty-six thousand bodies of water in the United States could be regarded as interstate waters under the 1956 act.[88] Amendments implemented in 1961, the subsequent *Water Quality Act,* 1965, which directed states to prepare standards for stream quality, and the *Clean Water Restoration Act,* 1966, which expanded federal grants and matching funds to participating states, alleviated only some of the difficulties.[89]

By 1970 a growing intolerance of unregulated pollution was evident at the federal level in presidential pronouncements, in the actions of the Corps of Engineers, and in renewed efforts to bring stringency to regulatory controls.[90] The *Clean Water Act* of 1972, in a call for comprehensive programs to prevent, reduce, or eliminate the pollution of navigable waters, invoked the importance of protecting fish and aquatic wildlife. The measures envisaged included water quality standards and implementation plans for interstate and intrastate waters. These were to be formulated at the state level for review by the Environmental Protection Agency, to which the previous Federal Water Quality Administration had been relocated in 1970. The legislation addressed the impact on water quality not only of "point" sources such as municipal and industrial discharges, but also of "non-point" sources of pollution such as agricultural runoff and forest management operations or silvacultural practices. For point sources at least, the ambitious objective of the *Clean Water Act* was captured in the concept of a National Pollutant Discharge Elimination System, which called for industrial dischargers to adopt the "best practicable control technology currently available."[91]

The *Clean Water Act*'s immediate contribution to water quality improvements was considerable. The extensive introduction of treatment facilities saw the proportion of the population served by wastewater treatment leap from 42 percent in 1970 to 74 percent by 1985.[92] Declining levels of contamination and rising oxygen levels constituted undeniable progress. But were the means employed suitable? One critique, committed to efficient flushing, accused the *Clean Water Act* of striking a "romantic pose," charging that its failure to adopt a "sophisticated" view of pollution would result in a very expensive effort generating only trivial benefits. In this view, nationwide standards were misguided; federal river-basin agencies could better generate the regional policies that would take distinctive regional needs and priorities into account. It seemed senseless, for example, to insist upon the best practicable treatment of wastewater along Lake Superior's "near-virgin shores." The *Clean Water Act* was charged with replacing "an overly fractionalized system" with "an overly centralized one."[93] Within the context of a well-established and successful federal framework for water quality in the United States, healthy debate about the appropriate relationship between national, state, and local initiatives has been ongoing.[94]

In Canada, measures directed toward water quality at the national level coincided with developments in the United States. Canadian authorities had been induced to support a sewage treatment initiative on a national scale, in part as a consequence of American pressure to implement remedial measures around the Great Lakes. In 1957, for example, the Ontario communities of Sarnia and Trenton received mandatory orders from the federal government to install sewage works. The provincial premier, after meeting with municipal representatives, told the legislature that he was dismayed at being forced to put such pressure on the municipalities "by directions from Ottawa, coming from a high diplomatic level."[95] Evidently, Washington officials had communicated to their Canadian counterparts concerns that had not reached Toronto at water level across Lake Ontario. The official rationale for Canada's federal government to venture into an area often seen as a local responsibility included mention of "the threat to national survival from indiscriminate and uncontrolled waste disposal" and a reminder of the importance of federal leadership, including financial support. Without an increase in expenditures for municipal sewage disposal, "no significant environmental improvement is possible."[96]

Water was among the subjects of discussion proposed for a national conference on conservation announced in 1958 by Canada's newly elected prime minister, John Diefenbaker. Several years of federal-provincial consultations and planning led to an important gathering in Montreal in the fall of 1961 where, from a range of perspectives, government officials and academic experts contemplated "Resources for Tomorrow."[97] Water was still on the agenda, and although the question of pollution had been discreetly placed in the twelfth spot on a list of a dozen related topics, the message conveyed by J.R. Menzies, chief of the Public Health Engineering Division of the Department of National Health and Welfare, was clear enough. Referring to studies by the World Health Organization and to conditions in parts of Canada, Menzies concluded disturbingly that "The sanitary control of the environment is slipping backward . . . The forces of deterioration have outstripped the forces of betterment."[98]

Like his professional counterparts elsewhere, Menzies pleaded for greater public awareness of pollution problems to reinforce the demands of medical and engineering personnel for greater expenditures on waste treatment. "Possibly the greatest present need is a better knowledge of pollution problems on the part of the taxpayer," he remarked. "So many demands are being made for increased expenditures in urban communities that sewage treatment gets little support in many areas."[99] Reference to the impact of insecticides such as DDT on fish life and municipal water supplies indicates that a few members of the scientific community were quietly humming the chorus when Rachel Carson's solo, *Silent Spring*, captured the public imagination in 1962.

Although they welcomed federal funding and its attendant benefits, Canadian provinces remained keenly sensitive to their perception of constitutional proprieties, especially insofar as environmental protection initiatives had implications for industrial activities. Thus, when introducing a new water statute to the Canadian parliament in late 1969, federal legislators were careful to ensure that their more forthright proposals were accompanied by conciliatory gestures and offers of ample funding. Let us cooperate, introductory provisions of the *Canada Water Act (CWA)* seemed to suggest, by providing for federal-provincial committees to consult about, advise upon, and coordinate policies and programs for comprehensive water resource management on a national, provincial, regional, lake, or river-basin scale.[100] Arguably, *CWA* funding facilitated a transition toward new objectives, including sustaining the functions of freshwater environments.[101] Yet, as monies flowed disproportionately into major structural works that were not inconsistent with the aspirations of earlier generations for large-scale energy and infrastructure, less sympathetic observers began to see environmental opportunities fall victim to "an inclement institutional environment." The influence of established agencies curtailed experimentation with promising and innovative measures that might otherwise have been fostered.[102]

The *Canada Water Act* contemplated mechanisms for water quality management, that is, "any aspect of water resource management that relates to restoring, maintaining or improving the quality of water" in parts of the country where water quality management had become "a matter of urgent national concern." The federal government, in conjunction with a province or provinces – or on a unilateral basis in the case of inter-jurisdictional waters where reasonable efforts had failed to secure agreement – might create agencies with specific responsibility to plan for the restoration, preservation, and enhancement of water quality levels.[103] Recommendations would address water quality standards, waste discharges and treatment procedures, waste treatment, sampling, and other aspects of a comprehensive plan for the area in question, even including the novel – to Canadians – possibility of effluent fees.[104]

Advocates portrayed pollution discharge fees as a means of creating incentives for polluters to identify alternatives to existing production arrangements or treatment practices. Yet to others, effluent fees represented "pay as you go pollution" amounting to a permissive approach to environmental contamination. Certain officials thought it odd that while provincial governments might be trying to impose treatment obligations on industry, federal authorities would be offering to relieve polluters from anything more onerous than a payment schedule. In such circumstances, one provincial minister suggested, industries might claim that if they had paid a fee, they couldn't be expected to clean up as well.[105] Predictable resistance to implementation left whatever theoretical promise this aspect of the *Canada Water Act* offered entirely unfulfilled. Instead, such water quality initiatives as the Canadian federal government was actually willing to pursue were taken in connection with its residual authority over fisheries.

In 1970, coincident with the *Canada Water Act,* changes to federal fisheries legislation enhanced its utility as a mechanism for environmental protection. The process leading to these amendments dated from the midsixties, when federal fisheries officials acknowledged the systematic nature of pollution problems affecting Canadian fisheries. However, the fisheries department's legislative solution differed significantly from the *Canada Water Act,* whose principles, revolving around energy and resources, embodied the priorities of its makers. The measures advocated, and eventually implemented, in fisheries legislation remained ad hoc rather than comprehensive. They were designed to address localized situations related to a single resource rather than to work within the total management approach associated with the *Canada Water Act*. But fisheries officials were not without good cause in advocating a less overarching approach to environmental quality. They felt more confident of the constitutional foundations of their initiative and wanted to reduce dependence on the vagaries of intergovernmental cooperation. In a wry memo addressed to his counterpart at Energy, Mines and Resources, the fisheries minister expressed a firm disinclination to relegate the protection of fisheries to the "tender mercies of jointly manned federal-provincial committees of civil servants."[106]

Jack Davis, the pragmatic federal minister of fisheries who oversaw the 1970 amendments, was keen to get on with the job and (unlike those who formulated the *Canada Water Act*) not particularly fussy about the theoretical framework. Davis viewed fish as a "first line of defence" against aquatic pollution. In his words, "Anything that harms fish . . . may be harmful to man himself." A healthy fish population would signal environmental health, so "a healthy environment and a healthy fishery is undoubtedly the best insurance policy we can buy in our battle against pollution in water."[107]

Seen in this light, the long-standing inclination to sacrifice "a few fish" in the interests of industrial advancement appeared much less defensible than in former times. Davis opted to contain environmental pollution by confining it to the industrial settings where it originated: "Pollution . . . must be stopped at the factory fence."[108] In contrast to the *Canada Water Act*, which espoused assimilation and flexibility, Davis envisaged uniform national standards that would override differences in the purported assimilative capacity of natural waterways. This approach, similar to that of the US *Clean Water Act*'s sectorally oriented endorsement of best available control technology, was specifically intended to avert the risk that some jurisdictions would sacrifice environmental protection for short-term economic advantage.[109] In addition, although new facilities would be expected to comply immediately, existing industries would be allowed to negotiate their implementation of standards. Although Davis appeared capable of firmness – "Industry . . . must keep its poisons to itself. Canada's cities and towns must do likewise"[110] – succeeding federal governments have generally declined to assert the limits of their environmental powers.[111]

The early 1970s also saw important developments in British pollution law. In 1973, national water legislation provided for an integrated system of river and water management to be administered by ten regional water authorities. A *Control of Pollution Act* (*COPA*) was enacted in 1974. Primarily known as a pioneering venture into the management of toxic and controlled wastes, *COPA* extended coverage of discharge control arrangements to encompass most inland, underground, tidal, and coastal waters. Significantly, the legislation promised to allow any member of the public to launch proceedings against polluters whose effluent did not comply with consent or licence conditions. Thus, anyone – and not only those with a proprietary interest in the water harmed – would have the right to take action against water authorities and other dischargers for wastewater or sewage treatment deficiencies. It appeared that the residual constraints on legal action to safeguard waterways traceable back to the 1876 *Rivers Pollution Prevention Act* might finally disappear.[112]

When the implementation of significant aspects of *COPA* was delayed in 1975, critics linked this to the fact that roughly 30 percent of discharges to rivers did not comply with existing standards. It was almost a decade before key provisions of *COPA* were fully implemented, by which time even the royal commission standard dating from the First World War era had been abandoned as unattainable. As of 1988, two thousand miles of British rivers were deemed polluted, and discharges from twenty-seven hundred sewage outfalls failed to meet the conditions of their operating permits or consents, despite successful and high-profile prosecutions against the Thames and Anglian water authorities, among others.

By this point, a new round of statutory reform was under way, leading not only to privatization of water and sewage services, but to the separation of the operational and enforcement responsibilities that had previously been combined in the hands of water authorities. A national inspectorate, the National Rivers Authority, exercised its independence after 1989 in a series of initiatives against pollution.[113] Meanwhile, the impact of historical attitudes to waste disposal and the risks associated with inadequately maintained sewerage systems were exposed in legal controversies over groundwater contamination and sewage flooding that very publicly worked their way to senior courts.[114]

The important advances of the 1970s in national measures directed toward sewage and wastewater pollution were partially attributable to the determined efforts of advocates and litigators, and to mounting public pressure for domestic reform. It is also apparent that in certain circumstances international opinion and external pressure exerted a significant and sustained influence against the tradition of discharging wastes into water.

International Spillovers

The International Joint Commission, following extensive inquiry, reported on Great Lakes pollution to the governments of the United States and Canada in 1970. The completion of the IJC study, with its finding that, despite the terms of the historic Boundary Waters Treaty, "grave deterioration of water quality on each side of the boundary" was "causing injury to health and property on the other side," coincided with a general period of intense environmental concern. Such sentiment encouraged Canada and the United States to sign the Great Lakes Water Quality Agreement, a vital landmark in the series of official attempts to reverse the deeply entrenched pattern of disregard for water quality along extensive lengths of the international boundary.[115]

For the negotiators of the agreement, national philosophies of flushing now took on diplomatic overtones. US diplomats pointed out that "the higher Canadian percentage of sewered population was predicated on a lower level of treatment (only primary), while the United States was moving more slowly, but in the direction of a higher level of treatment (including secondary)." On the other hand, effluent quality achieved through primary treatment supplemented with chemicals would often be equivalent to secondary treatment. An additional consideration was the prevalence of separate sanitary and storm sewers in Canada. As a consequence, "a higher percentage of contaminated flow is treated than in the U.S., where a high percentage of flow often totally bypasses treatment." Canadian contaminant loadings were arguably lower on a per capita basis. Canada's then recent decision to control phosphates in detergents on a national basis, in contrast with US reliance on treatment in municipal sewage facilities, was also relevant to the negotiations, for if those arrangements proved to be inadequate, there would be no controls on phosphorus loadings.[116] In other words, locally preferred measures of addressing the consequences of flushing – typically those involving lower expenditures and touted as offering comparative advantages to communities near large bodies of water – were emerging as liabilities when the broader impacts came to public attention.

The first Great Lakes Water Quality Agreement in 1972 established a five-year program focusing on the eutrophication challenges in Lakes Erie and Ontario. This was very much the product of the IJC's own proposals for sewage remediation, although measures agreed upon by the two national governments fell somewhat short of the recommendations. The actual works were to be carried out by the relevant jurisdictions, with coordination, evaluation, and verification by the commission itself. The original agreement set out water quality objectives in both general and specific terms, before proposing preventive and remedial programs.[117] The parties were authorized to implement necessary measures according to their own legislative regimes, and the IJC assumed certain functions relating to information gathering and exchange, coordination, monitoring of progress, and making recommendations.[118]

General water quality objectives called for Great Lakes waters to be freed from surface scum, oil, and debris, as well as from substances that might result in conditions constituting nuisance. Nutrients resulting from human activity, to which municipal flushing contributed heavily, were also to be removed, along with substances capable of forming putrescent matter or objectionable sludge deposits. It was further agreed to limit the entry of substances that were toxic or harmful to human, animal, or aquatic life.[119]

In 1978, while reaffirming their original objectives, the two national governments introduced several extremely important elements to the definition of the task ahead. First, the 1978 agreement extended the challenge to what was known as the "Great Lakes Basin Ecosystem,"[120] where ecosystem was understood as "the interacting components of air, land, water and living organisms, including humans." Second, with measures taking hold to deal with phosphates and municipal sewage treatment, the parties were prepared to single out toxic substances for more elaborate definition and treatment.[121] In 1978 a toxic substance was recognized as "a substance which can cause death, disease, behavioural abnormalities, cancer, genetic mutations, physiological or reproductive malfunctions or physical deformities in any organism or its offspring, or which can become poisonous after concentration in the food chain or in combination with other substances." Early concerns centred on such substances as mercury, DDT, mirex, PCBs, and dioxin. Eventually, scientists working with the IJC's Water Quality Board identified over 350 synthetic toxic chemicals in Great Lakes waters.[122] These had arrived from the atmosphere, overland, or via wastewater systems intended, like their predecessors, to flush contaminants into the aquatic environment.

By 1978 the parties had adopted an overall statement of purpose: "to restore and maintain the chemical, physical, and biological integrity of the waters of the Great Lakes Basin Ecosystem." For that purpose, they adopted a stringent policy calling for prohibitions against the discharge of toxic substances in toxic amounts, and for the virtual elimination of those toxic substances that persisted in the environment. Overall, then, the 1978 agreement carried forward the original agenda while extending the geographical scope of the assignment and placing greater emphasis on toxins, on surface runoff or "non-point" pollution, and on a broader ecosystems approach to basin management.[123]

The condition of the Great Lakes had declined to such an extent that mid-century promoters of continental water diversion schemes even invoked their cleansing benefits alongside water supply. For decades, mining engineer Tom Kierans had nurtured a proposal to "recycle" water from James Bay back into the continental heartland of North America. Kierans was finally inspired to share his vision in 1959 when yet another proposal to expand the Chicago diversion out of Lake Michigan caused Canada's ambassador in Washington to write to the US secretary of state advising that "Careful inquiry failed to reveal any sources of water in Canada which could be added to the present supplies of the basin to compensate for further withdrawals in the United States."[124] Kierans took exception to what he regarded as the limited scope of that assessment and began to campaign for a concept that came to be known as the Great Recycling and Northern Development (GRAND) Canal.

Kierans' GRAND Canal proposal called for the conversion of James Bay into a freshwater lake that would become the source of water pumped and channelled southward to reach Lake Huron via the French River. He expected to draw his water supply in the freshwater runoff of eleven million litres per second from the 700,000 km^2 of the James Bay drainage basin. The outcome of the scheme would be "a new mid-continent, water relay and replenishment transfer grid."[125]

US senator Frank Moss of Utah, an enthusiastic diversionist, shared some of Kierans' vision for northern waters. Indeed, he dedicated his book *The Water Crisis* to "those Utah pioneers whose first act in building a western empire was to divert a mountain stream to water their arid lands."[126] In a condensed account of another diversion scheme that held possibility for Utah's purposes, Moss described the proposal as "a continent-wide plan for collection, redistribution, and efficient utilization of waters now running off to the seas totally unused or only partially used." By means of a continental network of tunnels, canals, and improved natural channels linking chains of reservoirs, the collected surplus could be diverted southward and eastward to benefit, when distributed, one territory and seven provinces of Canada, thirty-five states of the United States, and three states of Mexico.[127] The estimated volume of water to be diverted annually was roughly equivalent to the overall flow of the St. Lawrence River.[128]

Among the benefits of continental water diversion that Moss touted were that "the level and purity of water in the Great Lakes would be restored and maintained, and the Lakes could be used as a distribution manifold, as . . . proposed specifically in the expanded Kierans' Plan."[129] Few were persuaded by the virtues of diversion, still less by the prospect of water quality improvements in the Great Lakes. A pair of American experts observed, "The added flow might be envisioned as flushing out the pollution of Lake Erie but, in fact, would be a minor increment to the existing flow of water through the Lake." Cities and industries wishing to postpone "the moments when the truth about pollution by effluent waters must be faced" would probably arrive at their day of reckoning well before continental water diversion schemes were implemented on the scale that Kierans and Moss had imagined.[130] Later estimates suggested that it might require 250 years "for pollution, or anything else to flush down the 3,700 kms from the head of Lake Superior to the Atlantic Ocean."[131]

The impact on the Atlantic, among other marine environments, of land-based pollution including sewage and industrial discharges was becoming an important late-twentieth-century concern at the international level. In the same year in which the Great Lakes Water Quality Agreement was signed by Canada and the United States, the European Community, newly expanded to include the United Kingdom, embarked formally on a set of environmental initiatives. Following a declaration agreed to by the heads of member states in October 1972, a series of environmental action programs was launched, leading eventually to the recognition of environmental protection in Europe's continually evolving constitutional framework. The quality of surface and underground waters, as well as recreational or bathing waters, has required continuing attention, including measures to regulate urban wastewater, industrial discharges, and the handling of agricultural wastes, among other ongoing sources of degradation. Legislative guidance in the form of directives has provided foundations for improvements on a variety of fronts, even if implementation has sometimes been hesitant.[132]

A four-year study of conditions at thirteen popular British coastal resort areas underlined the relationship between exposure to dirty beaches and a variety of infections, even as senior government officials sought to weaken the European Union's bathing water directive and to delay implementation of measures to raise treatment standards at sewage works.[133] Britain's somewhat indifferent approach to the directive was evident in its initial identification of only twenty-seven beaches to which the bacteriological standards would apply. Land-locked Luxembourg had designated more.[134] British reluctance to embrace the bathing water directive was undoubtedly attributable to the financial implications of remedying the consequences of the historical policy of discharging untreated wastes into the sea. In reliance on the "dilute and disperse" approach to coastal sewage, secondary treatment was available at only 2 percent of UK sewage outfalls into the sea in the early 1990s.

Britain also distinguished itself in this era as the only European country continuing to dump sewage sludge at sea. Roughly a third of the sewage sludge generated from treatment in the United Kingdom was still being shipped offshore – a practice dating back over a century – to sites five to ten miles distant from major coastal communities. In 1987 this amounted to around 5,077 thousand tonnes. In combination with international pressure, a European directive on urban wastewater treatment calling for secondary treatment as a minimum and proposing 1998 as the terminal date for dumping sewage in the sea, accelerated reconsideration of these deeply entrenched practices.[135]

A pair of international conventions negotiated among North Atlantic maritime nations in the early 1970s (Oslo 1972 and Paris 1974) set in motion the processes of regulating marine pollution, first from waste

Firsthand observation and exposure to the adverse effects of coastal sewage discharges and offshore dumping of wastes has encouraged public-interest protests in several regions of the world. Surfers Against Sewage has been active in Britain in drawing attention to the importance of safeguarding coastal waters.

dumping by ships and then from land-based sources. These agreements committed the signatories to act against pollution from substances that could pose human health hazards, harm marine life, or damage amenities. They identified and classified pollutants according to their capacity to cause harm, banning the most harmful and subjecting the dumping of others to national licensing systems under the supervision of international agencies.[136]

The effect of collective initiatives pursuant to the original conventions was supplemented by international pressure in a series of North Sea conferences bringing together environment ministers from the eight countries sharing the basin. Beginning in 1984, the North Sea ministerial conferences addressed both industrial waste discharges and Britain's continuing search for comparative advantage in the disposal of sewage sludge at sea. The cynical logic of environmental devastation, grounded in economic self-interest and justified by reference to the virtues of Britain's short, swift rivers and the power of tidal currents, was systematically exposed by critics. Many of Britain's most polluting industrial operations, noted Nigel Haigh of the Institute for European Environmental Policy, were situated around estuaries or coastal areas well downstream from drinking-water sources. That fact alone had been offered to support the argument that a significant national comparative advantage would be lost if effluents discharged into coastal waters were required to meet standards applied to waterways used to supply drinking water. As Haigh elaborated this increasingly objectionable proposition, "Britain for pollution purposes, it can be argued, is well favoured by geography just as for transport purposes or, more facetiously, for the purposes of growing lemons, it is disadvantaged by geography." Just as Italian lemon growers enjoy the benefits of their sunny climes, or German manufacturers profit from proximity to the European marketplace, so it was argued that Britain should quite properly take advantage of a national "ability to locate industries on estuaries or on the coast where acute pollution problems are less likely to arise and where the sea water can assimilate or destroy the pollutants."[137] But, as we had so often learned in the case of rivers, lakes, and streams, accumulating evidence cast doubt on the ability of oceans to immunize humans against their own wastes. In 1990, participants at the North Sea conference accepted 1998 as a terminal date for dumping sewage sludge.[138]

As 1998 drew to a close, Britain's final acceptance of the end of ocean sewage dumping was ultimately based on the importance of maintaining good relations around the basin rather than upon scientific or economic conviction. "Politics and a collective inclination to buttress a new morality aimed at protecting the North Sea on precautionary grounds have prevailed."[139] Two British specialists summed up the situation as the dumping era came to a close: "The need to repatriate this sludge has highlighted the dilemma of contamination. About one-half of the total is not disposable on land because it contains a sufficient quantity of residues to be classified as controlled waste and, hence, disposable only at approved sites."[140] Long after the delusion had been exposed, the convenient practice of using water to "assimilate" whatever wastes could be flushed into it had no further claim to reliability.

Arrangements for governing water had engaged the attention of policy makers in Britain, Canada, and the United States following the Second World War. Various efforts were made to integrate water-related functions of government or to coordinate the responsibilities of diverse professional organizations associated with water management. In several cases, the watershed model achieved new prominence. In others, responsibility shifted upwards on the national hierarchy. But insofar as water quality was concerned, no jurisdiction successfully provided the discipline of enforcement, and reluctance to establish standards or to identify fundamental objectives continued to compromise results. Not surprisingly, the fate of lakes and rivers reappears on legislative and judicial agendas. It was increasingly clear that the preconditions for improving water quality encompassed much more than resources and technology. They extended well beyond the realms of legislation, enforcement, and administration, and raised questions about political organization and the scale of effective operations as well as cultural values and attitudes toward the natural environment.

Conclusion

WATER QUALITY AND THE
FUTURE OF FLUSHING

CANADA, GREAT BRITAIN, AND THE United States stood second, fourth, and twelfth, respectively, in a 2003 international analysis ranking the water quality of 122 nations from Finland, in first place, to Belgium, which, thanks to agricultural runoff and other forms of contamination, managed to squeeze out Morocco for the distinction of having the worst water in the world.[1] The rankings, of course, assess current conditions in the participating nations against each other's results, rather than against external or more independent norms, for such evaluations have typically been satisfied to measure human performance against human performance.

Comprehensive assessments of water quality have been difficult to develop. It is easier to assess immediate local conditions than to confidently assign responsibility for factors or trends – or to conclude that such conditions accurately reflect broader norms.[2] Recent United Nations attempts to ascertain the impact of pollution on freshwater resources suggest that on the order of two million tons of human, industrial, chemical, and agricultural waste are disposed of daily into receiving waters. Acknowledging limitations in pollution data, the study gives one estimate of global wastewater production at about 1,500 km^3 a day. Working from the assumption that one litre of wastewater would contaminate eight litres of fresh water, the authors estimated the current global pollution burden at close to an unfathomable 12,000 km^3, roughly equivalent in volume to Lake Superior.[3]

Although it has been possible to identify entire rivers that have been transformed, lost forever, or dedicated to the assimilation or "translocation" of waste over the past two centuries, the full environmental consequences of urban development and industrialization – both highly reliant upon water-borne waste removal – are still to be assessed. Low water levels in North America's Great Lakes are increasing the risk that turbulence from vessels on the surface will disturb the long-dormant residue or legacy of earlier generations of civic and industrial wastes. Dredging projects attempting to clear harbours and canals frequently involve the unfortunate side effect of recirculating toxic municipal and industrial sediments. Residues of prescription drugs and pharmaceuticals, following routes pioneered by domestic sewage, organic industrial wastes, and then production chemicals and heavy metals, are the latest stowaways to appear unannounced in waterways around the world. Researchers are already reporting traces of antibiotics, hormones, antidepressants, and cancer drugs in surface waters. Personal-care products from deodorants to sunscreen are also now making their way into surface and groundwater supplies.

Nor are the oceans immune from the impacts of flushing, whether via inadequate or poorly maintained sewerage and wastewater arrangements or via direct dumping by coastal communities. A number of major urban centres still regularly discharge untreated or only partially treated sewage into coastal waters, and it is only recently that significant efforts have been made to control the sanitary practices of heavily populated ocean liners. Experts on the scientific aspects of marine environmental protection have documented widespread resulting illness alongside environmental impacts ranging from instances of eutrophication to the complete destruction of coral reefs.[4] The popularity of the beach as a recreational destination reached unpredicted heights in the past century.[5] However, beach closures and pollution advisories increased quite dramatically in the United States in the late 1990s,[6] and in Europe, where regulatory measures to safeguard bathing-water quality have been established and refined over many years, close vigilance remains essential. Through organizations such as Surfers against Sewage or Surfriders, coastal swimmers are demanding results that could not have been imagined from Justice Best's futile attempt to persuade his fellow judges of the merits of sea bathing in England two centuries ago. To remedy that original failure, desperate rearguard international actions are now under way.[7]

The legacy of flushing, a foundation of the civilization we have inherited, is psychological and cultural as well as physical. More than a century after the earliest remedial initiatives, the condition of New Jersey waterways has become a cultural reference point. In *The Bone Collector*, a riveting movie thriller with Denzel Washington as a quadriplegic detective, a taxi driver from Jersey remarks matter-of-factly, "They have toxins in the water." The opening story in Melissa Bank's bestseller *The Girls' Guide to Hunting and Fishing* recalls an animated debate over the comparative recreational attractions of Nantucket and Loveladies on the New Jersey shore. After her mother confidently asserted that a lagoon and canal on the latter were "absolutely" swimmable, the narrator, Jane Rosenal, reflected quietly, "I didn't want to acid rain on my mother's parade, but the lagoon had oil floating on the surface and the bottom was sewagey soft."[8]

Flushing and the widely available water supplies that make flushing possible have become conveniences that are largely taken for granted in Europe and North America. Science fiction writer Bruce Sterling, author of *A Good Old-Fashioned Future*, argues that we can look forward to a time when "the internet will become furniture, just like running water and the telephone."[9] A more complacent approach to water supply might seem difficult to imagine, but there are contenders. Peter Mayle, a celebrated observer of European idiosyncrasies, reflects – however charmingly – the same casual disinterest in *French Lessons: Adventures with a Knife, Fork, and*

Coastal waters remain vulnerable to both sewage spills and direct discharges. In March 2001, a tugboat pushing a barge ruptured a 72-inch pipe beneath Biscayne Bay, Florida, unleashing millions of gallons of untreated sewage and forcing officials to close a large part of the bay to swimmers.

Corkscrew. To accentuate the culinary passions of his subjects, Mayle draws attention to salt as a topic of intense interest among gourmets, contrasting their ardour with the more conventional diner's inclination to dismiss salt as "about as fascinating as a glass of tapwater."[10] The danger in such a casual disregard for the fundamentals was exposed in the 2002 report of a judicial inquiry into a severe experience of water contamination in Walkerton, Ontario: "We may have become victims of our own success, taking for granted our drinking water's safety. The keynote in the future should be vigilance."[11]

Although water supply and drinking-water quality occasionally engage public attention, it is less evident that significant changes have taken place in regard to wastewater. Sewage treatment remains, in the words of one well-placed environmental official, "a largely invisible basic service,"[12] at times kept intentionally out of sight. With a federal election campaign under way, aides to Paul Martin, then Canada's minister of finance and not yet prime minister, cancelled a scheduled visit to a plastics manufacturing plant in the city of Windsor because the facility, it had been learned, produced toilet seats. Journalists reported that Mr. Martin's political staff were "mortified, then livid" at the so-called optics of the situation.[13] They evidently saw no electoral benefits, nothing ministerial – and certainly not prime ministerial – at this particular venue. In their assessment that there could be no upside to a televised audience with the porcelain god, Martin's advisors may have been correct.

I noted a similar impulse toward avoidance when a youthful caller asked the hosts of a Calgary phone-in radio show why Tigger put his head in the toilet. Tension in the studio was unmistakable to listeners; the damage control impulse kicked in. The prominent local radio personalities were clearly stumped by the young lad's question, uncertain about what to expect, and very, very anxious not to put their feet in it. When the giggling six-year old produced the answer – "He was looking for Pooh" – relief was audible. Life went on.

For most of us, flushing just makes things disappear, and we have largely become accustomed to making this happen with the flip of a lever or the push of a button. To address the problem of those who are disinclined to expend even this much energy, electronic sensing devices have introduced a gratifying measure of automation. It is difficult to find any discernible interest in the flow of water into the tank – the storage, treatment, and delivery of water of drinkable quality – or in anything that might happen after pulling the chain, pushing the button, pressing the handle, or stepping away from the electronic beam – the treatment of wastewater and the discharge of effluent. The deceptive immediacy and finality of flushing discourages both forethought and afterthought about its economic and environmental consequences.

If there is any popular interest in plumbing today, it seems to be directed toward domestic innovation, as the cult of improvement stimulates continuing curiosity about the future of flushing. Where are we going from here? Perhaps a Japanese-designed model with a remote-controlled lid, a heated ergonomically designed seat, twin-nozzled warm water jets, and hot-air dryer will begin to attract a global following. Such "Washlet" toilets, ranging in price from about $1,000 to $5,000, have reportedly garnered a significant share of the Japanese market. Caution is advised, however, as worn-out wiring in earlier models has sparked at least four fires, placing patrons uncomfortably in the hot seat.[14] The self-closing toilet seat once promoted as a Father's Day gift has not won universal acceptance. Another innovation, the shuttle *Columbia*'s capacity to vacuum human waste into the black hole of outer space, seems even less likely to develop a market on Earth. But one never knows. The *Columbia* approach, as far as I am able to determine, has no direct impact on the water quality of the home planet, although some may find it disturbing that the "just get rid of it" approach is as appealing to those with a NASA budget as it has long been to those with more restricted purses.

As riparian landlords in Britain squabbled with fellow water users almost two centuries ago, judges seemed blissfully unaware of potential damage to the waterways. Such rivers as the Irwell, the Irk, and their North American counterparts the Merrimack and Hudson – lawfully, in Chief Justice Lord Ellenborough's opinion – became "corrupted in quality." It once seemed inconceivable that anything more was at stake than the allocation of an infinite supply of aquatic resources between contesting claimants whose expressed wants constituted the entire spectrum of imaginable goals.

A variety of contending uses, including irrigation, fishing, navigation, and waterpower production, figured prominently for much of the nineteenth century. In that purposeful age, expectations about natural water quality were adapted to accommodate "reasonable uses." Even as municipal water supply systems proliferated, the seemingly benign category of "reasonable use" soon encompassed the "reasonable" infusion of wastes. Indeed, with the encouragement of sanitary reformers, waste removal – both municipal and industrial – contended with ancient rights of fishing and navigation for recognition among the purposes of waterways, sometimes even threatening to usurp their place on a fluctuating hierarchy of water uses.

Some were willing to accept – or even to embrace with enthusiasm – the proposition that at least certain waterways were actually intended to serve as sewers, while powerful undercurrents of thought diverted public and professional attention from environmental degradation. Arguments and assumptions drawn from science, economics, public morality, and law typically ascribed waste-removal strategies to "necessity" and downgraded their impacts to the status of "inconvenience." Or, discounted environmental losses were balanced against enthusiastic projections of commercial prosperity.

At a somewhat more popular level, conventional morality also condoned water pollution. The list of excuses is long: "There is no point going after me when everyone else does the same thing; that will hardly improve the waterway." "It can't actually be wrong to pollute since everyone behaves this way." "It would be unfair to single out industrial and manufacturing polluters when municipalities discharge their effluent into the same waters." "Our activity is so beneficial that it more than compensates for the loss of a few fish." "I tried my best." "My pollution is good for the water; it feeds the fish or it cleans the water." "There's really no harm done. What is pollution anyway?" "My effluent is cleaner than the river. What more can I do?" "Of course we would like to do something, as soon as we are able to take care of more pressing demands." "If we really need clean water, we can produce it with filtration, with chlorination, or we can extract it from the sea. And there will soon be better and cheaper technology to do the job, so there is no reason to act now." The rhetoric used to deflect responsibility or to minimize its significance has been well developed to support the assumption that discharging "residuals" is an appropriate use of water resources. That is the culture of flushing, a downward spiral that has been difficult to reverse, the valiant efforts of critics and opponents notwithstanding.

Rivers, lakes, vulnerable aquifers, and coastal waters were sacrificed to waste through ignorance of the consequences and through misunderstanding and delusion about self-purification and the apparent infinity of the oceans. Though there is some truth in self-purification as applied to organic wastes, the convenient temptation to overtax natural capacity was generally irresistible.[15] Even when the limitations of self-purification and the vulnerability of the oceans became evident, these comforting misconceptions endured, eventually even securing official endorsement in such pronouncements as "Streams are nature's sewers." The impulse to retain these notions long after their reliability was in question was reinforced by the comparative advantage that uninhibited flushers of waste appeared to enjoy over those who were compelled by geographic circumstances, insistent neighbours, or good conscience to direct their revenues to waste handling, to treatment and disposal, or even to the prevention and reduction of contamination.

Legal observers of the evolution of environmental challenges typically identify the late 1960s or early 1970s as a point of origin for an ongoing series of legislative responses. The so-called first generation of legislation, characterized as remedial in nature, was directed toward the distinct concerns of air and water pollution. Said to have been largely symbolic, this range of legislative activity was also limited by its reliance upon executive discretion and a relative lack of public involvement. A second generation of legislative initiatives, emerging in the mid-1970s, was more comprehensive in acknowledgment of the complex and integrated quality of environmental issues. Although significantly more preventive in orientation, measures from this period were still criticized for continued reliance upon an adversarial, rights-oriented, and competitive approach to the environment. Greener measures, the so-called third generation of environmental regulation, were said to be grounded in the recognition of the need for a much more broadly based approach to environmental participation whereby both rights and obligations would be accommodated.[16] And now market instruments vie for consideration. With so many generations of environmental law spawned in little more than three decades, one imagines their progenitors breeding faster than fruit flies. But the generations were by no means as sharply delineated as these analyses might suggest;[17] nor do the 1960s represent their point of origin.

As the generally unanticipated and almost invariably underestimated consequences of flushing from municipal and industrial sources became apparent during the nineteenth century, a range of observers set out to challenge the processes of deterioration. Some wanted to catch fish, generally for sale and consumption; some needed enough water to float vessels across; others preferred not to smell the waterways whose shorelines they inhabited. By invoking long-standing legal principles to safeguard well-recognized human interests oriented around their own activities, fishing, navigation, and more traditional riparian interests served informally and indirectly as proxies or surrogates for a range of natural functions of water.[18] Joined by a vigorous community of public health officials and supported by regulatory initiatives tending to be more formidable on paper than in practice, these groups resisted continued deterioration of the waterways.

For an appreciation of some of the obstacles, environmental historian Arthur F. McEvoy's discussion of the perceived value of fish is instructive.

In market terms, he explains, a fish is a commodity whose price to consumers is established through the mechanism of supply and demand. But, in an industry that pays no rent for access to its resource base, the market price fails utterly to reflect the value of the captured fish's forgone contribution to sustaining the long-term supply: "As if they were gold nuggets, the industry ripped fish out of their environment with no thought to the long-term consequences of its actions." Yet a fish is part of a species, he continues, "a coherent, self-perpetuating entity that works: it accumulates nutrients from its environment according to a genetic code adapted to its particular environment."[19] Water, of course, was that environment, although its essential contribution to life processes, like the fisheries themselves, tended to be overlooked or under-appreciated. Its uses – generally private or commercial in nature – were widely encouraged, with little recognition of the possibility that other functions of water might be undermined in the process.

The degradation of waterways has taken place over the past two centuries within the context of a wider range of considerations concerning individual liberty, progress and social advancement, the relationship of the human species to the planet it currently inhabits, and so on, for the values of the law ultimately reflect the values of the community it is expected to serve. The law is also often expected to contribute to the advancement or realization of the aspirations and goals of the community. Those aspirations, which increasingly include water quality as an element of environmental protection, have not always been achieved, partly as a consequence of the limits of the law, including the law's accommodation of other competing priorities. All those other priorities are, of course, equally human, and reconciliation thus tends to be democratic and political. But as Oliver A. Houck, professor of environmental law, cautions, "Once people's wants and desires become the standard for the harvest – whether the subject is timber, grass, watersheds or tuna – there is no standard at all."[20]

Somewhat ironically, the latest routes back from this precipice depend equally upon people's wants and desires, the new assumption, however, being that those wants and desires now encompass environmental protection, and possibly even a willingness to pay for it. As observed by the author of a survey of pollution control, information-based strategies for environmental protection – among the latest alternatives – have been given increasing emphasis because "regulatory, economic and policy means of controlling pollution are ultimately only as effective as the public wants them to be, and the degree to which information is available and understood is a major influence on public opinion."[21]

The importance of information and understanding in societies driven by public opinion is highlighted by certain lacunae in leadership. The fact that politicians prepared to dismiss environmental protection with the remark that "You can buy salmon in a can" remain electable is more than a little discouraging.[22] The difficulty of getting beyond the equivalent conviction that clean water comes right out of a tap is illustrated by one mayor's pithy account of shortfalls in Canadian federal government funding for sewage treatment facilities: "The government was listening to the public and the public wasn't clamouring for sewage treatment plants."[23]

In the same way that a better appreciation is needed of the context, functions, value, and importance of salmon outside of a can, a broader understanding of the importance of water outside the tap is essential to the long-term success of pollution control initiatives dependent upon public information strategies. Recognition of the value of water to the Earth and to society, currently a subject of extensive research,[24] might support pricing reforms to promote conservation, or might more realistically indicate the significance of the losses that have customarily been disregarded when yet another proposal for economic development is being advanced.

Public interests in water, once largely consisting of rights to fishing and navigation, have come to be understood – at least by some – to encompass more than mere uses of waters delivered by providence for the exclusive enjoyment of current claimants. Language has emerged to identify those interests. Some have adopted the concepts of "in-stream" and "non-consumptive" uses to focus on the essential contributions of waterways. More recently, ecosystem functions and integrity, biodiversity, and sustainability have been appended to the list of goals, occasionally near the top of the hierarchy where, I believe, they belong.

There is no reason this late in the present volume to venture far into the realms of biodiversity, ecosystem integrity, or environmental sustainability, but their general relevance requires some consideration. Biological diversity has been understood to encompass variety in several forms – genetic, species, and ecological. Although these forms of biodiversity are interdependent, ecological diversity may be most relevant here, as it concerns "the variety of habitats, biotic communities, and ecological processes found in the biosphere and the great variety within ecosystems in terms of difference of habitat and ecological processes."[25]

Biodiversity performs critical roles. In the first instance, it provides the medium for the flow of both energy and material that establishes the functional properties of ecosystems. Moreover, it contributes to the resilience of those ecosystems. This latter attribute, according to expert opinion, "could turn out to be the leading service supplied by biodiversity insofar as all other services appear to depend on it to a sizable degree. As biodiversity is depleted, there is often – not always – a decline in the integrity of ecosystem processes that supply environmental services."[26]

The principles of ecosystem management are therefore also under scrutiny in light of the long-term benefits that might be secured. A number of central elements of ecosystem management have been identified, but its fundamental goals must be the preservation of biological diversity or the restoration of biodiversity where regional fauna and flora have been lost. Scientific principles and ecological research will need to be carefully incorporated into management procedures to ensure that policy initiatives and operational plans are formulated and adjusted so as to minimize or avoid disruption of natural processes. This is as true in relation to unseen flows of waste water as in relation to transportation corridors, residential development, and industrial siting. If viable communities and economic opportunities are to be ensured, ecosystem management must contribute to sustainable development of resources in ways that are compatible with prevailing natural processes. Given that ecosystems transcend typical jurisdictional boundaries, ecosystem management calls for structures and procedures that enhance cooperation between governments, agencies, and communities in ways that have often been lacking in our response to wastewater.[27] Legal and environmental scholar Robert B. Keiter has outlined the function of ecosystem-based management from the perspectives of science and politics: "As a scientific matter, it focuses management at the appropriate level to guard against species loss and to ensure sustainable resource systems; as a political matter, it conveys, powerfully, a sense of interconnectedness among species and the irrelevance of political boundaries in the face of natural processes."[28]

Much thought is being given to what these concepts mean and to how, in particular, watersheds and eco-regions – although these concepts are not without difficulty[29] – might offer considerable promise for improved water management, including quality objectives that encompass a range of enforceable indicators.[30] Simultaneously, water-source protection measures are being introduced as one means of safeguarding the quality of drinking water. Some researchers hopefully anticipate that "the recent blossoming of natural flow restoration projects may herald the beginning of efforts to undo some of the damage of past flow alterations. The next century holds promise as an era for renegotiating human relationships with rivers, in which lessons from past experience are used to direct wise and informed action in the future."[31]

This thoughtful reflection strikes me as just about right. We need a generation of transformation more than occasional international "water years," or even "water decades," however welcome these may be.[32] It would not hurt at all to renegotiate human relationships with waterways, for a number of assumptions underpinning human entitlements to water resources may be ready for serious re-examination.[33] Biodiversity, sustainability, and ecological integrity – although none of these concepts is free from controversy – are sufficiently well understood that their incorporation as norms into legislation is already widely under way. These emerging societal objectives will require explicit legal protection in order to reinforce the indirect environmental safeguards that have been intermittently provided by riparian rights claims, fisheries cases, and navigation.

Although this inquiry into the culture of flushing offers no clear policy prescription, let alone technical guidance to those now contemplating new treatment technologies or the prospect of greatly expanded wastewater recycling, reflection on past experience may cast some light upon the general context in which choices were made. When courts, legislatures, and often the communities they served historically accepted the adverse consequences of flushing, they tended to do so with reference to perceived municipal or industrial necessities, by discounting the significance of the impacts to the level of inconvenience, or in anticipation of compensatory economic advantage. Perhaps it would be worthwhile to imagine environmental necessities as a point of departure, to acknowledge that impacts will generally extend far beyond the realm of inconvenience, and to concede that for badly underestimating environmental losses there is a high price to pay.

Notes

Foreword

1 UNICEF, Statistics, Sanitation, http://childinfo.org/areas/sanitation/, and UN HABITAT, *Water and Sanitation in the World's Cities: Local Action for Global Goals* (London: Earthscan, 2003); see also S.A. Esrey, J.B. Potash, L. Roberts, and C. Shiff, "Effects of Improved Water Supply and Sanitation on Ascariasis, Diarrhoea, Dracunculiasis, Hookworm Infection, Schistosomiasis, and Trachoma," *Bulletin of the World Health Organization* 69, 5 (1991): 609-21, and L. Fewtrell, R.B. Kaufmann, D. Kay, W. Enanoria, L. Haller, and J.M. Colford, "Water, Sanitation and Hygiene Interventions to Reduce Diarrhoea in Less Developed Countries: A Systematic Review and Meta-analysis," *Lancet Infectious Diseases* 5 (2005): 42-52.
2 Quoted in Tom Koch, *Cartographies of Diseases: Maps, Mapping, and Medicine* (Redlands, CA: ESRI Press, 2005), 83.
3 Koch, *Cartographies*, 95-101, 105-14, 90-93, quotes on pp. 95, 98; see also W.H. Frost, ed., *Snow on Cholera: Being a Reprint of Two Papers by John Snow, M.D.* (New York: Commonwealth Fund, 1936), and Peter Vinten-Johansen, Howard Brody, Nigel Paneth, Stephen Rachman, Michael Rip, and Donald Zuck, *Cholera, Chloroform and the Science of Medicine: A Life of John Snow* (New York: Oxford University Press, 2003).
4 The "critical illness" idea derives from Susan Sontag, *Illness as Metaphor* (New York: Vintage Books, 1978).
5 Frederick Engels, *The Condition of the Working Class in England* (London: Electric Book, 2001), 113; Hugh Miller, *First Impressions of England and Its People* (Boston: Gould and Lincoln, 1851), 62.
6 Christopher Otter, "Cleansing and Clarifying: Technology and Perception in Nineteenth-Century London," *Journal of British Studies* 43, 1 (January 2004): 44-46.
7 See, for example, Theodore Steinberg, *Nature Incorporated: Industrialization and the Waters of New England* (Cambridge, UK, and New York: Cambridge University Press, 1991).

8 For more detail here, see the important article by Ken Cruikshank and Nancy B. Boucher, "Blighted Areas and Obnoxious Industries: Constructing Environmental Inequality on an Industrial Waterfront, Hamilton, Ontario, 1890-1960," *Environmental History* 9, 3 (July 2004): 464-96, quotes from p. 469.
9 Arn M. Keeling, "The Effluent Society: Water Pollution and Environmental Politics in British Columbia, 1889-1980" (PhD diss., University of British Columbia, 2004); Arn M. Keeling, "Sink or Swim: Water Pollution and Environmental Politics in Vancouver, 1889-1975," *BC Studies* 142-43 (Summer-Autumn 2004): 69-101; Sally Hermansen and Graeme Wynn, "Reflections on the Nature of an Urban Bog," *Urban History Review* 34,1 (Fall 2005): 9-27.
10 See City of Toronto website, http://www.toronto.ca/water/wastewater_treatment/history.htm.
11 See "Palace of Purification: The R.C. Harris Filtration Plant," *Toronto Star*, 2002, http://bumblenut.com/drawing/comics/harris_filtration_plant/index.shtml; City of Toronto website, http://www.toronto.ca/water/supply/filtration.htm; *Toronto Plus.ca Online Magazine*, http://www.torontoplus.ca/portal/profile.do?profileID=1025587; Toronto Green Community and Toronto Field Naturalists, "Archive of Past Walks," Lost River Walks, http://www.lostrivers.ca/Warch9899.htm.
12 Glen Lowry, "The Representation of 'Race' in Ondaatje's *In the Skin of a Lion*," *CLCWeb: Comparative Literature and Culture. A WWWeb Journal* 6,3 (September 2004), http://clcwebjournal.lib.purdue.edu/clcweb04-3/lowry04; Carol L. Beran, "Ex-centricity. Michael Ondaatje's *In the Skin of a Lion* and Hugh MacLennan's *Barometer Rising*," *Studies in Canadian Literature/Etudes en littérature canadienne* 18,1 (1993).
13 Michael Ondaatje, *In the Skin of a Lion* (Toronto: McClelland and Stewart, 1987), 105-6.
14 Maria Kaika and Eric Swyngedouw give this point postmodern expression in "Fetishizing the Modern City: The Phantasmagoria of Urban Technological Networks," *International Journal of Urban and Regional Research* 24,1 (March 2000): 120-38, where they begin with Deleuze and Guattari to establish that "the city is a circulatory conduit"; American historian Joel Tarr writes in a somewhat different register in "The Metabolism of the Industrial City: The Case of Pittsburgh," *Journal of Urban History* 28,5 (July 2002): 511-45.
15 Trudeau quote from George Hutchison and Dick Wallace, *Grassy Narrows* (Toronto: Van Nostrand Reinhold, 1977), 178. For Chief Seattle's speech, see Rudolf Kaiser, "'A Fifth Gospel, Almost.' Chief Seattle's Speech(es): American Origins and European Reception," in Christian F. Feest, ed., *Indians and Europe: An Interdisciplinary Collection of Essays* (Aachen: Ed. Herodot, Rader-Verl., 1987), 505-25, and Rudolf Kaiser, *"A Whole Religious Concept"? Chief Seattle's Speech(es): American Origins and European Reception: Almost a Detective Story* (Hildesheim: Nortorf, 1985).
16 Jamie Benidickson, *Idleness, Water, and a Canoe: Reflections on Paddling for Pleasure* (Toronto: University of Toronto Press, 1997); Bruce W. Hodgins and Jamie Benidickson, *The Temagami Experience: Recreation, Resources, and Aboriginal Rights in the Northern Ontario Wilderness* (Toronto: University of Toronto Press, 1989); Jamie Benidickson, "Ontario Water Quality, Public Health and the Law, 1880-1930," in G. Blaine Baker and Jim Phillips, eds., *Essays in the History of Canadian Law*, vol. 8 (Toronto: University of Toronto Press, 1999), 115-41.
17 See Toronto Water section of City of Toronto website, http://www.toronto.ca/water/index.htm.

18 Efraim Halfon and Don Poulton, "Distribution of Chlorobenzenes, Pesticides and PCB Congeners in Lake Ontario near the Toronto Waterfront," *Water Pollution Research Journal of Canada* 27,4 (1992): 751-72 (includes citations to other works; see the "Discussion: Spatial Patterns" section for this quote).
19 Primary treatment screens out large debris (rags, condoms) and usually a little more than half of suspended solids such as toilet paper, feces, and any other floatable substances carried down the sewers. Secondary treatment filters out 90 percent of suspended solids, and disposes of them on land.
20 Sierra Legal Defence Fund, "Canada's Sewage Report Card Results," media release, http://www.sierralegal.org/m_archive/pro4_09_08.html; Rhiannon Coppin, "When Flush Comes to Shove," *Vancouver Courier*, 6 July 2005, http://www.vancourier.com/issues05/071205/news/072105nn1.html.
21 David Goldblatt, *Social Theory and the Environment* (Cambridge: Polity Press, 1996), 167.
22 I derived this phrase and some early inspiration from William Leiss and Christina Chociolko, *Risk and Responsibility* (Montreal and Kingston: McGill-Queen's University Press, 1994).
23 Ulrich Beck, *Risk Society: Towards a New Modernity* (London: Sage, 1992); Ulrich Beck, *Ecological Politics in an Age of Risk* (Cambridge: Polity Press, 1995). My discussion of Beck has been influenced and guided by Goldblatt, *Social Theory and the Environment*, especially pp. 154-87; quote from p. 166.
24 Goldblatt, *Social Theory and the Environment*, 170-73.
25 Christopher Hume, "Toronto's Deco Days," *Toronto Saturday Star*, 20 September 2003; see also the autobiographical notes by Rick Bebout titled "Promiscuous Affections: A Life in the Bar 1969-2000," http://webhome.idirect.com/~rbebout/bar/citizen1.htm.
26 On the hollowing out of the state and its implications, see Jeremy Wilson, "For the Birds: Neoliberalism and the Protection of Diversity in British Columbia," *BC Studies* 142-43 (Summer-Autumn 2004): 241-77, and Anita Krajnc, "Wither Ontario's Environment? Neo-conservatism and the Decline of the Environment Ministry," *Canadian Public Policy* 26,1 (2000): 111-27; for the Beck quote, see Goldblatt, *Social Theory and the Environment*, 166.

INTRODUCTION: THE CULTURE OF FLUSHING

1 Joseph Sax, "Why I Teach Water Law," *Journal of Law Reform* 18,2 (1985): 273.
2 Ibid., 274.
3 Ibid., 273.
4 Joel A. Tarr, *The Search for the Ultimate Sink: Urban Pollution in Historical Perspective* (Akron, OH: University of Akron Press, 1996).
5 Richard S. Campbell et al., "Water Management in Ontario: An Economic Evaluation of Public Policy," *Osgoode Hall Law Journal* 12 (1974): 502.
6 Dennis R. O'Connor, Commissioner, *Report of the Walkerton Inquiry*, Part 1: *The Events of May 2000 and Related Issues* ([Toronto]: Ontario Ministry of Attorney General, 2002).
7 Irving Fox, "Institutions for Water Management in a Changing World," *Natural*

Resources Journal 16 (1976): 748-49, citing, among other things, J. Krutilla and O. Eckslem, "Multiple Purpose River Development," *Natural Resources Journal* 2 (1970): 52-70.

CHAPTER 1: THE ADVANTAGE OF A FLOW OF WATER

1 Adam Smith, *The Wealth of Nations* (1776, repr.; Chicago: Encyclopaedia Britannica, 1952), 12.
2 William Blackstone, *Commentaries on the Laws of England*, 4 vols. (Oxford: Clarendon Press, 1765-69). For an introduction to the vast literature on Blackstone's conception of property law and water, see Michael Taggart, *Private Property and Abuse of Rights in Victorian England: The Story of Edward Pickles and the Bradford Water Supply* (Oxford: Oxford University Press, 2002), 110-13.
3 Blackstone, *Commentaries on the Laws of England*, vol. 2, 403.
4 Earl Finbar Murphy, *Water Purity: A Study in Legal Control of Natural Resources* (Madison: University of Wisconsin, 1961), 7.
5 Antoine-Laurent Lavoisier, *Oeuvres de Lavoisier*, 6 vols., ed. J.B. Dumas and Edouard Grimaux (Paris: n.p., 1862-93), 2: 4, quoted in J.B. Gough, "Lavoisier's *Memoirs on the Nature of Water and Their Place in the Chemical Revolution*," *Ambix* 30, 2 (July 1983): 93.
6 John Gribbin, *Science: A History, 1543-2001* (Harmondsworth: Penguin Books, 2003), 271-83.
7 Emmanuel Le Roy Ladurie, Introduction to Jean-Pierre Goubert, *The Conquest of Water: The Advent of Health in the Industrial Age*, trans. Andrew Wilson (Cambridge: Polity Press, 1989), 2.
8 Terry Reynolds, *Stronger Than a Hundred Men: A History of the Vertical Water Wheel* (Baltimore: Johns Hopkins University Press, 1983), 276.
9 Shaw actually returned water to the Irwell, but the configuration of the river was such that the point of re-entry was downstream from Bealey's facilities.
10 *Bealey v. Shaw* (1805), 6 East 207, 102 E.R. 1267.
11 Ibid., 1270.
12 Ibid., 1269. For a detailed analysis of opinion in this case set against earlier legal developments, see Joshua Gertzler, *A History of Water Rights at Common Law* (Oxford: Oxford University Press, 2004), 207-12.
13 Mr. Francis to F.O. Ward, 29 March 1854, quoted in *Chicago Sewerage: Report of the Results of Examinations Made in Relation to Sewerage in Several European Cities in the Winter of 1856-57* (Chicago: Chicago Board of Sewerage, 1858), 10.
14 W.P. Hazeldine, "Textile and Trade Effluents and Their Treatment," *Journal of the Textile Institute* 40 (1949): P1, 090-6, quoted in Louis Klein, *River Pollution II Causes and Effects* (London: Butterworths, 1962), 117. (A fellmonger is a dealer in fells, or sheepskins, who separates the wool from the pelts.)
15 Frederick Engels, *The Condition of the Working Class in England*, trans. and ed. W.O. Henderson and W.O. Chaloner (Oxford: Basil Blackwell, 1958), 60.
16 *Nuttall v. Bracewell* (1866), 2 L.R. Ex. 1 at 9.
17 Louis C. Hunter, *History of Industrial Power in the United States, 1780-1930*, 3 vols. (Charlottesville: University of Virginia Press, 1979), 1: 1-6.

18 Ibid., 184.
19 Francis to Ward, in *Chicago Sewerage*, 10.
20 Murphy, *Water Purity*, 50-52.
21 George Derby, Secretary, Massachusetts State Board of Health, "Mill Dams and Other Water Obstructions," in Massachusetts State Board of Health, *Third Annual Report* (Boston, 1872), 60.
22 Marshall Ora Leighton, *Sewage Pollution in the Metropolitan Area near New York City and Its Effect on Inland Water Resources*, US Geological Survey Water-Supply and Irrigation Paper No. 72 (Washington, DC: Government Printing Office, 1902), 10.
23 Alice Outwater, *Water: A Natural History* (New York: Basic Books, 1996), 105.
24 E.C. Pielou, *Fresh Water* (Chicago: University of Chicago Press, 1998), 209-13.
25 George Eliot, *The Mill on the Floss* (New York: Norton, 1994), 130.
26 John Sutcliffe, quoted in Gertzler, *A History of Water Rights*, 40.
27 Joseph K. Angell, *The Law of Watercourses*, 2nd ed. (Boston: Little and Brown, 1833), vii, quoted in Morton J. Horwitz, *The Transformation of American Law, 1760-1860* (Cambridge, MA: Harvard University Press, 1977), 40.
28 C.M. Haar, "Legislative Change of Water Law in Massachusetts," in D. Haber and S.W. Bergen, eds., *The Law of Water Allocation in the Eastern United States* (New York: Ronald Press, 1958), 9.
29 Carol Rose, "Energy and Efficiency in the Realignment of Common Law Water Rights," *Journal of Legal Studies* 19 (1990): 261; Horwitz, *Transformation of American Law*.
30 Horwitz, *Transformation of American Law*, 36.
31 Theodore Steinberg, *Nature Incorporated: Industrialization and the Waters of New England* (Cambridge: Cambridge University Press, 1991), 23.
32 Ibid., 16.
33 This is in contrast to other interpretations that regard law as "norms laid down by officials in power, secular embodiments of natural law, or social phenomena with a distinctive kind of past." Robert Summers, *Instrumentalism and American Legal Theory* (Ithaca, NY: Cornell University Press, 1982), 20.
34 Tony Scott and Georgina Coustalin, "The Evolution of Water Rights to 1850," in *The Evolution of Property Rights over Natural Resources* (manuscript version, September 1996), 12.
35 Samuel Weil, "Waters: American Law and French Authority," *Harvard Law Review* 33 (1919): 133–39.
36 *Tyler v. Wilkinson*, 4 Mason 397; 24 F. Cas. 472 (1827).
37 Rose, "Energy and Efficiency," 289.
38 *Tyler v. Wilkinson*, 474. On the general significance of "reasonable use" as a mediating device in cases of conflict over nineteenth-century property rights, see Robert G. Bone, "Normative Theory and Legal Doctrine in American Nuisance Law: 1850 to 1920," *Southern California Law Review* 59 (1986): 1148-50.
39 James Kent, *Commentaries on American Law*, 439.
40 Ibid.
41 Ibid., 441.
42 Ibid., 440.
43 Ibid., 441.
44 *Wright v. Howard* (1823), 1 Sim. & St. 190, 57 E.R. 76 (Ch.).
45 *Wood v. Waud* (1849), 3 Ex. 748 at 781, 154 E.R. 1047.

46 Parke B., in *Embrey v. Owen* (1851), 155 E.R. 579 at 585, referring to 3 Kent's Commentaries, Lecture 52, 439-45. Robert Bone helpfully refers to nineteenth-century riparian rights as relative rather than absolute: "The content of the riparian water right was partly defined by reference to the consequences which the exercise of the right might have on other riparian owners." Bone, "Normative Theory," 1132.
47 *Embrey v. Owen*, 587.
48 John Langton, *Early Days in Upper Canada*, ed. W.A. Langton (Toronto: Macmillan, 1926), 47-48.
49 Nicol H. Baird, "Report on the Overflowing of the Scugog River and Lake," 10-18 October 1835, quoted in Neil S. Forkey, "Damning the Dam: Ecology and Community in Ops Township, Upper Canada," *Canadian Historical Review* 79 (1998): 83.
50 Ibid., 84.
51 Ibid., 94.
52 This local retaliation against mill owners was by no means unique: Several decades later, a mill on Wisconsin's Crawfish River mysteriously burned one night, even though a state health study of the river, a particularly sluggish waterway, had failed to establish a clear link between the miller's dam and the prevalence of malaria. The mill owner sold out to neighbouring farmers, who removed the dam. Malaria disappeared from the vicinity. (Murphy, *Water Purity*, 78). The malicious destruction of a dam might even involve death: *Hudson v. Napanee River Improvement Co.* (1914), 31 O.L.R. 47.
53 Forkey, "Damning the Dam," 94.
54 Ibid., 84.
55 J. Willard Hurst, *Law and Economic Growth: The Legal History of the Lumber Industry in Wisconsin, 1836-1915* (Cambridge, MA: Harvard University Press, 1964), 216.
56 Hunter, *Industrial Power*, 1: 140.
57 *Strickler v. Todd*, 10 Sergeant and Rawle (Pennsylvania, 1823), 63 at 68, quoted in ibid., 148.
58 Hurst, *Law and Economic Growth*, 173.
59 Haar, "Legislative Change," 9.
60 Christopher Hamlin, *A Science of Impurity: Water Analysis in Nineteenth Century Britain* (Bristol: Adam Hilger, 1990); and Steinberg, *Nature Incorporated*.
61 *Blair and Sumner v. Deakin; Eden and Thwaites v. Deakin* (1887), 57 Law Times 522.
62 For a fascinating and detailed account of water supply controversy in the Bradford district, see Michael Taggart, *Private Property and Abuse of Rights in Victorian England* (Oxford: Oxford University Press, 2002).
63 Steinberg, *Nature Incorporated*, 46.
64 Haar, "Legislative Change," 9.
65 Hunter, *Industrial Power*, 1: 156.
66 *Miner v. Gilmour*, [1858] 14 E.R. 861 at 870. Miner, on this analysis, was in no position to insist that the waters be held back to his advantage and to the detriment of Gilmour's downstream gristmill, for Miner's rights derived from Horner: the latter, having sold the gristmill site unconditionally, had abandoned whatever opportunity might once have existed to impose limitations on the water rights of the property, which eventually came into Gilmour's hands.
67 Gertzler, *A History of Water Rights*, 292-94.
68 *Red River Roller Mills v. Wright*, 30 Minn. 249 at 253, 15 N.W. 167 at 169 (1883).
69 Robert Abrams, "Water Allocation by Comprehensive Permit Systems in the East:

Considering a Move away from Orthodoxy," *Virginia Environmental Law Journal* 9 (1990): 257-58, noting variations in the reasonable use rule.
70 Hunter, *Industrial Power*, 1: 141.
71 N. William Hines, "Nor Any Drop to Drink: Public Regulation of Water Quality, Part I: State Pollution Control Programs," *Iowa Law Review* 52 (1966): 196-201.

CHAPTER 2: NAVIGATING AQUATIC PRIORITIES

1 John Baddeley, *The London Angler's Book, or Waltonian Chronicle* (London: G. Parsonage, 1834), cited in Carleton B. Chapman, "The Year of the Great Stink," *Pharos* 35 (July 1972): 96.
2 *Rex v. Medley et al.* 172 E.R. 1246 (K.B. 1834).
3 Ibid.
4 Ibid. See also *Weekly Dispatch*, 10 February 1834; *London Times*, 8 February 1834.
5 *Rex v. Medley et al.*
6 Ibid.
7 Ibid., 1250.
8 Ibid.
9 William Cronon, *Changes in the Land: Indians, Colonists, and the Ecology of New England* (New York: Hill and Wang, 1983), 155-56; Louis C. Hunter, *Industrial Power in the United States, 1780-1930*, 3 vols. (Charlottesville: University of Virginia Press, 1979), 1: 144-46.
10 Henry D. Thoreau, *The Illustrated "A Week on the Concord and Merrimack Rivers,"* ed. Carl F. Hovde, William L. Howarth, and Elizabeth Witherell (Princeton, NJ: Princeton University Press, 1983), 88, quoted in Theodore Steinberg, *Nature Incorporated: Industrialization and the Waters of New England* (Cambridge: Cambridge University Press, 1991), 4.
11 Steinberg, *Nature Incorporated*, 207-10.
12 Margaret Beattie Bogue, *Fishing the Great Lakes: An Environmental History, 1783-1933* (Madison: University of Wisconsin Press, 2000), 17.
13 Ibid., 141.
14 G.H. Benzenberg, "The Sewerage System of Milwaukee and the Milwaukee River Flushing Works," *Transactions of the American Society of Civil Engineers* 30 (December 1893): 373-74.
15 Bogue, *Fishing the Great Lakes*, 143.
16 As Arthur F. McEvoy has observed in an important study of California fisheries, the fishermen's environmental and resource problems evolved through the interaction between ecology, legal process, and harvesting practices, each of which could change independently of that interaction. Arthur F. McEvoy, *The Fisherman's Problem: Ecology and Law in the California Fisheries, 1850-1980* (Cambridge: Cambridge University Press, 1986), 13.
17 Quoted in Derek Fraser, "The Politics of Leeds Water," *Publications of the Thoresby Society*, vol. 53, part 1, no. 116, 50.
18 Charles Milnes Gaskell, "On the Pollution of Our Rivers," *Nineteenth Century*, July 1903, 86.
19 Hussey Vivian, MP, 3 Hansard 164: 770, quoted in Roy M. MacLeod, "Government

and Resource Conservation: The Salmon Acts Administration, 1860-1886," *Journal of British Studies* 7 (1968): 119.
20 William Ffennell, *Fifth Annual Report of the Inspectors of Salmon Fisheries (England and Wales)*, 1866, 4, quoted in MacLeod, "Government and Resource Conservation," 125.
21 MacLeod, "Government and Resource Conservation," 125.
22 Buckland, *Sixth Annual Report of the Inspectors of Salmon Fisheries (England and Wales)*, 1867, 4, quoted in MacLeod, "Government and Resource Conservation," 128.
23 MacLeod, "Government and Resource Conservation," 129.
24 *Salmon Act, 1861* (U.K.), 24 & 25 Vict., c. 109, s. 5. See also J.O. Taylor, *The Law Affecting River Pollution* (Edinburgh: W. Green and Son, 1928), 43, 63-64.
25 MacLeod, "Government and Resource Conservation," 147.
26 Ibid., 147.
27 *Seventeenth Annual Report of the Inspectors of Salmon Fisheries (England and Wales)*, 1878, 53, quoted in MacLeod, "Government and Resource Conservation," 146.
28 Buckland's 1878 address to the Sanitary Institute (delivered at the Parkes Museum of Hygiene, London, 26 February 1885), quoted in Henry Robinson, "Address on River Pollution" 2.
29 See Cronon, *Changes in the Land;* Graeme Wynn, *Timber Colony* (Toronto: University of Toronto Press, 1980), 93.
30 Charlotte Whitton, *A Hundred Years A-Fellin* (Braeside: Gillies Brothers, 1943), 157.
31 Proudfoot V.C. did not deliver a written judgment when he granted the injunction at the trial level.
32 *McLaren v. Caldwell* (1882), 8 S.C.R. 435 at 440 per Chief Justice William J. Ritchie, and at 467 per Justice John W. Gwynne; (1884), 9 App. Cas. 392.
33 J. Linton, *Beneath the Surface: The State of Water in Canada* (Ottawa: Canadian Wildlife Federation, 1997), 9-11.
34 Wynn, *Timber Colony*, 94.
35 Ibid., 94.
36 *Palmer v. Mulligan*, 3 Caines R. 307 (1805).
37 Carol Rose, "Energy and Efficiency in the Realignment of Common Law Water Rights," *Journal of Legal Studies* 19 (1990): 261-96, 283.
38 *Snow v. Parsons*, 28 Vt. 459 (1856).
39 *Hayes v. Waldron*, 44 N.H. 580 (1863). Steinberg, *Nature Incorporated*, 146-47, argues that the flexible rule opened the way for abuse in the interests of the community. For further interpretation, see Robert G. Bone, "Normative Theory and Legal Doctrine in American Nuisance Law: 1850 to 1920," *Southern California Law Review* 59 (1986): 1206-8.
40 See, e.g., *Red River Roller Mills v. Wright*, 30 Minn. 249, 15 N.W. 167 (1883).
41 *Austin v. Snider* (1861), 21 U.C.Q.B. 299; *Mitchell v. Barry* (1867), 26 U.C.Q.B. 416.
42 Canada, *House of Commons Debates* (28 February 1871), 193.
43 Hamilton H. Killaly, *Report on the Commission into the Condition of Navigable Streams* (Ottawa: I.B. Taylor, 1873), 2.
44 *Austin v. Snider* (1861), 21 U.C.Q.B. 299 at 303-4.
45 R. Peter Gillis, "Rivers of Sawdust: The Battle over Industrial Pollution in Canada, 1865-1903," *Journal of Canadian Studies* 21 (1986): 87-88.
46 Hamilton H. Killaly, *Report on the Commission into the Condition of Navigable Streams* (Ottawa: I.B. Taylor, 1873), 8.

47 Ibid., 7.
48 Ibid., appendix 25.
49 Ibid., 12.
50 Ibid., 14, 12.
51 *An Act for the Better Protection of Navigable Streams and Rivers,* S.C. 1873, c. 65, s. 4.
52 John P.S. McLaren, "The Tribulations of Antoine Ratté: A Case Study of the Environmental Regulation of the Canadian Lumbering Industry in the Nineteenth Century," *University of New Brunswick Law Journal* 33 (1984): 221.
53 On the basis of these efforts, Ottawa operators obtained exemptions from the act by Order-in-Council 23 June 1880. Ibid., and Gillis, "Rivers of Sawdust," 90-91.
54 *Ratté v. Booth* (1890), 15 A.C. 188 at 190.
55 Ibid.
56 "Farmer Drowned," *Ottawa Citizen,* 12 November 1897.
57 *Ratté v. Booth* (1885), 10 O.R. 351; (1886), 11 O.R. 491; (1887), 14 O.A.R. 419 (C.A.); (1890), 15 A.C. 188 (P.C.); see also *Globe* (Toronto), 24 March 1885; *An Act Respecting Saw Mills on the Ottawa River,* S.O. 1885, c. 24. Further reference to federal concern with lumber refuse as an interference with navigation may be found in Judith Tulloch, *The Rideau Canal: Defence, Transportation, Recreation* (Ottawa: Parks Canada, 1981), 25.
58 Canadian federal government powers over "Navigation and Shipping" and "Sea Coast and Inland Fisheries" are found in the *Constitution Act, 1867* (U.K.), 30 & 31 Vict., c. 3, ss. 91(10) and 91(12).
59 "Annual Report of the Department of Fisheries, 1889," Canada, *Sessional Papers,* 1890, vol. 23, no. 12, s. 17, quoted in Gilbert Allardyce, "'The Vexed Question of Sawdust': River Pollution in Nineteenth-Century New Brunswick," *Dalhousie Review* 52 (1972): 177-90.
60 Vict., c. 36, s. 7 quoted in Allardyce, "'Vexed Question of Sawdust,'" 123.
61 Quoted in Allardyce, "'Vexed Question of Sawdust,'" 124.
62 Gillis, "Rivers of Sawdust," 94.
63 Canada, *House of Commons Debates* (19 June 1894), 4566.
64 Ontario, Department of Fisheries, *Annual Report,* 1899, 40; *Annual Report,* 1900, 18; *Annual Report,* 1902, 37; *Annual Report,* 1903, 15. Pollution from sawdust, industrial waste, and municipal sewage remained ongoing concerns to fisheries officials.
65 Hiram Robinson, Managing Director, to Sir Charles Tupper, Minister of Marine and Fisheries, 10 August 1894, Hawkesbury Lumber Company Papers, Archives of Ontario.
66 Hiram Robinson to J.R. Booth, 18 September 1901, Hawkesbury Lumber Company Papers, Archives of Ontario.
67 Gillis, "Rivers of Sawdust," 100.
68 *Hunter v. Richards* (1913), O.W.N. 854; for two previous decisions, see (1911) 2 O.W.N. 855, and (1912) 3 O.W.N. 1432 where Clute J. considers the *Navigable Waters Protection Act,* R.S.C. 1906 ch. 115, sec.19.
69 George Perkins Marsh, *Man and Nature,* ed. David Lowenthal (Cambridge, MA: Harvard University Press, 1965), 107-8.
70 *Forest and Stream* 22 (3 July 1884), 441, quoted in Donald J. Pisani, "Fish Culture and the Dawn of Concern over Water Pollution in the United States," *Environmental Review* 8 (1984): 121.
71 Albert E. Cowdrey, "Pioneering Environmental Law: The Army Corps of Engineers and the Refuse Act," *Pacific Historical Review* 44 (1975): 331.

72 John Opie, *Nature's Nation: An Environmental History of the United States* (New York: Harcourt Brace and Company, 1998), 308-9.
73 Circumstances on the Pacific coast differed in many respects, as California and other Western states incorporated and developed alternative water law arrangements. By 1884, the so-called Debris Cases finally began to bring hydraulic mining under control. See Samuel Weil, "Fifty Years of Water Law," *Harvard Law Review* 50 (1936): 254.
74 E.G. Blackford, quoted in Pisani, "Fish Culture," 122. As John H. Fertig explains, "The state is . . . concerned with the domestic food supply. It is the owner of the fish in the streams, and under its police power can refuse to allow the entrance into streams of any matter deleterious to the propagation of fish. The duty of preserving the fish and game from extinction is as clear as its duty to secure for its citizens a supply of wholesome food." John H. Fertig, "The Legal Aspects of the Stream Pollution Problem," *American Journal of Public Health* 16 (1926): 785.
75 U.S. Statutes, 1899, c. 425, ss. 10 and 13; 30 Stat. 1152.
76 William H. Rodgers Jr., "Industrial Water Pollution and the Refuse Act: A Second Chance for Water Quality," *University of Pennsylvania Law Review* 119 (April 1971): 766.
77 On the *Refuse Act*, see James M. Fallows, *The Water Lords* (New York: Grossman, 1971), 204-16; and Ray M. Druley, "The Refuse Act of 1899," 3 *BNA Environ. Rep.* 11 (1972) 28.
78 U.S. Statutes, 1899, c. 425, s. 13; 30 Stat. 1152.
79 Rodgers, "Industrial Water Pollution and the Refuse Act," 777-78.
80 *United States v. Republic Steel*, 362 U.S. 482 at 506n26 (1960). The exemption originated in 1894 and was retained in the 1899 legislation. It was the opinion of the judge advocate general in 1907 that "the control of waters for drinking and sanitary purposes, and the regulation of the flow and of the deposit of sewage . . . are fully within state control as an incident of their police power." Exempted from state control were structures that would affect navigation. US War Department, Office of the Judge Advocate General, *A Digest of the Opinions of the Judge Advocates General of the Army* (Washington, DC: Government Printing Office, 1912), 756.
81 Quote and summary from Paul Johnson, *The Birth of the Modern, World Society, 1815-1830* (London: Weidenfeld and Nicolson, 1991), 755.
82 *Blundell v. Catterall* (1821), 5 B. & Ald. 315, 106 E.R. 1190.
83 Ibid., 1207, 1206.
84 Ibid., 1197.
85 Ibid., 1196.
86 Ibid., 1204.
87 Ibid., 1203.
88 Ibid., 1204, 1205.
89 Ibid., 1202 per Holroyd J.
90 Ibid., 1193.
91 Lawrence Wright, *Clean and Decent: The Fascinating History of the Bathroom and the Water Closet* (Toronto: University of Toronto Press, 1967), 95.
92 *Blundell v. Catterall* (1821), 1194.
93 Ibid., 1195.
94 Ibid., 1194.
95 Ibid., 1205.

96 Ibid., 1204.
97 Ibid., 1197.
98 Ibid., 1207.
99 Leonard R. Jaffee, "State Citizen Rights Respecting Greatwater Resource Allocation: From Rome to New Jersey," *Rutgers Law Review* 25 (1971): 612-13.

CHAPTER 3: A SOURCE OF CIVIC PRIDE

1 Joel A. Tarr et al., "Water and Wastes: A Retrospective Assessment of Wastewater Technology in the United States, 1800-1932," *Technology and Culture* 25 (1984): 230.
2 J.A. Hassan, "The Growth and Impact of the British Water Industry in the Nineteenth Century," *Economic History Review*, 2nd ser., 38 (1985): 532.
3 Christopher Hibbert, *London: The Biography of a City* (Harmondsworth: Penguin, 1980), 47-48.
4 John Graham-Leigh, *London's Water Wars: The Competition for London's Water Supply in the Nineteenth Century* (London: Francis Boutle, 2000), 9.
5 Bernard Rudden, *The New River: A Legal History* (Oxford: Clarendon Press, 1985), 32.
6 Ibid., 11-12.
7 Ibid., 50.
8 Ibid.
9 Graham-Leigh, *London's Water Wars*, 11.
10 Ibid., 36.
11 Quoted in Thomas C. Keefer, *Report on a Water Supply for the City of Toronto* (Toronto, 1857), 17.
12 Graham-Leigh, *London's Water Wars*, 15.
13 Keefer, *Report on a Water Supply for Toronto*, 14.
14 John D. Bell, "Report on the Importance and Economy of Sanitary Measures to Cities," in *Proceedings and Debates of the Third National Quarantine and Sanitary Convention* (New York: Edmund Jones, 1859), 576-77, cited in Tarr et al., "Water and Wastes," 228.
15 Graham-Leigh, *London's Water Wars*, 13-14.
16 Rudden, *The New River*, 138.
17 [John Wright], *The Dolphin*, 1827, quoted in Richard Trench and Ellis Hillman, *London under London: A Subterranean Guide* (London: John Murray, 1993), 88; Christopher Hamlin, *A Science of Impurity: Water Analysis in Nineteenth Century Britain* (Bristol: Adam Hilger, 1990), 81, identifies the author as John Wright and explains his outrage at the Grand Junction operation.
18 Keefer, *Report on a Water Supply for Toronto*, 18; Hassan, "Growth and Impact," 534; Christopher Hamlin, "Muddling in Bumbledom: On the Enormity of Large Sanitary Improvements in Four British Towns, 1855-1885," *Victorian Studies* 32 (1988): 55-83. For a case study, see Michael Taggart, *Private Property and Abuse of Rights in Victorian England: The Story of Edward Pickles and the Bradford Water Supply* (Oxford: Oxford University Press, 2002), 14-20.
19 The operational assets of the private water companies were transferred to the London Metropolis Water Board in 1904.
20 Graham-Leigh, *London's Water Wars*, 98.

21 Bill Luckin, *Pollution and Control: A Social History of the Thames in the Nineteenth Century* (Bristol: Adam Hilger, 1986), 38-39. The Chelsea Company had introduced filtration in the wake of a royal commission established in 1828 in response to Wright's denunciation of the Dolphin facility. Hamlin, *A Science of Impurity*, 95.
22 Rudden, *The New River*, 157.
23 Graham-Leigh, *London's Water Wars*, 99-100.
24 The *Metropolis Water Act*, 1852, 15 & 16 Vict., c. 84., lists East London Water Works, Southwark and Vauxhall Water Company, West Middlesex Waterworks Company, Grand Junction Water Works Company, Kent Waterworks, and Hampstead Waterworks Company, in addition to the New River, Chelsea, and Lambeth companies.
25 Nelson M. Blake, *Water for the Cities: A History of the Urban Water Supply Problem in the United States* (Syracuse, NY: Syracuse University Press, 1956), 4.
26 Ibid., 24-35.
27 Iron pipes had been available as early as 1746, but were not used extensively before 1817, when legislation in England required iron mains in ten years. Lewis Mumford, *The City in History: Its Origins, Its Transformations, and Its Prospects* (New York: Harcourt Brace and World, 1961). The advent of extruded lead tubes, used in England after 1820, and the mass production after midcentury of cast-iron, wrought-iron, and stoneware pipes encouraged consistency in plumbing. Ellen Lupton and J. Abbot Miller, *The Bathroom, the Kitchen, and the Aesthetics of Waste: A Process of Elimination* (Cambridge, MA: MIT List Visual Arts Center, 1992), 22.
28 Blake, *Water for the Cities*, 82-83.
29 Ibid., 44-62; Edwin G. Burrows and Mike Wallace, *Gotham: A History of New York City to 1898* (New York: Oxford University Press, 1998), 360-62.
30 Quoted in Blake, *Water for the Cities*, 51.
31 Ibid., 100-5.
32 Ibid., 110.
33 Ibid., 114-15.
34 Ibid., 120.
35 Burrows and Wallace, *Gotham*, 589.
36 Allen Hazen, *Clean Water and How to Get It* (New York: John Wiley and Sons, 1907), 2.
37 W.H.G. Armytage, *A Social History of Engineering* (Cambridge, MA: MIT Press, 1961), 434.
38 Hazen, *Clean Water and How to Get It*, 1.
39 Blake, *Water for the Cities*, 164-65.
40 Ibid., 167-68. See also Richard L. Bushman and Claudia L. Bushman, "The Early History of Cleanliness in America," *Journal of American History* 74 (1988): 1213-38.
41 Ann Durkin Keating, "Many Systems: Water Provision in Nineteenth Century Chicago," in Howard Rosen and Ann Durkin Keating, eds., *Water and the City: The Next Century* (Chicago: Public Works Historical Society, 1991), 91.
42 See, for example, on Boston, Fern L. Nesson, *Great Waters: A History of Boston's Water Supply* (Hanover and London: University Press of New England, 1983). Ellis L. Armstrong, ed., *History of Public Works in the United States, 1776-1976* (Chicago: American Public Works Association, 1976) presents a valuable survey.
43 Maureen Ogle, "Redefining 'Public' Water Supplies, 1870-1890: A Study of Three Iowa Cities," *Annals of Iowa* 50 (Spring 1990): 507-30.

44 *Report of the Water Committee Submitting the Reports of the Engineers of the New Water Works of Montreal* (Montreal, 1854), 3-5.
45 John Irwin Cooper, *Montreal: A Brief History* (Kingston: McGill-Queen's University Press, 1969), 27.
46 T.C. Keefer, *Report on a Preliminary Survey for the Water Supply of the City of Montreal*, (Montreal, 1852), 6.
47 Keating, "Many Systems," 92.
48 W. Atherton, *Montreal*, 2 vols. (reprinted University Microfilms International, 1982), 2: 409.
49 Chris Armstrong and H.V. Nelles, *Monopoly's Moment: The Organization and Regulation of Canadian Utilities* (Philadelphia: Temple University Press, 1986; paperback, Toronto: University of Toronto Press, 1988), 19.
50 Elwood Jones and Douglas McCalla, "Toronto Waterworks, 1840-77: Continuity and Change in Nineteenth Century Toronto Politics," *Canadian Historical Review* 60 (1979): 302-3.
51 Furniss on two occasions sold his Toronto operations to other private interests who were unable to fulfill their mortgage obligations to him, with the result that the facilities returned to his ownership and control.
52 Jones and McCalla, "Toronto Waterworks," 304-5.
53 *Proceedings of the Standing Committee on Fire, Water and Gas of the City of Toronto in Connexion with the Supply of Water to the City* (Toronto, 1854), 49.
54 Ibid.
55 Ibid., 14.
56 Jones and McCalla, "Toronto Waterworks," 311.
57 *Toronto Globe*, 24 May 1872, quoted in Jones and McCalla, "Toronto Waterworks," 316.
58 *Toronto Water Works Act*, S.O. 1872, c. 79, Preamble.
59 Jones and McCalla, "Toronto Waterworks," 320.
60 Hassan, "Growth and Impact," 534.
61 Joel A. Tarr, "The Separate vs. Combined Sewer Problem: A Case Study in Urban Technology Design Choice," *Journal of Urban History* 5 (1979): 310; Earl Finbar Murphy, *Water Purity: A Study in Legal Control of Natural Resources* (Madison: University of Wisconsin Press, 1961), 27.
62 Stuart Galishoff, "Triumph and Failure: The American Response to the Urban Water Supply Problem, 1860-1923," in Martin V. Melosi, ed., *Pollution and Reform in American Cities, 1870-1930* (Austin and London: University of Texas Press, 1980), 51.
63 Galishoff, "Triumph and Failure."
64 Leo G. Denis, *Water Works and Sewerage Systems* (Ottawa: Commission of Conservation, 1916), 136-37.
65 Maureen Ogle, "Water Supply, Waste Disposal and the Culture of Privatism in the Mid-Nineteenth-Century American City," *Journal of Urban History* 25 (1999): 328.
66 Henry Bixby Hemenway, *Legal Principles of Public Health Administration* (Chicago: T.H. Flood, 1914), 687-88.
67 Ibid., 689.
68 Armstrong and Nelles, *Monopoly's Moment*, 32.
69 Murphy, *Water Purity*, 24.
70 See Martin V. Melosi, *The Sanitary City: Urban Infrastructure in America from Colonial*

Times to the Present. Creating the North American Landscape (Baltimore: Johns Hopkins University Press, 2000), 119-26.
71 M.N. Baker, *The Quest for Pure Water*, 2nd ed. (New York: American Waterworks Association, 1981); Hamlin, *A Science of Impurity.*
72 Rudden, *The New River*, 131-32; Hamlin, *A Science of Impurity*, 120; Graham-Leigh, *London's Water Wars;* Letty Anderson, "Hard Choices: Supplying Water to New England Towns," *Journal of Interdisciplinary History* 15 (1984): 221-23.
73 H.V. Nelles, Introduction to T.C. Keefer, *Philosophy of Railroads and Other Essays* (Toronto: University of Toronto Press, 1972), xxv.
74 Ibid.
75 Keefer, *Philosophy of Railroads* (Montreal and Toronto, 1850), quoted in Nelles, Introduction to Keefer, *Philosophy*, xxv.
76 *Proceedings of the Standing Committee*, 71.
77 T.C. Keefer, *Report on a Preliminary Survey for the Water Supply of the City of Montreal* (Montreal, 1852), 4.
78 T.C. Keefer, *Report on a Preliminary Survey for the Water Supply of the City of Montreal* (Montreal, 1854), 37.
79 Josiah Quincy, quoted in Blake, *Water for the Cities*, 175.
80 Quoted in Blake, *Water for the Cities*, 144.
81 Keefer, *Water Supply of the City of Montreal* (1852), 6.
82 Ibid., 9.
83 *Proceedings of the Standing Committee*, 43.
84 Keefer, *Water Supply of the City of Montreal* (1854), 51-52.
85 Keefer, *Water Supply of the City of Montreal* (1852), 21.
86 Tarr et al., "Water and Wastes," 231; Lupton and Miller, *Aesthetics of Waste*, 22; Donald J. Pisani, "Fish Culture and the Dawn of Concern over Water Pollution in the United States," *Environmental Review* 8 (1984): 124.
87 James Mansergh, *The Water Supply of the City of Toronto, Canada* (Westminster, 1896). As a general rule, throughout this book consumption levels have simply been reported as they appear in original sources, and no attempt has been made to convert or standardize references to imperial (4.5 litres per gallon) and American (3.8 litres per gallon) water volume measures.
88 Allen Hazen, *Clean Water and How to Get It*, 2nd ed. (New York: John Wiley and Sons, 1914), 151.
89 Hassan, "Growth and Impact," 538. F.W. Robins, *The Story of Water Supply* (Oxford: Oxford University Press, 1946), 198, contains details on rates and distribution levels for eight London water supply companies in 1843.
90 Hassan, "Growth and Impact," 539.
91 Jean-Pierre Goubert, *The Conquest of Water: The Advent of Health in the Industrial Age*, trans. Andrew Wilson (Cambridge: Polity Press, 1989), 51, 112.
92 Ibid., 196-97.
93 Hazen, *Clean Water and How to Get It*, 152.
94 Mansergh, *Water Supply*, 25.
95 Ibid., 26.
96 Denis, *Water Works*, 135.
97 Keefer, *Water Supply of the City of Montreal* (1852), 6.
98 Ibid.

99 Benjamin Henry Latrobe, quoted in Blake, *Water for the Cities,* 26.
100 Keefer, *Water Supply of the City of Montreal* (1852), 9-10.
101 John C. Weaver and Peter de Lottinville, "The Conflagration and the City: Disaster and Progress in British North America during the Nineteenth Century," *Social History* 13 (1980): 447; Burrows and Wallace, *Gotham,* 594-95.
102 Charles M. Holt, *Insurance Law of Canada* (Montreal: C. Theoret, 1898), 6.
103 (U.K.), 6 & 7 Vict., c. 109.
104 (U.K.), 6 & 7 Vict., c. 75.
105 Keefer, *Report on a Water Supply for Toronto,* 15.
106 *Proceedings of the Standing Committee,* 55. The figures are reported in pounds as originally recorded because Canada did not adopt the decimal system until the late 1850s.
107 Ibid., 56.
108 Anderson, "Hard Choices," 217.

CHAPTER 4: THE WATER CLOSET REVOLUTION

1 Stephen Halliday, *The Great Stink of London: Sir Joseph Bazalgette and the Cleansing of the Victorian Metropolis* (Stroud, Gloucestershire: Sutton Publishing, 1999); Roy Palmer, *The Water Closet: A New History* (Newton Abbot: David and Charles, 1973).
2 May Stone, "The Plumbing Paradox: American Attitudes toward Late Nineteenth-Century Domestic Sanitary Arrangements," *Winterthur Portfolio* 14 (1979): 284-85.
3 Joel A. Tarr et al., "Water and Wastes: A Retrospective Assessment of Wastewater Technology in the United States, 1800-1932," *Technology and Culture* 25 (1984): 231.
4 Lucinda Lambton, *Temples of Convenience* (New York: St. Martin's Press, 1978), 8.
5 John Graham-Leigh, *London's Water Wars: The Competition for London's Water Supply in the Nineteenth Century* (London: Francis Boutle, 2000), 20; Ian McNeill, *Joseph Bramah: A Century of Invention* (New York: Augustus M. Kelley, 1968), 26-30.
6 Janet Roebuck, *Urban Development in 19th-Century London: Lambeth, Battersea and Wandsworth 1838-1888* (London and Chichester: Phillimore, 1979), 161-62.
7 Lawrence Wright, *Clean and Decent: The Fascinating History of the Bathroom and the Water Closet* (Toronto: University of Toronto Press, 1967), 148.
8 Carleton B. Chapman, "The Year of the Great Stink," *Pharos* 35 (July 1972): 92.
9 Lewis Mumford, *The City in History: Its Origins, Its Transformations, and Its Prospects* (New York: Harcourt Brace and World, 1961), 462;
10 Anthony S. Wohl, *Endangered Lives: Public Health in Victorian Britain* (London: J.M. Dent, 1983), 107-9; Jean-Pierre Goubert, *The Conquest of Water: The Advent of Health in the Industrial Age,* trans. Andrew Wilson (Cambridge: Polity Press, 1989), 96.
11 Maureen Ogle, *All the Modern Conveniences: American Household Plumbing, 1840-1890* (Baltimore and London: Johns Hopkins University Press, 1996), 61.
12 Palmer, *The Water Closet,* 39.
13 Ogle, *All the Modern Conveniences,* 71-83.
14 Richard L. Bushman and Claudia L. Bushman, "The Early History of Cleanliness in America," *Journal of American History* 74 (1988): 1225.
15 Ellen Lupton and J. Abbot Miller, *The Bathroom, the Kitchen, and the Aesthetics of Waste: A Process of Elimination* (Cambridge, MA: MIT List Visual Arts Center, 1992), 22.
16 Chapman, "The Year of the Great Stink," 93.

17 Joel A. Tarr, "The Separate vs. Combined Sewer Problem: A Case Study in Urban Technology Design Choice," *Journal of Urban History* 5 (1979): 336n6.
18 Ogle, *All the Modern Conveniences*, 4.
19 Tarr, "The Separate vs. Combined Sewer Problem," 336n6.
20 Ogle, *All the Modern Conveniences*, 17.
21 Ibid., 35; see also H. Wayne Morgan, "America's First Environmental Challenge, 1865-1920," in Margaret Francine Morris, ed., *Essays on the Gilded Age* (Austin and London: University of Texas Press, 1973), 87.
22 Ogle, *All the Modern Conveniences*, 15, 92.
23 Bushman and Bushman, "The Early History of Cleanliness," 1225.
24 See Ogle, *All the Modern Conveniences*, 52-60, concerning local ordinances.
25 Stat. (U.K.), 23 Hen. VIII., c. 5 (1532-33), quoted in Edith G. Henderson, *Foundations of English Administrative Law* (Cambridge, MA: Harvard University Press, 1963), 28.
26 Henry Jephson, *The Sanitary Evolution of London* (London: T. Fisher Unwin, 1907), 16.
27 Leslie B. Wood, *The Restoration of the Tidal Thames* (Bristol: Adam Hilger, 1982), 27. The districts overseen by the commissioners were the City of London, Westminster and part of Middlesex, Holborn and Finsbury, Tower Hamlets, St. Katherine, Poplar and Stockwell, Greenwich, and Surrey and Kent.
28 Roebuck, *Urban Development*, 42.
29 Palmer, *The Water Closet;* Halliday, *The Great Stink of London*.
30 J. Toulmin Smith, *Metropolis Local Management Act, 1855* (London, 1857), 1.
31 *Metropolis Local Management Act, 1855* (U.K.), 18 & 19 Vict., c. 120, s. 81.
32 Wright, *Clean and Decent*, 94.
33 Bernard Rudden, *The New River: A Legal History* (Oxford: Clarendon Press, 1985), 155.
34 Graham-Leigh, *London's Water Wars*, 79.
35 Roebuck, *Urban Development*, 162.
36 Battersea, Clapham, Putney, Wandsworth, Streatham, and Tooting.
37 *Tinkler v. Board of Works for the Wandsworth District* (1857-58), 27 Law J. Ch. 342 at 346.
38 Ibid., 349.
39 Wandsworth was certainly not alone in suffering this particular neglect at the hands of the building trades. See the *Builder*, "Proceedings under Metropolitan Building Act," 22 February 1862, 138; "Cases under the Building Act," 9 August 1862, 574; "Cases under the Metropolitan Building Act," 20 December 1862, 917.
40 Wandsworth District Board of Works, *Fifth Annual Report* (1861), 14.
41 "Demolition of a House: Metropolitan Management Act," *Builder*, 1 November 1862, 794.
42 Wandsworth District Board of Works, *Seventh Annual Report* (1863), 15.
43 The initial reports of the case appear in the *Times* for 18 and 22 April 1863 at pp. 13 and 11, respectively.
44 "Demolition of a House," 794.
45 *Metropolis Local Management Act, 1855*, s. 76.
46 *Times*, 22 April 1863, 11.
47 *Cooper v. Board of Works for the Wandsworth District* (1863), 143 E.R. 414.
48 Wandsworth District Board of Works, Minutes, 22 October 1862, Wandsworth Borough Council, Battersea Library, London.
49 A Select Committee in 1866 recommended abolition of the water companies, in large

part because of problems associated with the transition from intermittent to constant supply. Consumers and ratepayers demonstrated considerable reluctance to undertake the plumbing changes necessitated by the shift in supply arrangements, and it was thought that only a public body could appropriately exercise the powers required to compel London residents to put and maintain their domestic plumbing in proper order to receive a constant flow of water. Graham-Leigh, *London's Water Wars*, 112.
50 Wandsworth Board of Works, *Seventh Annual Report* (1863), 15.
51 The London authority is s. 44 of the *Public Health (London) Act*, amending 18 & 19 Vict., c. 120, s. 88; see also *Public Health Act, 1875*, ss. 39 and 45; and *Public Health Acts Amendment Act, 1907*, s. 47; Alexander MacMorran and E.J. Naldrett, *The Public Health (London) Act, 1891*, 2nd ed. (London: Butterworth, 1910), 101.
52 Michael Leapman, *The World for a Shilling* (London: Headline Books, 2001), 93.
53 Ibid., 94-95; Wright, *Clean and Decent*, 200; Lambton, *Temples of Convenience*, 10.
54 *Biddulph v. Vestry of St. George, Hanover Square* (1863), 33 L.J. Ch. 411, 8 L.T. (N.S.) 558.
55 MacMorran and Naldrett, *The Public Health (London) Act, 1891*, 101-3. Gabriel Chevallier's novel *Clochemerle*, a charming account of the installation of a urinal in a conservative French village, exemplifies the sensitivities engaged.
56 Ross Cranston, *Legal Foundations of the Welfare State* (London: Weidenfeld and Nicolson, 1985), 144.
57 See cases in MacMorran and Naldrett, *The Public Health (London) Act, 1891*, 87, note (i).
58 *Wood v. Widnes Corporation*, [1898] 1 Q.B. 463, 67 L.J.Q.B. 254, 77 L.T. 779.
59 *St. John's, Hackney (Vestry of) v. Hutton*, [1897] 1 Q.B. 210, 66 L.J.Q.B. 74.
60 Henry Bixby Hemenway, *Legal Principles of Public Health Administration* (Chicago: T.H. Flood, 1914), 697, citing *Harrington v. Providence*, 20 R.I. 223.
61 F. Garvin Davenport, "The Sanitation Revolution in Illinois, 1870-1900," *Journal of the Illinois State Historical Society* 66 (1973): 307-8.
62 Nancy Tomes, *The Gospel of Germs: Men, Women and the Microbe in American Life* (Cambridge: Harvard University Press, 1998), 185.
63 Ogle, *All the Modern Conveniences*, 27.
64 Suellen Hoy, *Chasing Dirt: The American Pursuit of Cleanliness* (New York and Oxford: Oxford University Press, 1995), 65.
65 Ogle, *All the Modern Conveniences*, 124.
66 *Public Health (London) Act, 1891* (U.K.), c. 76, s. 45.
67 Although the policy-making enthusiast might hail the introduction of financial incentives such as this user fee as a desirable attempt to manage demand for environmental amenities, a moment's sympathetic reflection will suggest that in these matters the demand side (to borrow the economist's term) is inelastic – sometimes excruciatingly so. In France it had been a long-standing practice to impose charges for the use of latrines, although by 1849 the French were entertaining the idea of free public facilities for both men and women, at least throughout Paris. Goubert, *The Conquest of Water*, 94; Alain Corbin, *The Foul and the Fragrant: Odor and the French Social Imagination* (Cambridge, MA: Harvard University Press, 1986), 117.
68 Ogle, *All the Modern Conveniences*, 128-43.
69 Ibid., 145.
70 Boston Medical Commission, *Sanitary Condition of Boston* (1875, repr., New York: Arno Press, 1977), 173-74.
71 *Encyclopaedia Britannica*, 11th ed., s.v. "Sewerage," 738.

72 Palmer, *The Water Closet*, 96-97.
73 "The Sewers and Sewage Farms of Berlin," *Engineering News and American Railway Journal*, 27 August 1896, 139-41.
74 Stone, "The Plumbing Paradox," 294.
75 Lupton and Miller, *Aesthetics of Waste*, 23.
76 *Encyclopaedia Britannica*, 11th ed., s.v. "Sewerage," 740.
77 William Spinks, *The Drainage of Villages* (London: Sanitary Publishing, 1895), 19-20.
78 Charles V. Chapin, *Municipal Sanitation in the United States* (Providence, RI: Snow and Farnham, 1901), 115.
79 Edwin G. Burrows and Mike Wallace, *Gotham: A History of New York City to 1898* (New York: Oxford University Press, 1998), 1194-96. Martin V. Melosi, "Pragmatic Environmentalist: Sanitary Engineer George E. Waring Jr.," in *Essays in Public Works History* (Washington, DC: Public Works Historical Society, 1977), 4: 12-18; Morgan, "America's First Environmental Challenge," 93-94.
80 Chapin, *Municipal Sanitation*, 132-33.
81 Hoy, *Chasing Dirt*, 105.
82 Ontario Provincial Board of Health (hereafter Ontario PBH), *Annual Report, 1884*, 22.
83 Ontario PBH, *Annual Report, 1889*, 23, 25.
84 Ontario PBH, *Annual Report, 1893*, 100e. It is assumed that complaints not otherwise accounted for remained outstanding as the year ended.
85 Ontario PBH, *Annual Report, 1894*, 115-16.
86 Heather A. MacDougall, "The Genesis of Public Health Reform in Toronto, 1869-1890," *Urban History Review* 10, 3 (1982): 6. Similar measures were implemented in other communities such as Kingston, Ontario.
87 Ontario PBH, *Annual Report, 1891*, 101.
88 Ibid., 102, 103, 105.
89 Tarr, "The Separate vs. Combined Sewer Problem," 309; H.H. Stanbridge, *History of Sewage Treatment in Britain* (Maidstone, Kent: Institute of Water Pollution Control, 1976), 1: 43.
90 Tarr et al., "Water and Wastes," 234. By 1899, over two hundred municipalities had purchased private water companies in the United States. Letty Anderson, "Hard Choices: Supplying Water to New England Towns," *Journal of Interdisciplinary History* 15 (1984): 221.
91 Jephson, *The Sanitary Evolution of London*, 96-97. The *Plumber and Sanitary Engineer* (later the *Sanitary Engineer* and eventually the *Engineering Record*) began publication in 1877. Stone, "The Plumbing Paradox," 291. Four years later, in 1881, the New York City Board of Health began to supervise plumbers and to require registration. Ibid., 295.

CHAPTER 5: MUNICIPAL EVACUATION

1 Joel A. Tarr, "The Separate vs. Combined Sewer Problem: A Case Study in Urban Technology Design Choice," *Journal of Urban History* 5 (1979): 310.
2 Henry Bixby Hemenway, *The Legal Principles of Public Health Administration* (Chicago: T.H. Flood, 1914), 681.
3 Martin V. Melosi, *The Sanitary City: Urban Infrastructure in America from Colonial*

Times to the Present. Creating the North American Landscape (Baltimore: Johns Hopkins University Press, 2000), 91.
4 Edinburgh, Commissioners of Police, Papers Relating to the Noxious Effects of the Fetid Irrigation around the City of Edinburgh (Edinburgh, 1839), 8, quoted in P.J. Smith, "The Foul Burns of Edinburgh: Public Health Attitudes and Environmental Change," *Scottish Geographical Magazine* 91 (1975): 28.
5 Melosi, *The Sanitary City*, 47; on the linkage of drainage to miasmas, see Maureen Ogle, "Water Supply, Waste Disposal, and the Culture of Privatism in the Mid-Nineteenth-Century American City," *Journal of Urban History* 25 (1999): 335.
6 Stanley K. Schultz, *Constructing Urban Culture: American Cities and Urban Planning, 1800-1920* (Philadelphia: Temple University Press, 1989), 125-27.
7 Christopher Hamlin, *A Science of Impurity: Water Analysis in Nineteenth Century Britain* (Bristol: Adam Hilger, 1990), 74.
8 John P.S. McLaren, "Nuisance Law and the Industrial Revolution – Some Lessons from Social History," *Oxford Journal of Legal Studies* 3 (1983): 190-94.
9 William Lambe, *An Investigation of the Properties of Thames Water* (London: Butcher and Underwood, 1828), 52, quoted in Hamlin, *A Science of Impurity*, 73.
10 Hamlin, *A Science of Impurity*, 83; Bill Luckin, *Pollution and Control: A Social History of the Thames in the Nineteenth Century* (Bristol: Adam Hilger, 1986), 38-39; Jean-Pierre Goubert, *The Conquest of Water: The Advent of Health in the Industrial Age*, trans. Andrew Wilson (Cambridge: Polity Press, 1989), 110.
11 *Rees's Cyclopedia*, v. 38, s.v. "Water," quoted in Hamlin, *A Science of Impurity*, 78-79.
12 Thomas Spencer, *Report on the Quality of the New River Company's Water* (London, 1855), 53.
13 Hamlin, *A Science of Impurity*, 100.
14 Ibid., 108-9.
15 Ibid., 110-16.
16 Ibid., 104.
17 See ibid., 127-29.
18 *Proceedings of the Standing Committee on Fire, Water and Gas of the City of Toronto in Connexion with the Supply of Water to the City* (Toronto, 1854), 21.
19 Henry Croft, "The Mineral Springs of Canada," *Canadian Journal* 1 (1853): 154.
20 *Proceedings of the Standing Committee*, 21.
21 Ibid., 48.
22 Chris Armstrong and H.V. Nelles, *Monopoly's Moment: The Organization and Regulation of Canadian Utilities* (Philadelphia: Temple University Press, 1986; paperback, Toronto: University of Toronto Press, 1988), 31.
23 *Proceedings of the Standing Committee*, 19.
24 Reverend Charles Kingsley, *The Water Babies*, quoted in Lucinda Lambton, *Temples of Convenience* (New York: St. Martin's Press, 1978), 10.
25 Schultz, *Constructing Urban Culture*, 114.
26 Melosi, *The Sanitary City*, 44-52.
27 Earl Finbar Murphy, *Water Purity: A Study in Legal Control of Natural Resources* (Madison: University of Wisconsin Press, 1961), 33; Stephen Halliday, *The Great Stink of London: Sir Joseph Bazalgette and the Cleansing of the Victorian Metropolis* (Stroud, Gloucestershire: Sutton Publishing, 1999), 39.

28 Melosi, *The Sanitary City*, 48.
29 W.R. Cornish and G. de N. Clark, *Law and Society in England 1750-1950* (London: Sweet and Maxwell, 1989), 160.
30 The death rate plummeted from fifty-five per thousand in 1851 under the former system to thirteen per thousand in 1853 with the introduction of new arrangements. Murphy, *Water Purity*, 33.
31 Leslie B. Wood, *The Restoration of the Tidal Thames* (Bristol: Adam Hilger, 1982), 21; Murphy, *Water Purity*, 33.
32 Francis Sheppard, *London 1808-1870: The Infernal Wen* (London: 1971), 101.
33 Anthony S. Wohl, *Endangered Lives: Public Health in Victorian Britain* (London: J.M. Dent, 1983), 234.
34 Halliday, *The Great Stink of London*, 92.
35 Henry Mayhew, quoted in Lawrence Wright, *Clean and Decent: The Fascinating History of the Bathroom and the Water Closet* (Toronto: University of Toronto Press, 1967), 154. The atmosphere is effectively re-created in Clare Clark's recent work *The Great Stink: A Novel of Corruption and Murder beneath the Streets of Victorian London* (Orlando, FL: Harcourt Books, 2005).
36 J.M. Chapman and Brian Chapman, *The Life and Times of Baron Haussmann: Paris in the Second Empire* (London: Weidenfeld and Nicolson, 1957), 121.
37 Felix Barker and Peter Jackson, *London: 2000 Years of a City and Its People* (London: Castle and Company, 1974), 279.
38 Quoted in Chapman and Chapman, *Baron Haussmann*, 121.
39 Halliday, *The Great Stink of London*.
40 Wood, *The Restoration of the Tidal Thames*, 32.
41 Wohl, *Endangered Lives*, 107.
42 For construction details, including the embankment, see Stephen Inwood, *A History of London* (New York: Carroll and Graf, 1998), 434; Roy Porter, *London: A Social History* (Cambridge, MA: Harvard University Press, 1995), 164. Goubert, *The Conquest of Water*, 61-62, provides further descriptive detail.
43 *AG v. Metropolitan Board of Works* (1863), 1 H. & M. 298; 298, 71 E.R. 131 at 138.
44 Ibid.
45 Halliday, *The Great Stink*, 1-5.
46 Carleton B. Chapman, "The Year of the Great Stink," *Pharos* 35 (July 1972): 97.
47 Wohl, *Endangered Lives*, 234.
48 *Public Health Act, 1848* (U.K.), 11 & 12 Vict., c. 63.
49 Richard J. Evans, *Death in Hamburg: Society and Politics in the Cholera Years, 1830-1910* (Oxford: Clarendon Press, 1987), 133-34.
50 Alain Corbin, *The Foul and the Fragrant: Odor and the French Social Imagination* (Cambridge, MA: Harvard University Press, 1986), 225.
51 Ogle, "Water Supply, Waste Disposal," 336.
52 Edwin G. Burrows and Mike Wallace, *Gotham: A History of New York City to 1898* (New York: Oxford University Press, 1998), 588-89.
53 John H. Griscom, *The Uses and Abuses of Air: Showing Its Influence in Sustaining Life and Producing Disease*, 3rd ed. (New York: Redfield, 1854), 183, quoted in May Stone, "The Plumbing Paradox: American Attitudes toward Late Nineteenth-Century Domestic Sanitary Arrangements," *Winterthur Portfolio* 14 (1979): 292. See also Suellen Hoy, *Chasing Dirt: The American Pursuit of Cleanliness* (New York and Oxford: Oxford University Press, 1995), 26-27.

54 Ogle, "Water Supply, Waste Disposal," 325. See also H. Wayne Morgan, "America's First Environmental Challenge, 1865-1920," in Margaret Francine Morris, ed., *Essays on the Gilded Age* (Austin and London: University of Texas Press, 1973), 87.
55 *Journal of the American Society of Civil Engineers* (15 November 1889): 162, quoted in Louis P. Cain, "Raising and Watering a City," *Technology and Culture* 13 (1972): 363.
56 Joanne Abel Goldman, *Building New York's Sewers: Developing Mechanisms of Urban Management* (West Lafayette, IN: Purdue University Press, 1997), 70-75.
57 Boston Medical Commission, *Sanitary Condition of Boston* (1875, repr., New York: Arno Press, 1977), 171, 172.
58 Murphy, *Water Purity*, 60.
59 Before the introduction of reinforced concrete around 1905, vitrified clay pipe predominated in small-diameter sewers, with larger conduits constructed from brick. In 1909, 18,361 miles of combined sewers transporting human wastes and stormwater in the same pipe constituted 74 percent of the total sewerage network; an additional 8,361 miles of separate sanitary sewers represented 21 percent of the system, and 1,352 miles of storm sewers constituted the remainder. Joel A. Tarr et al., "Water and Wastes: A Retrospective Assessment of Wastewater Technology in the United States, 1800-1932," *Technology and Culture* 25 (1984): 237.
60 Winslow and Phelps, *Purification of Boston Sewage*, 96.
61 Ibid., 11. In the case of Boston, the resulting overall flow of 46 billion gallons of sewage per year contained 1.5 million kilograms of nitrogen in the form of free ammonia alone.
62 Wohl, *Endangered Lives*, 107; see, overall, Wohl, *Endangered Lives*, 80-116. See also Christopher Hamlin, "Muddling in Bumbledom: On the Enormity of Large Sanitary Improvements in Four British Towns, 1855-1885," *Victorian Studies* 32 (1988): 55-83; and H.H. Stanbridge, *History of Sewage Treatment in Britain* (Maidstone, Kent: Institute of Water Pollution Control, 1976), 1: 22, 34.
63 Winslow and Phelps, *Purification of Boston Sewage*, 11. Around the turn of the century, Germany was regarded as being ahead of much of Europe in terms of sewerage, but still lagging behind England. Sewerage connections in Berlin were not widespread until the 1880s, and much of the Prussian countryside continued to rely on privy middens and cesspits into the twentieth century. Evans, *Death in Hamburg*, 132-33; "Recent Tendencies in Sewage Disposal Practice," *Canadian Engineer* 15 (2 October 1908): 700.
64 W.R. Nichols and George Derby, "Sewerage; Sewage; The Pollution of Streams; The Water-Supply of Towns," in Massachussetts State Board of Health, *Fourth Annual Report* (1873), 40, quoted in Theodore Steinberg, *Nature Incorporated: Industrialization and the Waters of New England* (Cambridge: Cambridge University Press, 1991), 205.
65 Massachussets State Board of Health, *Eighth Annual Report* (1877), 62, 63, quoted in Steinberg, *Nature Incorporated*, 211.
66 Goldman, *New York's Sewers*, 22-23.
67 Tarr, "The Separate vs. Combined Sewer Problem," 315.
68 As explained by Theodore Steinberg, "An especially large quantity of organic pollution can deplete the stream's supply of dissolved oxygen. In addition, inorganic pollutants, such as acids and alkalis present in industrial waste, can destroy the bacteria needed to carry out self-purification. What remains of the organic matter is broken down by organisms called anaerobic bacteria that do not need free oxygen. An exact understanding of the self-purification of rivers . . . was largely unknown until late in the nineteenth century." Steinberg, *Nature Incorporated*, 206-7.

69 Ibid., 205.
70 L. Frederick Rice, "Engineer's Report," Report on Proposed Water Works at Lawrence, Mass. (Lawrence, 1872), 26, quoted in ibid., 224.
71 "Annual Report of the Public Health Officer, Journals of the Nova Scotia House of Assembly," 1917, quoted in J. Nedelsky, "From Common Law to Commission: Development of Water Law in Nova Scotia," in *Water and Environmental Law* (Halifax: Institute for Resources and Environmental Studies, Dalhousie University, 1981), 102.
72 A.D. Weston, "Disposal of Sewage into the Atlantic Ocean," in Langdon Pearse, ed., *Modern Sewage Disposal: Anniversary Book of the Federation of Sewage Works Associations* (New York: Federation of Sewage Works Associations, 1938), 211; Charles V. Chapin, *Municipal Sanitation in the United States* (Providence, RI: Snow and Farnham, 1901), 300.
73 Winslow and Phelps, *Purification of Boston Sewage*, 20.
74 Steinberg, *Nature Incorporated*, 235.
75 City of Baltimore, Sewerage Commission, *Annual Report*, 1899, 15-16, quoted in John Capper, Garrett Power, and Frank R. Shivers Jr., *Chesapeake Waters: Pollution, Public Health, and Public Opinion, 1607-1972* (Centreville, MD: Tidewater Publishers, 1983), 88-89.
76 A.K. Warren and A.M. Rawn, "Disposal of Sewage into the Pacific Ocean," in Pearse, ed., *Modern Sewage Disposal*, 204.
77 Winslow and Phelps, *Purification of Boston Sewage*, 23.
78 Chapin, *Municipal Sanitation*, 300.
79 Ibid., 699.
80 Margaret Beattie Bogue, *Fishing the Great Lakes: An Environmental History, 1783-1933* (Madison: University of Wisconsin Press, 2000), 144-45.
81 Steinberg, *Nature Incorporated*, 215.
82 Tarr, "The Separate vs. Combined Sewer Problem," 311-12.
83 Cuthbert W. Johnson, *On Fertilizers*, 3rd ed. (London: J. Ridgway, 1851), 117.
84 Albert E. Lauder, *The Municipal Manual: A Description of the Constitution and Functions of Urban Local Authorities* (London: P.S. King, 1907), 46
85 Justus von Liebig, *Familiar Letters*, 473, quoted in Christopher Hamlin, "Providence and Putrefaction: Victorian Sanitarians and the Natural Theology of Health and Disease," *Victorian Studies* 28 (1985): 385.
86 J.J. Rowley, *The Difficult and Vexed Question of the Age: Sewage of Towns: How to Dispose of It, by Making None* (Sheffield, 1884), 11; Johnson, *On Fertilizers*, 120-24, provides further detail on pricing.
87 Tarr, "The Separate vs. Combined Sewer Problem," 313; John Graham-Leigh, *London's Water Wars: The Competition for London's Water Supply in the Nineteenth Century* (London: Francis Boutle, 2000), 97.
88 Michael Leapman, *The World for a Shilling: How the Great Exhibition of 1851 Shaped a Nation* (London: Headline Books, 2002), 92.
89 Henry Moule's pamphlets on the subject include *The Impossibility Overcome: Or the Inoffensive, Safe, and Economical Disposal of the Refuse of Towns and Villages* (1870); *Town Refuse, the Remedy for Local Taxation* (1872); and *National Health and Wealth Promoted by the General Adoption of the Dry Earth System* (1873). See also his *Earth Sewage versus Water Sewage, or, National Health and Wealth Instead of Disease and Waste* (Ottawa, 1868).

90 *Encycloapedia Britannica*, 11th ed., s.v. "Sewerage," 737.
91 Schultz, *Constructing Urban Culture*, 134.
92 Rowley, *The Difficult and Vexed Question of the Age*, 8.
93 Sandra Gwyn, *The Private Capital: Ambition and Love in the Age of Macdonald and Laurier* (Toronto: McClelland and Stewart, 1984), 55.
94 Willis Chapman, *How to Do It: Some Suggestions on House Sanitation – Being a Paper Prepared for the Association of Executive Health Officers of Ontario* (Toronto: Warwick and Sons, 1891), 10.
95 Goldman, *Building New York's Sewers*, 24-25; Donald J. Pisani, "Fish Culture and the Dawn of Concern over Water Pollution in the United States," *Environmental Review* 8 (1984): 117, 124.
96 *Encyclopaedia Britannica*, 11th ed., s.v. "Sewerage," 737.
97 R. Baumeister, *The Cleaning and Sewerage of Cities*, adapted by J.M. Goodell (New York: Engineering News, 1895), 214.
98 Ibid., 216.
99 Ibid.
100 Lauder, *The Municipal Manual*, 46.
101 Henry Robinson, *Sewerage and Sewage Disposal* (London: Biggs, 1896), 55-56; S.E. Finer, *The Life and Times of Sir Edwin Chadwick* (London: Methuen, 1980), 223-24; J.C. Wylie, *The Wastes of Civilization* (London: Faber and Faber, 1959), 55.
102 Quoted in W.H. Corfield, *On the History of Sewage Disposal Inquiries* (London, 1888), 9.
103 Quoted in F.T.K. Pentelow, *River Purification* (London: Edward Arnold, 1953), 5-6.
104 Quoted in Corfield, *Sewage Disposal Inquiries*, 16.
105 Lord Ravensworth, *Parliamentary Debates*, 179 (1865), 374-78, quoted in John Sheail, "Town Wastes, Agricultural Sustainability and Victorian Sewage," *Urban History* 23 (1996): 193-94.
106 Wohl, *Endangered Lives*, 110.
107 Sheail, "Town Wastes," 197.
108 Winslow and Phelps, *Purification of Boston Sewage*, 28.
109 G.H. Benzenberg, "The Sewerage System of Milwaukee and the Milwaukee River Flushing Works," *Transactions of the American Society of Civil Engineers* 30 (December 1893): 370.
110 J.S. Billings, "Sewage Disposal in Cities," *Harper's Magazine*, September 1885, 581.
111 Winslow and Phelps, *Purification of Boston Sewage*, 28, 96.
112 "A Famous Sewer Farm," *Canadian Engineer* 10 (September 1903): 248.
113 Charles Gilman Hyde, "A Review of Progress in Sewage Treatment during the Past Fifty Years in the United States," in Pearse, ed., *Modern Sewage Disposal*, 8; for early chronology, see "Recent Tendencies in Sewage Disposal Practice," *Canadian Engineer* 15 (2 October 1908): 701.
114 Winslow and Phelps, *Purification of Boston Sewage*, 23.
115 Ibid., 96.
116 Tarr et al., "Water and Wastes," 239.
117 Of the 279 systems, 155 were combined and 124 separate. Leo G. Denis, *Water Works and Sewerage Systems of Canada* (Ottawa: Commission of Conservation, 1916), 176.
118 "Sewage Treatment in Great Britain and Some Comparisons with Practice in the United States," *Engineering News*, 6 October 1904, 310-11.

CHAPTER 6: LEARNING TO LIVE DOWNSTREAM

1. *Oldaker v. Hunt* (1854), 6 De G. M. & G. 376.
2. John P.S. McLaren, "The Common Law Nuisance Actions and the Environmental Battle – Well-Tempered Swords or Broken Reeds?" *Osgoode Hall Law Journal* 10 (1972): 537; and McLaren, "Nuisance Law and the Industrial Revolution – Some Lessons from Social History," *Oxford Journal of Legal Studies* 3 (1983): 170.
3. *Attorney-General v. Birmingham* (1858), 70 E.R. 220 at 222.
4. For a succinct discussion of the mid-nineteenth-century process of court reform, see W.R. Cornish and G. de N. Clark, *Law and Society in England, 1750-1950* (London: Sweet and Maxwell, 1989), 38-45.
5. *Attorney-General v. Birmingham* at 224.
6. Ibid., 224.
7. Ibid., 226.
8. Asa Briggs, *Victorian Cities* (Harmondsworth, UK: Pelican Books, 1968), 211-16.
9. E.P. Hennock, *Fit and Proper Persons: Ideal and Reality in Nineteenth-Century Urban Government* (London: Edward Arnold, 1973), 108.
10. Anthony S. Wohl, *Endangered Lives: Public Health in Victorian Britain* (London: J.M. Dent, 1983), 110.
11. Sewage Utilization Act, *1865* (U.K.), 28 & 29 Vict., c. 75; Hennock, *Fit and Proper Persons*, 108-9.
12. C.-E.A. Winslow and Earle B. Phelps, *Purification of Boston Sewage*, US Geological Survey Water-Supply and Irrigation Paper No. 185 (Washington, DC: Government Printing Office, 1906), 990.
13. Wohl, *Endangered Lives*, 159.
14. S.G. Checkland, *The Rise of Industrial Society in England, 1815-1885* (London: Longman, 1964), 255.
15. Hennock, *Fit and Proper Persons*, 109.
16. A number of the leading examples are listed and described in Edwin B. Goodell, *A Review of the Laws Forbidding Pollution of Inland Waters in the United States*, US Geological Survey, Department of the Interior, Water-Supply and Irrigation Paper No. 152 (Washington, DC: Government Printing Office, 1905), 16-20, 25.
17. Briggs, *Victorian Cities*, 216-19.
18. Birmingham Sewage Inquiry Committee, 1871, quoted in H.H. Stanbridge, *History of Sewage Treatment in Britain* (Maidstone, Kent: Institute of Water Pollution Control, 1976), 1: 35.
19. Winslow and Phelps, *Purification of Boston Sewage*, 29.
20. Wicksteed's correspondence with Francis Hincks on this initiative is reproduced in T.C. Keefer, *Report on a Preliminary Survey for the Water Supply of the City of Montreal* (Montreal, 1854).
21. John Sheail, "Town Wastes, Agricultural Sustainability and Victorian Sewage," *Urban History* 23 (1996): 195.
22. Earl Finbar Murphy, *Water Purity: A Study in Legal Control of Natural Resources* (Madison: University of Wisconsin Press, 1961), 39.
23. Christopher Hamlin, "Providence and Putrefaction: Victorian Sanitarians and the Natural Theology of Health and Disease," *Victorian Studies* 28 (1985): 392-94.
24. H.H. Stanbridge, quoted in Sharon Beder, "From Sewage Farms to Septic Tanks: Trials

and Tribulations in Sydney," *Journal of the Royal Australian Historical Society* 79 (1993): 72-95, note 34.
25 Arthur J. Martin, *The Sewage Problem: A Review of the Evidence Collected by the Royal Commission on Sewage Disposal* (London: Sanitary Publishing, 1905), 54-63.
26 Christopher Hamlin, *A Science of Impurity: Water Analysis in Nineteenth Century Britain* (Bristol: Adam Hilger, 1990), 183; see also Beder, "From Sewage Farms to Septic Tanks," note 15; Stephen Halliday, *The Great Stink of London: Sir Joseph Bazalgette and the Cleansing of the Victorian Metropolis* (Stroud, Gloucestershire: Sutton Publishing, 1999), 19; Henry Robinson, *Sewerage and Sewage Disposal* (London: Biggs, 1896), 110. For reaction from the agricultural and scientific communities to irrigation farming, see Sheail, "Town Wastes," 198-203.
27 Wohl, *Endangered Lives*, 110.
28 Winslow and Phelps, *Purification of Boston Sewage*, 32.
29 Wohl, *Endangered Lives*, 110.
30 Beder, "From Sewage Farms to Septic Tanks," note 56.
31 Eliot C. Clarke, "Report to a Commission Appointed to Consider a General System of Drainage for the Valleys of the Mystic, Blackstone and Charles Rivers, Mass.," quoted in Kivas Tully and W.J. McAlpine, "Report on Intercepting and Outfall Sewers and Final Disposal of the Sewage of the City of Toronto," (1886), 39.
32 Royal Commission on Rivers Pollution, 1867, quoted in F.T.K. Pentelow, *River Purification: A Legal and Scientific View of the Last 100 Years* (London: Edward Arnold, 1953), 44-45.
33 Quoted in Tully and McAlpine, "Report on Intercepting and Outfall Sewers," 40.
34 Operating by means of anaerobic putrefaction, septic tanks were reasonably effective by the late 1890s, although their antecedents might be traced to an earlier period. They reached the height of popularity in England in the early 1900s, when it was initially hoped that they could eliminate sludge entirely. But when this expectation proved illusory, municipal interest declined dramatically. Stanbridge, *History of Sewage Treatment in Britain*, vol. 4, *Removal of Suspended Matter*, 42-44. Refinements of the septic concept resulted in a two-storey tank, known after its German designer as the Imhoff tank. These won immediate popularity for primary treatment throughout Europe and the United States, where roughly seventy-five communities had installations of this type by the First World War. Martin V. Melosi, *The Sanitary City: Urban Infrastructure in America from Colonial Times to the Present. Creating the North American Landscape* (Baltimore: Johns Hopkins University Press, 2000), 171.
35 Hennock, *Fit and Proper Persons*, 110; Briggs, *Victorian Cities*, 216-17.
36 Winslow and Phelps, *Purification of Boston Sewage*, 31, 49, 52, 58, 79, 80, 90.
37 Wohl, *Endangered Lives*, 109.
38 Joseph Chamberlain, quoted in Briggs, *Victorian Cities*, 224.
39 Briggs, *Victorian Cities*, 217.
40 William S. Childe-Pemberton, *The Life of Lord Norton* (London: John Murray, 1909), 283.
41 Royal Commission on Rivers Pollution, 1867, quoted in William Howarth, *Water Pollution Law* (London: Shaw, 1988), 119-21.
42 Ibid.
43 Royal Commission on Rivers Pollution, 1867, quoted in Wohl, *Endangered Lives*, 245.
44 "Rivers Pollution Bill," *Field*, 5 August 1876, 155; see also "The Rights and Obligations

of Riparian Proprietors," *American Law Register* 7 (1859): 705, cited in Robert G. Bone, "Normative Theory and Legal Doctrine in American Nuisance Law: 1850 to 1920," *Southern California Law Review* 59 (1986): 1132n78.
45 *Pennington v. Brinsop Hall Coal Company*, [1877] Law Reports, Ch. D. 769 at 771.
46 Bone, "Normative Theory and Legal Doctrine," 1101; Joel Franklin Brenner, "Nuisance Law and the Industrial Revolution," *Journal of Legal Studies* 3 (1973): 403; D.M. Provine, "Balancing Pollution and Property Rights: A Comparison of the Development of English and American Nuisance Law," *Anglo-American Law Review* 7 (1978): 31.
47 *Hole v. Barlow* (1858), 140 E.R. 1113; Brenner, "Nuisance Law," 410-13.
48 *Bamford v. Turnley* (1862), 122 E.R. 25.
49 Ibid., 31.
50 Ibid., 28.
51 Ibid., 31.
52 Ibid., 33.
53 Ibid., 32.
54 *Stockport Waterworks v. Potter* (1861), 158 E.R. 433.
55 McLaren, "Nuisance Law and the Industrial Revolution," 155; see also Brenner, "Nuisance Law and the Industrial Revolution"; Provine, "Balancing Pollution and Property Rights."
56 McLaren, "Nuisance Law and the Industrial Revolution"; Brenner, "Nuisance Law and the Industrial Revolution," 329-31.
57 Sheail, "Town Wastes."
58 Bill Luckin, *Pollution and Control: A Social History of the Thames in the Nineteenth Century* (Bristol: Adam Hilger, 1986), 164.
59 T.E.P., "Great Destruction of Fish in the Ribble," *Field*, 10 July 1875, 34.
60 Edward Frankland, quoted in David Knight, *The Age of Science: The Scientific World View in the Nineteenth Century* (Oxford: Basil Blackwell, 1986), 147. The message, according to Knight, was "a confidence that problems can be solved, and do not just have to be lived with" (148).
61 Pentelow, *River Purification*, 7-8.
62 Royal Commission on Rivers Pollution, 1867, quoted in Higgins, *A Treatise on the Law: Relating to the Pollution and Obstruction of Watercourses* (London: Stevens and Haynes, 1877), 2-3. The commissioners published proposed standards for a series of substances. Higgins, "A Treatise on the Law," 3-7.
63 Higgins, *A Treatise on the Law*, 8-9.
64 Luckin, *Pollution and Control*, 163-69.
65 On the early use of related language, see Brenner, "Nuisance Law and the Industrial Revolution," 428; Eric Ashby and Mary Anderson, *The Politics of Clean Air* (Oxford: Clarendon Press, 1981). Ashby and Anderson also give an interesting account of British alkali inspectors.
66 Hubert Stuart Moore, *The Salmon and Freshwater Fisheries Act, 1923, and the Rivers Pollution Prevention Acts* (London: Sweet and Maxwell, 1924), 147.
67 J.O. Taylor, *The Law Affecting River Pollution* (Edinburgh: W. Green and Son, 1928), 82-83.
68 Circular Letter Issued by the Local Government Board, 6 August 1877, John Lambert, Secretary, cited in J.V.V. Fitzgerald, *The Law Affecting Pollution of Rivers* (London: Knight, 1902), 153.

69 Ibid.
70 "The Rivers Pollution Bill," *Field,* 29 July 1876, 140.
71 Ibid., 5 August 1876, 155.
72 C.E. Saunders, "Legislation for the Purification of Rivers and Its Failures," *Transactions of the Society of Medical Officers of Health* (1886-87): 81.
73 Ibid., 81-83.
74 Luckin, *Pollution and Control,* 169.
75 Frank Spence, "How to Stop River Pollution," *Contemporary Review,* September 1893, 428.
76 Ibid., 430
77 Harry Arthurs, *Without the Law* (Toronto: University of Toronto Press, 1985).
78 Spence, "How to Stop River Pollution," 430.
79 Lamorock Flower, "The Fouling of Streams," *Transactions of the Sanitary Institute of Great Britain* 9 (1887-88): 267.
80 *Local Government Act* (1888), s. 14.
81 Brenner, "Nuisance Law and the Industrial Revolution," 429.
82 See Moore, *The Salmon and Freshwater Fisheries Act,* 150, also referring to the *Rivers Pollution (Border Councils) Act,* 1898, dealing with rivers partly in England and partly in Scotland.
83 Wohl, *Endangered Lives,* 251-54, contains a general discussion.
84 Flower, "The Fouling of Streams," 261.
85 Wohl, *Endangered Lives,* 252, identifies only three significant differences between the Mersey and Irwell act and the 1876 original. Nonetheless, apparent progress did occur, there and in the West Riding of Yorkshire. Ibid., 242.
86 Charles Milnes Gaskell, "On the Pollution of Our Rivers," *Nineteenth Century,* July 1903, 92; Eighteen of these board members were chosen by the West Riding County Council; Leeds selected four, Bradford and Sheffield three each, and Huddersfield and Halifax one each.
87 Wohl, *Endangered Lives,* 253.
88 Gaskell, "On the Pollution of Our Rivers," 95.
89 Ibid.
90 *Yorkshire West Riding C.C. v. Holmfirth Sanitary Authority,* [1894] 2 Q.B. 846.
91 Luckin, *Pollution and Control,* 171, lists a number of important initiatives occurring at the same time and reports on their accomplishments.
92 Gaskell, "On the Pollution of Our Rivers," 94. A similar story applied to the Dearne, once grossly polluted. There, by 1902, fifteen new sewage works had been added to the twenty-eight operating in 1896. Trade effluents were treated by twenty-four out of thirty facilities, in contrast with only ten out of twenty-eight six years before.
93 Gaskell, "On the Pollution of Our Rivers," 93-94; David Kinnersley, *Troubled Water: Rivers, Politics and Pollution* (London: Hilary Shipman, 1988), 67, indicates continued improvements through to the 1930s.
94 Wohl, *Endangered Lives,* 253.
95 Taylor, *The Law Affecting River Pollution,* 74.
96 Ibid., 76.
97 C.F. Folsom, "On the Disposal of Sewage," in *Seventh Annual Report of the State Board of Health of Massachusetts* (Boston: Wright and Potter, 1876), 350-51, quoted in Hamlin, "Providence and Putrefaction," 393.

98 George W. Fuller, "Administrative Problems in the Control of Pollution of Streams," *American Journal of Public Health* 16 (1926): 779.
99 Ibid.
100 Charles F. Folsom, "The Pollution of Streams," Massachussetts State Board of Health, *Eighth Annual Report* (1877), 73, quoted in Theodore Steinberg, *Nature Incorporated: Industrialization and the Waters of New England* (Cambridge: Cambridge University Press, 1991), 228.
101 Ibid., 229-30.
102 Winslow and Phelps, *Purification of Boston Sewage*, 13.
103 *Encyclopaedia Britannica*, 11th ed., s.v. "Sewerage," 735.
104 Quoted in Wohl, *Endangered Lives;* Pentelow, in *River Purification,* questions this analysis.
105 Taylor, *The Law Affecting River Pollution*, 69.
106 Wohl, *Endangered Lives*, 256.
107 Winslow and Phelps, *Purification of Boston Sewage*, 91.
108 John H. Fertig, "The Legal Aspects of the Stream Pollution Problem," *American Journal of Public Health* 16 (1926): 786.
109 *Nolan v. New Britain* (1897), 38 A. 703; *Morgan v. City of Danbury* (1896), 35 A. 499; *Weber v. Town of Berlin* (1904), 13 O.L.R. 302.
110 Carleton B. Chapman explains how such a nuisance was created: "When dissolved oxygen is totally utilized in the decomposition of organic materials certain inorganic salts, including sulphates, begin to give up oxygen. As a consequence, hydrogen sulphide is formed and the stream becomes malodorous." "The Year of the Great Stink," *Pharos* 35 (July 1972): 94.
111 J.J. Rowley, *The Difficult and Vexed Question of the Age: Sewage of Towns: How to Dispose of It, by Making None* (Sheffield, 1884), 4.

Chapter 7: The Bacterial Assault on Local Government

1 J.A. Hassan, "The Growth and Impact of the British Water Industry in the Nineteenth Century," *Economic History Review*, 2nd ser., 38 (1985): 543.
2 Edward Frankland, 1868, quoted in Theodore Steinberg, *Nature Incorporated: Industrialization and the Waters of New England* (Cambridge: Cambridge University Press, 1991), 214. See also Christopher Hamlin, *A Science of Impurity: Water Analysis in Nineteenth Century Britain* (Bristol: Adam Hilger, 1990), ch. 7.
3 *A.G. v. Cockermouth Local Board* (1874), 18 L.R. Eq. 174 at 175.
4 *Final Report of the Rivers Pollution Commission, June 1874,* quoted in J.V.V. Fitzgerald, *The Law Affecting Pollution of Rivers* (London: Knight, 1902), 32.
5 For details, see Gavin Thurston, *The Great Thames Disaster* (London: Allen and Unwin, 1965).
6 *Times*, 14 October 1878, 7, col. 6.
7 Carleton B. Chapman, "The Year of the Great Stink," *Pharos* 35 (July 1972): 100.
8 Christopher Hamlim, "William Dibdin and the Idea of Biological Sewage Treatment," *Technology and Culture* 29 (1988): 198-99.
9 *Bamford v. Turnley* (1862), 122 E.R. 25.
10 W.H. Corfield, *On the History of Sewage Disposal Inquiries* (London, 1888), 27-28.

11 Leslie B. Wood, *The Restoration of the Tidal Thames* (Bristol: Adam Hilger, 1982), 41-42; Royal Commission on Metropolitan Sewage Discharge, *First Report 1884; Second and Final Report 1884.*
12 Chapman, "The Year of the Great Stink," 100; Stephen Inwood, *A History of London* (New York: Carroll and Graf, 1998), 434.
13 Henry Robinson, *Sewerage and Sewage Disposal* (London: Biggs, 1896), 84.
14 *Encyclopaedia Britannica*, 11th ed., s.v. "Sewerage," 743.
15 Wood, *The Restoration of the Tidal Thames*, 49. Dibdin's findings were of great interest on both sides of the Atlantic and contributed to significant developments in sewage treatment. C.-E.A. Winslow and Earle B. Phelps, *Purification of Boston Sewage*, US Geological Survey Water-Supply and Irrigation Paper No. 185 (Washington, DC: Government Printing Office, 1906), 18, 30, 54-56. See also Martin V. Melosi, *The Sanitary City: Urban Infrastructure in America from Colonial Times to the Present. Creating the North American Landscape* (Baltimore: Johns Hopkins University Press, 2000), 169.
16 Major Greenwood, *Epidemics and Crowd Diseases: An Introduction to the Study of Epidemiology* (New York: Macmillan, 1937), 146-56.
17 "Sawdust Survey," P.H. Bryce, Secretary, Item 32, RG 62, Series B4, Scrapbooks, Provincial Board of Health, 1884, Archives of Ontario; "Survey of Industries and Nuisances," P.H. Bryce, Secretary, Item 56, RG 62, Series B4, Scrapbooks, 10 May 1886, Archives of Ontario.
18 Quoted in C.-E.A. Winslow, "Pioneers of Sewage Disposal in New England," in Langdon Pearse, ed., *Modern Sewage Disposal: Anniversary Book of the Federation of Sewage Works Associations* (New York: Federation of Sewage Works Associations, 1938), 279.
19 Donald J. Pisani, "Fish Culture and the Dawn of Concern over Water Pollution in the United States," *Environmental Review* 8 (1984): 124.
20 Marshall Ora Leighton, *Sewage Pollution in the Metropolitan Area near New York City and Its Effect on Inland Water Resources*, US Geological Survey Water-Supply and Irrigation Paper No. 72 (Washington, DC: Government Printing Office, 1902), 9-10.
21 Massachusetts State Board of Health, Lunacy and Charity, *First Annual Report* (1880), quoted in Barbara Gutmann Rosenkrantz, *Public Health and the State: Changing Views in Massachusetts, 1842-1936* (Cambridge, MA: Harvard University Press, 1972), 81.
22 Ontario, Provincial Board of Health, *Annual Report, 1892*, 50.
23 Rosenkrantz, *Public Health and the State*, 104.
24 John Duffy, *The Sanitarians: A History of American Public Health* (Urbana and Chicago: University of Illinois Press, 1990), 176, 194, 201-2.
25 "Report of Sedgewick's presentation to the Lowell Water Board," 9 January 1891, available through the University of Massachussetts at Lowell, Center for Lowell History: http://library.uml.edu/clh/Typh/Typ2.html.
26 William Thompson Sedgwick, quoted in Rosenkrantz, *Public Health and the State*, 105.
27 Winslow and Phelps, *Purification of Boston Sewage*, 37.
28 Ibid.
29 Hiram F. Mills of the Massachusetts State Board of Health, reporting to the Board in 1890, as quoted in Winslow and Phelps, *Purification of Boston Sewage*, 37.
30 Ibid., 39.
31 Keith Vernon, "Pus, Sewage, Beer and Milk: Microbiology in Britain, 1870-1940," *History of Science* 28 (1990): 289.

32 Joel A. Tarr, "Industrial Wastes and Public Health: Some Historical Notes, Part 1, 1876-1932," *American Journal of Public Health* 75 (1985): 1060.
33 Duffy, *The Sanitarians*, 129, 187.
34 Earl Finbar Murphy, *Water Purity: A Study in Legal Control of Natural Resources* (Madison: University of Wisconsin, 1961), 29-30.
35 Henry Bixby Hemenway, *Legal Principles of Public Health Administration* (Chicago: T.H. Flood, 1914), 684.
36 *Public Health Act (Ont.)*, 1882, 45 Vict., c. 29.
37 J.E. Hodgetts, *From Arm's Length to Hands-On: The Formative Years of Ontario's Public Service, 1867-1940* (Toronto: University of Toronto Press for the Ontario Historical Studies Series, 1995), 20-21.
38 Ibid., 14.
39 Rosenkrantz, *Public Health and the State*, 52.
40 Ontario, Provincial Board of Health, *Annual Report, 1884*, 15, 22, 26.
41 Ibid., 22.
42 P.H. Bryce, "Inspection of Muskoka Health Resorts," in Ontario, Provincial Board of Health, *Annual Report, 1899*, 69.
43 Charles A. Hodgetts and John A. Amyot, "Report on the Sanitary Conditions of the Muskoka and Kawartha Districts," in Ontario, Provincial Board of Health, *Sanitary Papers, 1904*, 143.
44 R.W. Bell, "Report re Sanitary Conditions at French River," in Ontario, Provincial Board of Health, *Annual Report, 1905*, 183.
45 Edwin B. Goodell, *A Review of the Laws Forbidding Pollution of Inland Waters in the United States*, US Geological Survey, Department of the Interior, Water-Supply and Irrigation Paper No. 152 (Washington, DC: Government Printing Office, 1905), 32.
46 Ibid., 33.
47 Ibid., 73.
48 Ibid., 107, 109.
49 See Ontario, Provincial Board of Health, *Annual Report, 1906*, 8, 111, 161, for expressions of the board's interest in legislative reform.
50 *Toronto Waterworks Act*, S.O. 1872, c. 79, s. 16.
51 The prohibition was introduced to Ontario law by the *Statute Law Amendment Act*, S.O. 1906, c. 19, assented to 14 May 1906.
52 S.O. 1912, c. 58, s. 8(o).
53 Ibid., s. 73.
54 Dr. J.W.S. McCullough, "Memorandum for Attorney General," 5 December 1912, Attorney General's Records, 1912, RG 4-32, no. 1678, Archives of Ontario.
55 *Re Waterloo Local Board of Health. Campbell's Case* (1918), 44 O.L.R. 338.
56 Ibid., 346.
57 Ibid., 348.
58 F.A. Dallyn to J.W.S. McCullough, 17 June 1919 and 2 August 1919, Provincial Board of Health Records, RG 62, Series G, Division of Sanitary Engineering, Archives of Ontario.
59 Ibid.
60 Allen Hazen, *Clean Water and How to Get It*, 2nd ed. (New York: John Wiley, 1914), 32.
61 Winslow and Phelps, *Purification of Boston Sewage*, 96.
62 Ibid., 11.
63 Hazen, *Clean Water and How to Get It*, 31.

64 Margaret Beattie Bogue, *Fishing the Great Lakes: An Environmental History, 1783-1933* (Madison: University of Wisconsin Press, 2000), 144-45.
65 George C. Whipple, *The Value of Pure Water* (New York: John Wiley and Sons, 1907), 36.
66 Joel A. Tarr, Terrie Yosie, and James McCurley, "Disputes over Water Quality Policy: Professional Cultures in Conflict, 1900-1917," *American Journal of Public Health* 70 (1980): 427; "Mr Roosevelt and the People," *Outlook* 96 (1910): 1; "The Pollution of Lakes and Rivers," *Outlook* 96 (1910): 144-45.
67 T. Aird Murray, *The Prevention of the Pollution of Canadian Surface Waters* (Ottawa: Commission of Conservation, 1912). The three articles reproduced in this pamphlet originally appeared in the *Toronto Globe*, 30 December 1911 and early January 1912.
68 Leighton, *Sewage Pollution in the Metropolitan Area*, 10.
69 International Joint Commission, "Progress Report of the IJC on the Reference by the United States and Canada," in *The Pollution of Boundary Waters*, 16 January 1914, 2-7.
70 For a description of the use of "indicator" organisms to detect contamination, see Dan Krewski et al., "Managing Health Risks from Drinking Water," *Journal of Toxicology and Environmental Health* 65 (2002): 1692-95.
71 IJC, "Progress Report of the IJC on the Reference by the United States and Canada," in *The Pollution of Boundary Waters*, 16 January 1914, 14. During at least one of the three years preceding the reference (1910-12), a death rate (per 100,000) of 196 was registered in Port Huron, Michigan; of 194 in Niagara Falls, New York; of 190 in Erie, Pennsylvania; of 109 in Marquette, Michigan; and above 50 in the Michigan communities of Alpena, Bay City, and Sault Ste. Marie, as well as in Duluth, Minnesota, and Sandusky, Ohio. The Ontario side produced equally alarming numbers - 330 in Sault Ste. Marie, 179 in Port Arthur, 134 in Sarnia, 86 in Niagara Falls, 63 in Brockville, and 57 and 55 in Walkerville and Windsor, respectively. Even comparatively fortunate Great Lakes cities such as Detroit, whose rate had never fallen below 15, and was above 30 in 1913, experienced levels of typhoid that would not be tolerated in Europe.
72 IJC, "Progress Report," 12.
73 The exception was Lackawanna, which discharged its water through Smoke's Creek, a mile and a half above the head of the river.
74 IJC, "Progress Report," 8.
75 Ibid., 45.
76 James Mansergh, *The Water Supply of the City of Toronto, Canada* (Westminster, 1896), 19.
77 IJC, "Progress Report," 21.
78 Ibid., appendix, 354.
79 Ibid., appendix, 353.
80 Ibid., appendix, 356.
81 Ibid., 3.
82 Ibid., 7-8.
83 Ibid., 12.
84 Leighton, *Sewage Pollution in the Metropolitan Area*, 35.
85 IJC, "Progress Report," 383.
86 Allan J. McLaughlin, in the US Public Health Service Annual Report 41-41 (1913), quoted in L.B. Dworsky, "Assessing North America's Management of Its Transboundary Waters," *Natural Resources Journal* 33 (1993): 427.
87 John Capper, Garrett Power, and Frank R. Shivers Jr., *Chesapeake Waters: Pollution, Public Health, and Public Opinion, 1607-1972* (Centreville, MD: Tidewater Publishers,

1983), 97; William L. Andreen, "The Evolution of Water Pollution Control in the United States: State, Local and Federal Efforts, 1789-1972: Part II," *Stanford Environmental Law Journal* 22, 2 (June 2003): 222-23; N. William Hines, "Nor Any Drop to Drink: Public Regulation of Water Quality, Part III: The Federal Effort," *Iowa Law Review* 52 (1967): 804-5.

88 The initial US Treasury Department Standard specified a maximum limit of 2 B. coli per 100 cc. George W. Fuller, "Relations between Sewage Disposal and Water Supply Are Changing," *Engineering News-Record*, 5 April 1917, 12.

89 B. Grover and D. Zussman, *Safeguarding Canadian Drinking Water* (Ottawa: Inquiry on Federal Water Resources, 1985); Frank Quinn, "The Evolution of Federal Water Policy," *Canadian Water Resources Journal* 10 (1985): 25.

90 John P.S. McLaren, "The Tribulations of Antoine Ratté: A Case Study of the Environmental Regulation of the Canadian Lumber Industry in the Nineteenth Century," *University of New Brunswick Law Journal* 33 (1984): 203.

91 John H. Taylor, "Fire, Disease and Water in Ottawa: An Introduction," *Urban History Review* 8 (1979): 7.

92 Allen Hazen, "Report on Improvement of the Ottawa Water Supply" (prepared for Ottawa City Council, October 1910), 3.

93 Ibid., 41.

94 *Re City of Ottawa and Provincial Board of Health* (1914), 33 O.L.R. 1.

95 Chris Warfe, "The Search for Pure Water in Ottawa: 1910-1915," *Urban History Review* 8 (1979): 90.

96 Canada, *Senate Debates* (2 March 1910), 349.

97 The measure was designed to prohibit the introduction into any navigable water or water flowing into navigable water any sewage matter, whether solid or liquid, or any other matter, refuse, or waste that was poisonous, noxious; putrid, or decomposing, unless it was disposed of according to regulations or pursuant to a permit.

98 Canada, *Senate Debates* (2 March 1910), 335.

99 The Boundary Waters Treaty had put domestic uses at the top of its hierarchy, and distinguished sanitation.

100 Canada, *Senate Debates* (2 March 1910), 342.

101 Murray, *The Prevention of the Pollution*, 7.

102 Ibid., 8.

103 Ibid., 6.

104 Canada, *Senate Debates* (2 March 1910), 341 (Senator McSweeney).

105 Canada, *Senate Debates* (3 March 1910), 352-53 (Senator Casgrain).

106 *Sanitary Review*, quoted in *Canadian Engineer* 16 (16 April 1909): 527.

107 Canada, *Senate Debates* (3 March 1901), 370.

108 Ibid.

109 Canada, *Senate Debates* (3 March 1910), 371-72.

110 Quoted in Jennifer Read, "Water Pollution Management in the Great Lakes Basin, 1900-1930" (paper presented to Themes and Issues in North American Environmental History Conference, April 1998), 9. See also Jennifer Read, "'A Sort of Destiny': The Multi-jurisdictional Response to Sewage Pollution in the Great Lakes, 1900-1930," *Scientia Canadensis* 55 (1999): 22-23, 103-29.

111 IJC, *Final Report of the International Joint Commission on the Pollution of Boundary Waters Reference* (Ottawa and Washington, DC: Government Printing Bureau, 1918).

112 Read, "Water Pollution Management," 14-15.
113 Ontario, Provincial Board of Health, *Annual Report, 1929*, 49.
114 Ibid.

CHAPTER 8: THE DILUTIONARY IMPULSE AT CHICAGO

1 Isaac Donaldson Rawlings, *The Rise and Fall of Disease in Illinois* ([Springfield, IL:] Illinois State Department of Public Health, 1927), 101-4, 113.
2 Quoted in William Cronon, *Nature's Metropolis: Chicago and the Great West* (New York and London: W.W. Norton and Company, 1991), 33.
3 Cronon, *Nature's Metropolis*, 301
4 Ibid., 33.
5 Ibid., 58.
6 Ellis Sylvester Chesbrough, *Report and Plan of Sewerage for the City of Chicago, Illinois, adopted by the Board of Sewerage Commissioners*, 31 December 1855, quoted in Louis P. Cain, "Raising and Watering a City: Ellis Sylvester Chesbrough and Chicago's First Sanitation System," *Technology and Culture* 13 (1972): 358.
7 Louis P. Cain, "Unfouling the Public's Nest: Chicago's Sanitary Diversion of Lake Michigan Water," *Technology and Culture* 15 (1974): 595.
8 Bruce Barker, "Lake Diversion at Chicago," *Case Western Reserve Journal of International Law* 18 (1986): 207.
9 Stanley A. Changnon and Mary E. Harper, "History of the Chicago Diversion," in Stanley A. Changnon, ed., *The Lake Michigan Diversion at Chicago: Past, Present and Future Regional Impacts and Responses to Global Climate Change* (Ann Arbor, MI: Great Lakes Environmental Research Laboratory, 1994), 17-19.
10 Cronon, *Nature's Metropolis*, 345.
11 W.H.G. Armytage, *A Social History of Engineering* (Cambridge, MA: MIT Press, 1961), 171.
12 Rawlings, *The Rise and Fall of Disease in Illinois*, 151.
13 Ibid., 221.
14 Libby Hill, *The Chicago River: A Natural and Unnatural History* (Chicago: Lake Claremont Press, 2000), 115.
15 Changnon and Harper, "History of the Chicago Diversion," 19-20; Herbert H. Naujoks, "The Chicago Water Diversion Controversy," *Marquette Law Review* 30 (1946): 152.
16 Department of Public Works, "International Waterways Commission, Proceedings and Report," Canada, *Sessional Papers*, 1913, no. 19a, 1184 (hereafter IWC Proceedings).
17 Isham Randolph, IWC Proceedings, 1167.
18 Psalm 107:23-30 (King James Version). Quoted in G.P. Brown, *Drainage Channel and Waterway: A History of the Effort to Secure an Effective and Harmless Method for the Disposal of the Sewage of the City of Chicago, and to Create a Navigable Channel between Lake Michigan and the Mississippi River* (Chicago: R.R. Donnelley and Sons, 1894), 429-30.
19 Changnon and Harper, "History of the Chicago Diversion," 20.
20 Cain, "Unfouling the Public's Nest," 596n6.
21 *The Standard Guide to Chicago* (Chicago, 1892), 109.
22 Ibid., 552.

23 William H. Wilson, *The City Beautiful Movement* (Baltimore: Johns Hopkins University Press, 1989), 57.
24 C.-E.A. Winslow and Earle B. Phelps, *Purification of Boston Sewage*, US Geological Survey Water-Supply and Irrigation Paper No. 185 (Washington, DC: Government Printing Office, 1906), 30.
25 *Standard Guide*, 111-14.
26 IWC Proceedings, 1183.
27 Ibid., 1188.
28 Albert E. Cowdrey, "Pioneering Environmental Law: The Army Corps of Engineers and the Refuse Act," *Pacific Historical Review* 44 (1975): 337-41.
29 Quoted in Cain, "Unfouling the Public's Nest," 598-99, citing C. Arch Williams, *The Sanitary District of Chicago* (Chicago: District, 1919).
30 Cain, "Unfouling the Public's Nest," 602.
31 Randolph, IWC Proceedings, 1168; IWC Proceedings, 265.
32 Hill, *The Chicago River*, 129-33.
33 "The Chicago Canal Opened," *New York Times*, 18 January 1900.
34 "Water in Chicago River," *New York Times*, 14 January 1900.
35 David R. Forgan of the Chicago Commercial Association, IWC Proceedings, 1182.
36 John Duffy, *The Sanitarians: A History of American Public Health* (Urbana and Chicago: University of Illinois Press, 1990), 144.
37 Upton Sinclair, *The Jungle*, ed. James Barrett (Urbana and Chicago: University of Illinois Press, 1988), 92.
38 Richard Yates, *Sanitary Investigations of the Illinois River and Tributaries* (1901), quoted in Rawlings, *The Rise and Fall of Disease in Illinois*, 167. See also Edwin O. Jordan, "The Self-Purification of Streams,",in *Chicago Decennial Publications*, 1st ser. (Chicago: University of Chicago Press, 1903), 10: 81-89.
39 As reported in the Supreme Court decision, *Missouri v. Illinois*, 200 U.S. 496 (1900), the statistics were as follows: 1891 (165), 1892 (441), 1893 (215), 1894 (171), 1895 (106), 1896 (106), 1897 (125), 1898 (95), 1899 (131), 1900 (154), 1901 (181), 1902 (216), 1903 (281).
40 Ibid., 521-22.
41 Ibid., 522-23.
42 Ibid., 521.
43 IWC Proceedings, 1170.
44 Ibid., 1171.
45 Ibid., 1172.
46 Ibid., 1177.
47 Ibid., 1178.
48 Ibid., 1179.
49 Ibid., 1185, 1195, 1193, 1196.
50 Ibid., 1174.
51 Ibid., 1197-98.
52 Ibid., 1189.
53 Ibid., 1191.
54 Naujoks, "The Chicago Water Diversion Controversy," 160.
55 R.A. Alger, quoted in Cain, "Unfouling the Public's Nest," 599.
56 Ibid., 600.

57 T. Aird Murray, *The Prevention of the Pollution of Canadian Surface Waters* (Ottawa: Commission of Conservation, 1912), 8.
58 James Simsarian, "The Diversion of Waters Affecting the United States and Canada," *American Journal of International Law* 32 (1938): 499.
59 IWC Proceedings, 1192.
60 Changnon and Harper, "History of the Chicago Diversion," 24-25.
61 IWC Proceedings, 1195.
62 "Return of Correspondence Protesting against the Illegal Diversion of the Waters of the Great Lakes by the Chicago Drainage Commission," Canada, *Sessional Papers,* 1924, no. 180, 108-9.
63 *Sanitary District of Chicago v. The United States,* 266 U.S. 405 at 430 (1924).
64 In "Report on the Diversion of Water from the Great Lakes and Niagara River, 1921," Colonel Warren of the United States Corps of Engineers indicated that "The diversion through the Chicago Sanitary Canal averaged 8,800 cubic feet per second in 1917, although some daily averages were 10,000 cubic feet per second or more." See Canada, *Sessional Papers,* 1924, no. 180, 128.
65 J.L.P. O'Hanly, "An Interim Report on the Effect of the Chicago Drainage Canal on the Level of the Great Lakes," Canada, *Sessional Papers,* 1896, no. 82 (hereafter O'Hanly Report).
66 O'Hanly Report, 7.
67 Ibid.
68 Memorandum contained in Canada, *Sessional Papers,* 1924, no. 180, 120.
69 An unpublished version of Poe's report is reproduced as Appendix A in the O'Hanly Report.
70 IWC Proceedings, 1191.
71 Thomas C. Keefer, "The Chicago Drainage Canal," *Canadian Engineer* 3 (August 1895): 91; C. Baillairgé, "An Invasion of Our Riparian Rights," *Canadian Engineer* 2 (April 1895): 349; Peter H. Bryce, "The Chicago Canal, and Some Sanitary Problems Connected Therewith," *Canadian Engineer* 3 (January 1896): 229.
72 Cain, "Unfouling the Public's Nest," 596, states, "While the Sanitary District initially described the excess capacity as having been accidental, they later claimed that this was the designed capacity and that the southern section's capacity was to have been increased to 14,000 cfs when Chicago's population warranted this increase."
73 Quoted in Canada, *Sessional Papers,* 1924, no. 180, 121.
74 J.Q. Dealey Jr., "The Chicago Drainage Canal and St. Lawrence Development," *American Journal of International Law* 23 (1929): 312.
75 Quoted in Canada, *Sessional Papers,* 1924, no. 180, 111.
76 Ibid.
77 Ibid.
78 Ibid.
79 Stimson's decision on the Sanitary District of Chicago application for permission to divert 10,000 cfs from Lake Michigan, 8 January 1913, is reproduced in Canada, *Sessional Papers,* 1924, no. 180, 113-18.
80 U.S. Statutes, 3 March 1899, c. 425, s. 10; 30 Stat. 1152.
81 Cain, "Unfouling the Public's Nest," 600n13.
82 Holmes in *Sanitary District of Chicago v. The United States,* 266 U.S. 405 at 432 (1924).
83 Naujoks, "The Chicago Water Diversion Controversy," 158.

84 Holmes in *Sanitary District of Chicago v. The United States*, 432.
85 Ibid., 424-25.
86 Dealey, "The Chicago Drainage Canal," 209.
87 These proceedings began in 1922 at the suggestion of the judge advocate general. Naujoks, "The Chicago Water Diversion Controversy," 162.
88 A. Dan Tarlock, "Global Climate Change and the Law of the Great Lakes Diversion," in Changnon, ed., *The Lake Michigan Diversion at Chicago*, 111.
89 For Canadian opinion and indications of Ontario's inclination to forego active protest in order not to interfere with provincial hydro ambitions, see "Re U.S. Litigation on Great Lakes Diversion," 1926, RG 4-32, no. 396, Archives of Ontario.
90 Joel A. Tarr et al., "Water and Wastes: A Retrospective Assessment of Wastewater Technology in the United States, 1900-1932," *Technology and Culture* 25 (1985): 230.
91 Attorney General of the United States to the Secretary of War, 13 February 1925, Appendix to the Defendants' Statement in *Wisconsin et al. v. Illinois and the Sanitary District of Chicago* in the Supreme Court of the United States, October Term, 1925 (hereafter Defendants' Statement).
92 Defendants' Statement, 66.
93 "Your predecessors, having in mind the various elements of an extremely complicated economic and sanitary problem, reached the conclusion that there could be – at least until Congress provided otherwise – a permit to use the amount indicated, but no more; and this could only proceed upon the theory that, while Congress had made a general prohibition of all impairment of navigable waters, it had delegated to the War Department the duty of deciding the degree of impairment which would bring a specific case within the general prohibition." Defendants' Statement, 66-67.
94 Ibid., 68.
95 Ibid., 68-69.
96 Actually, the 1889 legislation creating the SDC had anticipated argument along these lines by declaring in s. 24 that upon completion of the Main Channel, the same is declared a navigable stream.
97 Defendants' Statement, 46-47.
98 *Wisconsin v. Illinois*, 278 U.S. 367 at 418 (1929).
99 See *Wisconsin v. Illinois*, 281 U.S. 696 (1930). For the second Hughes report and another Holmes opinion, see *Wisconsin v. Illinois*, 281 U.S. 179 (1930).
100 Simsarian, "The Diversion of Waters Affecting the United States and Canada," 913.
101 Dealey, "The Chicago Drainage Canal," 310.
102 Meat packers successfully excluded SDC inspectors for years and resisted sewage-related taxes until intervention by the federal Public Works Administration in the 1930s. Martin V. Melosi, *The Sanitary City: Urban Infrastructure in America from Colonial Times to the Present. Creating the North American Landscape* (Baltimore: Johns Hopkins University Press, 2000), 241. During Prohibition, bootleggers disrupted sewage treatment operations with clandestine discharges of fermentation wastes. F.W. Mohlman, "Twenty-five Years of Activated Sludge," in Langdon Pearse, ed., *Modern Sewage Disposal: Anniversary Book of the Federation of Sewage Works Associations* (New York: Federation of Sewage Works Associations, 1938), 73.
103 Changnon and Harper, "History of the Chicago Diversion," 26; Robert Isham Randolph, "A History of Sanitation in Chicago," *Journal of the Western Society of Engineers* 44 (October 1939): 238; Illinois' decision to reduce actual diversion to 1,500 cfs, with

sewage effluent used to bring the total flow up to 3,200 cfs, was eventually challenged by neighbouring Great Lakes states who wanted treated effluent returned to Lake Michigan. A further US Supreme Court decree in 1967 had the effect of allowing Illinois to continue to divert treated effluent and stormwater out of the lake, although subsequent refinements to the order have imposed additional conditions and restrictions. See Barker, "Lake Diversion at Chicago."

104 Edward M. Martin, "Conviction of Chicago Sanitary District Officials," *National Municipal Review* 21 (March 1932): 202-3; "Four Found Guilty in Chicago Graft," *New York Times*, 6 February 1932, 6.

105 Wallace J. Laut, *A Canadian's Plan to Restore the Water Levels of the Great Lakes – The Campbell Plan for the Restoration of the Great Lakes Water Levels: The Mighty Albany – St. Lawrence of the Northland* (Toronto: n.p., 1925).

CHAPTER 9: SEPARATING WATER FROM THE WATERWAYS

1 Joel A. Tarr, "The Separate vs. Combined Sewer Problem: A Case Study in Urban Technology Design Choice," *Journal of Urban History* 5 (1979): 310; Earl Finbar Murphy, *Water Purity: A Study in Legal Control of Natural Resources* (Madison: University of Wisconsin Press, 1961), 27; Stuart Galishoff, "Triumph and Failure: The American Response to the Urban Water Supply Problem, 1860-1923," in Martin V. Melosi, ed., *Pollution and Reform in American Cities, 1870-1930* (Austin and London: University of Texas Press, 1980), 51; J.A. Hassan, "The Growth and Impact of the British Water Industry in the Nineteenth Century," *Economic History Review*, 2nd ser., 38 (1985): 534; Leo G. Denis, *Water Works and Sewerage Systems of Canada* (Ottawa: Printed by Mortimer Co., 1916), 137.

2 Robert Gottlieb, *Forcing the Spring: The Transformation of the American Environmental Movement* (Washington, DC: Island Press, 1993), 53-54.

3 George W. Fuller, "Relations between Sewage Disposal and Water Supply Are Changing," *Engineering News-Record*, 5 April 1917, 11. The broader effort to safeguard New York's water supply from pollution is summarized in Matthew Gandy, *Concrete and Clay: Reworking Nature in New York City* (Cambridge, MA: MIT Press, 2002), 43-44.

4 Christopher Hamlin, *A Science of Impurity: Water Analysis in Nineteenth Century Britain* (Bristol: Adam Hilger, 1990), 154.

5 John Burns, quoted in "Devastation by Reservoir," *Spectator*, 25 February 1911, 279.

6 James Mansergh, *The Water Supply of the City of Toronto, Canada* (Westminster, 1896), 16.

7 Edwin G. Burrows and Mike Wallace, *Gotham: A History of New York City to 1898* (Oxford and New York: Oxford University Press, 1999), 1228-30.

8 Marshall Ora Leighton, *Sewage Pollution in the Metropolitan Area near New York City and Its Effect on Inland Water Resources*, US Geological Survey Water-Supply and Irrigation Paper No. 72 (Washington, DC: Government Printing Office, 1902), 66.

9 Gerald T. Koeppel, *Water for Gotham: A History* (Princeton: Princeton University Press, 2000), 289.

10 By the mid-1880s Toronto's population exceeded 120 thousand. Twenty-one thousand water services supplied twelve million gallons daily to eighty-three thousand water takers who owned eleven thousand water closets. Forty-two thousand people continued

to rely on wells and the fourteen thousand remaining privies. Kivas Tully and W.J. McAlpine, "Report on Intercepting and Outfall Sewers and Final Disposal of the Sewage of the City of Toronto," (1886), 21.

11 Elwood Jones and Douglas McCalla, "Toronto Waterworks, 1840-77: Continuity and Change in Nineteenth Century Toronto Politics," *Canadian Historical Review* 60 (1979): 321.
12 *Toronto Globe,* 27 December 1892.
13 Mansergh, *Water Supply,* 5.
14 Ibid., 21.
15 Ibid.
16 Ibid.
17 Ibid., 52.
18 Arthur J. Martin, *The Sewage Problem: A Review of the Evidence Collected by the Royal Commission on Sewage Disposal* (London: Sanitary Publishing, 1905), 9-10; H.H. Stanbridge, *History of Sewage Treatment in Britain,* vol. 5, *Land Treatment* (Maidstone, Kent: Institute of Water Pollution Control, 1976), 25.
19 *City of Milwaukee v. Activated Sludge,* 69 F.2d. 577 at 583 (1934).
20 Ibid., 583.
21 Ibid.
22 Ibid., 584.
23 Ibid., 583.
24 Ibid.
25 Ibid., Patent 729.
26 Ibid., 75.
27 Langdon Pearse, ed., *Modern Sewage Disposal: Anniversary Book of the Federation of Sewage Works Associations* (New York: Federation of Sewage Works Associations, 1938), 72.
28 Ibid., 68.
29 William Howarth, *Water Pollution Law* (London: Shaw, 1988), 42.
30 Ibid., 43-44.
31 Ibid.
32 F.T.K. Pentelow, *River Purification* (London: Edward Arnold, 1952), 13-14.
33 Sharon Beder, "From Sewage Farms to Septic Tanks: Trials and Tribulations in Sydney," *Journal of the Royal Australian Historical Society* 79 (1993): 72-95.
34 H.W. Streeter, "Disposal of Sewage in Inland Waterways," in Pearse, ed. *Modern Sewage Disposal,* 200.
35 Beder, "From Sewage Farms to Septic Tanks."
36 Ibid.
37 John Sheail, "Town Wastes, Agricultural Sustainability and Victorian Sewage," *Urban History* 23 (1996): 207.
38 Denis, *Water Works and Sewerage Systems,* 136.
39 Ibid., 176.
40 Ibid., 5.
41 T. Aird Murray, *The Prevention of the Pollution of Canadian Surface Waters* (Ottawa: Commission of Conservation, 1912), 11.
42 "The Pollution of Streams," *Engineering Record* 60 (7 August 1909): 159.
43 Murray, *The Prevention of the Pollution,* 23-24.

44 Joel A. Tarr, *The Search for the Ultimate Sink: Urban Pollution in Historical Perspective* (Akron, OH: University of Akron Press, 1996), 46.
45 Leighton, *Sewage Pollution in the Metropolitan Area near New York City*, 9.
46 Joel A. Tarr, "Industrial Wastes and Public Health," *American Journal of Public Health* 75 (1985): 1059; Gottlieb, *Forcing the Spring*, 53. See also Edwin B. Goodell, *A Review of the Laws Forbidding Pollution of Inland Waters in the United States*, US Geological Survey, Department of the Interior, Water-Supply and Irrigation Paper No. 152 (Washington, DC: Government Printing Office, 1905), 30, 109.
47 Leighton, *Sewage Pollution in the Metropolitan Area near New York City*, 25.
48 Ibid., 26.
49 Ibid., 29.
50 Ibid., 33.
51 Ibid., 26, 29.
52 *Van Cleve v. Passaic Valley Sewerage Commissioners*, 60 A. 214 (1905).
53 Ibid.
54 Goodell, *Pollution of Inland Waters*, 121.
55 C.G. Hyde, "A Review of Progress in Sewage Treatment during the Past Fifty Years in the United States," in Pearse, ed., *Modern Sewage Disposal*, 3; A.D. Weston, "Disposal of Sewage into the Atlantic Ocean," in Pearse, ed., *Modern Sewage Disposal*, 213.
56 Leighton, *Sewage Pollution in the Metropolitan Area near New York City*, 63-64, 60, 35, 37, 61.
57 Seth G. Hess, "Interstate Action to Control Pollution," *State Government* 23 (1950): 204.
58 Burrows and Wallace, *Gotham*, 950.
59 Albert E. Cowdrey, "Pioneering Environmental Law: The Army Corps of Engineers and the Refuse Act," *Pacific Historical Review* 44 (1975): 336-37.
60 Tarr, *The Search for the Ultimate Sink*, 57.
61 Hess, "Interstate Action to Control Pollution."
62 Hyde, "A Review of Progress," in Pearse, ed., *Modern Sewage Disposal*, 3.
63 Hess, "Interstate Action to Control Pollution," 204.
64 Tarr, *The Search for the Ultimate Sink*, 58-59.
65 Cowdrey, "Pioneering Environmental Law," 341.
66 W.E. Adeney, *The Principles and Practice of the Dilution Method of Sewage Disposal* (Cambridge, 1928), 78; John Duffy, *The Sanitarians: A History of American Public Health* (Chicago: University of Illinois Press, 1992), 177.
67 Richard J. Evans, *Death in Hamburg: Society and Politics in the Cholera Years, 1830-1910* (Oxford: Clarendon Press, 1987), 146.
68 Ellis L. Armstrong, ed., *History of Public Works in the United States, 1776-1976* (Chicago: American Public Works Association, 1976), 235-36; "Filtered vs Unfiltered Water," *Canadian Engineer* 5 (October 1897): 165; Evans, *Death in Hamburg*.
69 George Chandler Whipple, *The Value of Pure Water* (New York: John Wiley and Sons, 1907), 10.
70 Murphy, *Water Purity*, 29.
71 Leighton, *Sewage Pollution in the Metropolitan Area near New York City*, 64.
72 Burrows and Wallace, *Gotham*, 1198.
73 Melosi, *The Sanitary City*, 45; Armstrong, ed., *History of Public Works in the United States*, 239.

74 James C. Connell, *Chicago's Quest for Pure Water* (Washington, DC: Public Works Historical Society, 1976), 15-18, cited in Duffy, *The Sanitarians,* 202.
75 Letty Anderson, "Water and the Canadian City," in Howard Rosen and Ann Durkin Keating, eds., *Water and the City: The Next Century* (Chicago: Public Works Historical Society, 1991), 40-41.
76 Melosi, *The Sanitary City,* 162.
77 Leighton, *Sewage Pollution in the Metropolitan Area near New York City,* 64.
78 "Sewage Pollution of Water Supplies," *Engineering Record* 48 (1903): 117, quoted in Joel A. Tarr, Terrie Yosie, and James McCurley, "Disputes over Water Quality Policy: Professional Cultures in Conflict, 1900-1917," *American Journal of Public Health* 70 (1980): 429.
79 Allen Hazen, *Clean Water and How to Get It* (New York: John Wiley and Sons, 1907), 35, 34.
80 Ibid., 35-36.
81 Ibid., 36-37.
82 Whipple, *The Value of Pure Water,* 7-9, 45, 51.
83 C.-E.A. Winslow and Earle B. Phelps, *Purification of Boston Sewage,* US Geological Survey Water-Supply and Irrigation Paper No. 185 (Washington, DC: Government Printing Office, 1906), 12.
84 Donald J. Pisani, "Fish Culture and the Dawn of Concern over Water Pollution in the United States," *Environmental Review* 8 (1984): 124; Clayton R. Koppes and William P. Norris, "Ethnicity, Class, and Mortality in the Industrial City: A Case Study of Typhoid Fever in Pittsburgh, 1890-1910," *Journal of Urban History* 11 (1985): 259; see also Mark J. Tierno, "The Search for Pure Water in Pittsburgh: The Urban Response to Water Pollution, 1893-1914," *Western Pennsylvania Historical Magazine* 70 (January 1977): 23-36.
85 Whipple, *The Value of Pure Water,* 37.
86 Ibid.; *State Sanitation: A Review of the Work of the Massachusetts State Board of Health,* 2 vols. (Cambridge, MA: Harvard University Press, 1917); Whipple with E.O. Jordan and C.-E.A. Winslow, *A Pioneer of Public Health: William Thompson Sedgwick* (New Haven: Yale University Press, 1924); Whipple, *The Microscopy of Drinking Water* (John Wiley, 1927).
87 Tarr et al., "Disputes over Water Quality," 430-31.
88 Gottlieb, *Forcing the Spring,* 57-59. See also Jennifer Read, "'A Sort of Destiny': The Multi-jurisdictional Response to Great Lakes Pollution, 1900-1930," *Scientia Canadensis* 55 (1999): 122; William L. Andreen, "The Evolution of Water Pollution Control in the United States: State, Local and Federal Efforts, 1789-1972: Part I," *Stanford Environmental Law Journal* 22, 1 (January 2003): 173.
89 Gene Desfor, "Planning Urban Waterfront Industrial Districts: Toronto's Ashbridge's Bay, 1889-1910," *Urban History Review* 17 (1988): 88; R.E. Riendeau, "Servicing the Modern City, 1900-1930," in Victor L. Russell, ed., *Forging a Consensus: Historical Essays on Toronto* (Toronto: University of Toronto Press, 1984), 162, 177; Catherine Brace, "Public Works in the Canadian City: The Provision of Sewers in Toronto, 1870-1913," *Urban History Review* 23 (March 1995): 33.
90 Files on the Fieldhouse litigation are located in the Archives of Ontario, Attorney General's Records, 1916, RG 4-32, no. 1578, "Fieldhouse v. City of Toronto" (Fieldhouse File), and in City of Toronto Legal Records Office (Toronto, Fieldhouse File).

91 Archives of Ontario, Fieldhouse File, "Statement of Claim."
92 Ibid., Commissioner of Works to Mayor Hocken and the Board of Control, 16 June 1914.
93 Ibid., R.M. Bell, Provincial Inspector of Health, to Provincial Board of Health, "Report re Nuisance Main Sewage Works, Toronto," 14 May 1915.
94 Ibid., Commissioner of Works and Medical Officer of Health to Mayor Church and Board of Control, 21 July 1916; "Memorandum for the Honourable the Attorney General," 5 December 1916.
95 *Toronto Daily Star*, 14 December 1920, 20.
96 *Fieldhouse v. Toronto* (1918), 43 O.L.R. 491 at 494.
97 Toronto, Fieldhouse File, William Johnston, City Solicitor, to Mayor Church and the Board of Control, 18 July 1918.
98 Ibid., Toronto City Council, Minutes, 1919, vol. 1, 27 January 1919 and 24 February 1919.
99 *Local Government Amendment Act, 1861* (U.K.), 24 & 25 Vict., c. 61, s. 4.
100 *A.G. v. Cockermouth Local Board* (1874), 18 L.R. Eq. 172 at 177. Other cases in which the legislation was examined include *Hainesworth v. West Riding Rivers Board* (1902), 5 L.G.R 356; *Staffordshire CC v. Seisdon RDC* (1907), 96 L.T. 328; *Hesketh v. Birmingham*, [1924] 1 K.B. 260.
101 *Public Health Act, 1875* (U.K.), 38 & 39 Vict., c. 55, s. 17.
102 See also *A.G. v. Halifax*, (1869) 39 L.J. (N.S.) 129.
103 *Attorney General v. Birmingham, Tame and Rea District Drainage Board*, [1908] 2 Ch. 551 at 553-54.
104 Ibid., 554.
105 Ibid., 555-56.
106 *Attorney General v. Birmingham, Tame, and Rea District Drainage Board*, [1910] 1 Ch. 48 at 55 (C.A.).
107 Ibid., 58.
108 Ibid., 56-57.
109 Ibid., 59.
110 Ibid., 50.
111 Ibid., 50-51. These findings were eventually accepted by the House of Lords in *Attorney-General v. Birmingham, Tame and Rea District Drainage Board*, [1912] H.L. 788.
112 David Kinnersley, *Troubled Water: Rivers, Politics and Pollution* (London: Hilary Shipman, 1988), 67.
113 James A. Alleman and T.B.S. Prakasam, "Reflections on Seven Decades of Activated Sludge History," *Journal of the Water Pollution Control Federation* 55 (1985): 436.
114 *City of Milwaukee v. Activated Sludge, Inc.*, 69 F.2d 577 at 593 (1934).
115 Alleman and Prakasam, "Reflections on Seven Decades of Activated Sludge History"; *City of Milwaukee v. Activated Sludge;* F.W. Mohlman, "Twenty-five Years of Activated Sludge," in Pearse, ed., *Modern Sewage Disposal*, 68-83. See also Pearse, ed., 8-11, 13, 283.

Chapter 10: Streams Are Nature's Sewers

1 John Duffy, *The Sanitarians: A History of American Public Health* (Urbana and Chicago: University of Illinois Press, 1990), 176.

2 "Tribune Topics," *Lawrence Evening Tribune,* 26 January 1891, quoted in Theodore Steinberg, *Nature Incorporated: Industrialization and the Waters of New England* (Cambridge: Cambridge University Press, 1991), 235.
3 Sir George Mackenzie, in *Mayor of Berwick v. Laird of Haining* (1661, Mor. Dict. 12772), quoted in J.O. Taylor, *The Law Affecting River Pollution* (Edinburgh: W. Green and Son, 1928), 21.
4 Taylor, *The Law Affecting River Pollution,* 22-25. On the early Edinburgh experience, see P.J. Smith, "The Foul Burns of Edinburgh: Public Health Attitudes and Environmental Change," *Scottish Geographical Magazine* 91 (1975): 25-37.
5 *Wood v. Waud* (1849), 154 E.R. 1047 at 1060-61.
6 *Merrifield v. Worcester* (1872), 110 Mass. 216 at 219-20.
7 *Ormerod v. Todmorden Mill* (1883), 11 Q.B. 155 at 168 (Brett M.R.).
8 *Earl of Harrington v. Corporation of Derby* (1905), 92 Law Times 153.
9 Ibid., 156.
10 Ibid., 158.
11 *McIntyre v. McGavin,* [1893] A.C. 268 at 277; See also *West Riding of Yorkshire v. Yorkshire Indigo, Scarlet & Colour Dyers, Ltd.* (1903), 67 J.P. 80.
12 Henry Bixby Hemenway, *Legal Principles of Public Health Administration* (Chicago: T.H. Flood, 1914), 701.
13 Joel A. Tarr, Terrie Yosie, and James McCurley, "Disputes over Water Quality Policy: Professional Cultures in Conflict, 1900-1917," *American Journal of Public Health* 70 (1980): 428.
14 *Wheatcroft v. Matlock Local Board* (1885), 52 Law Times 356.
15 *Falconar v. South Shields Corporation* (1895), 11 T.L.R. 223.
16 The sea and designated tidal waters might equally be classified as streams under the legislation, although watercourses mainly used as sewers prior to the *Rivers Pollution Prevention Act,* 1876, and flowing directly into the sea were excluded from stream status.
17 *West Riding of Yorkshire Rivers Board v. Gaunt (Reuben) and Sons, Ltd.* (1902), 67 J.P. 183, 19 T.L.R. 140 at 141.
18 *Newcastle-upon-Tyne v. Houseman* (1898), 63 J.P. 85; Bertram Jacobs, *A Manual of Public Health Law* (London: Sweet and Maxwell, 1912), 27.
19 *Airdrie Magistrates v. Lanark County Council,* [1910] A.C. 286.
20 Ibid., 289.
21 Ibid., 290.
22 Ibid., 291.
23 *George Legge & Son v. Wenlock Corp,* [1938] 1 All E.R. 37, [1938] A.C. 204.
24 Ibid., 40-41.
25 Ibid., 42.
26 *West Riding of Yorkshire Rivers Board v. Gaunt (Reuben) & Sons, Ltd.* (1902), 19 T.L.R. 140; *West Riding of Yorkshire Rivers Board v. Preston & Sons* (1904), L.T. 24.
27 *Butterworth v. West Riding of Yorkshire Rivers Board,* [1909] A.C. 45. A similar result was reached in *Kirkheaton Local Board v. Ainley,* [1892] 2 Q.B. 274.
28 *Young v. Bankier Distillery,* [1893] A.C. 691, 69 L.T. 838, 58 J.P. 100.
29 *Butterworth v. West Riding of Yorkshire Rivers Board,* [1909] A.C. 45.
30 London's Fleet is among the best known of these. See Nicholas Barton, *The Lost Rivers of London,* rev. ed. (London: Historical Publications, 1992).
31 Christine Rosen, "Differing Perceptions of the Value of Pollution Abatement across

Time and Place: Balancing Doctrine in Pollution Nuisance Law, 1840-1906," *Law and History Review* 11 (1993): 303.
32 *Sanderson v. Pennsylvania Coal,* 113 Pa. St. 126. Rosen, "Differing Perceptions," 334, 355, 364.
33 Quoted in *John Young v. Bankier,* [1891-94] All E.R. 439 (House of Lords) at 443.
34 Rosen, "Differing Perceptions," 303.
35 Ibid., 315.
36 Ibid., 319.
37 Ibid., 317-19.
38 Ibid., 322.
39 Ibid., 339.
40 Ibid., 340.
41 *Reideman v. Mr. Morris Electric Light Co.,* 56 App. Div. 23, 67 N.Y.S. 391 at 394 (1900), quoted in Rosen, "Differing Perceptions," 309.
42 *Nepisiguit Real Estate and Fishing Company v. Canadian Iron Corporation* (1913), 42 N.B.R. 387 at 394.
43 Rosen, "Differing Perceptions," 367.
44 Ibid., 369.
45 William L. Andreen, "The Evolution of Water Pollution Control in the United States: State, Local and Federal Efforts, 1789-1972: Part I," *Stanford Environmental Law Journal* 22, 1 (January 2003): 166.
46 Ibid., 380.
47 Clement Higgins, *Pollution and Obstruction of Watercourses* (London: Stevens and Haynes, 1877), 4.
48 *John Young v. Bankier,* 443.
49 Martin V. Melosi, *The Sanitary City: Urban Infrastructure in America from Colonial Times to the Present. Creating the North American Landscape* (Baltimore: Johns Hopkins University Press, 2000), 4.
50 Ibid., 9.
51 Taylor, *The Law Affecting River Pollution,* 9.
52 E.B. Besselièvre, "Statutory Regulation of Stream Pollution and the Common Law," *American Institute of Chemical Engineers, Transactions* 16 (1924): 217.
53 Marshall Ora Leighton, *Sewage Pollution in the Metropolitan Area near New York City and Its Effect on Inland Water Resources,* US Geological Survey Water-Supply and Irrigation Paper No. 72 (Washington, DC: Government Printing Office, 1902), 61.
54 Besselièvre, "Statutory Regulation of Stream Pollution," 217.
55 Pinchot, quoted in W.L. Stevenson, "The State vs. Industry, or the State with Industry," *American Institute of Chemical Engineers, Transactions* 16 (1924): 201.
56 Ibid.
57 Ibid., 205.
58 Besselièvre, "Statutory Regulation of Stream Pollution," 218.
59 Stevenson, "The State vs. Industry," 205.
60 W.L. Stevenson, "The Sanitary Conservation of Streams by Co-operation," *American Institute of Chemical Engineers, Transactions* 27 (1931): 9; Joel A. Tarr, "Industrial Wastes and Municipal Water Supplies in the 1920s," in Howard Rosen and Ann Durkin Keating, eds., *Water and the City: The Next Century* (Chicago: Public Works Historical Society, 1991), 261. In 1928, Pennsylvania industries agreed to implement

and maintain the protective measures then under development, to conduct regular testing programs, to notify the board in case of any accident threatening to interfere with the elimination or treatment of wastewaters, and, in the event of any such accident, to undertake emergency measures, including expeditious repair. In turn, Pennsylvania's Sanitary Water Board assumed responsibility for alerting waterworks and health officials downstream of any accident so that they could take precautions.

61 Stevenson, "The State vs. the Industry," 202.
62 *Water Pollution in the United States: Third Report of the Special Advisory Committee on Water Pollution* (Washington, DC: Government Printing Office, 1939), 56-57.
63 *First Report of the Joint Advisory Committee on River Pollution* (London, 1928), 4; Ministry of Health, Rivers Pollution Prevention Sub-committee of the Central Advisory Water Committee, *Prevention of River Pollution* (London, 1949), 14-17.
64 A.V. Delaporte, "Discussion," following Besselièvre, "Statutory Regulation of Stream Pollution," 228.
65 Ibid.
66 Ibid., 229.
67 Ibid., 229-30.
68 John Capper, Garrett Power, and Frank Shivers Jr., *Chesapeake Waters: Pollution, Public Health, and Public Opinion, 1607-1972* (Centreville, MD: Tidewater Publishers, 1983), 92-96.
69 Commissioners of Fisheries of Virginia, *Annual Report*, Fiscal Years 1915-1916, 12, quoted in ibid., 93.
70 *City of Hampton v. Watson*, 89 S.E. (Virginia 1916) 81-83, quoted in Capper, Power, and Shivers, *Chesapeake Waters*, 94-95.
71 *Darling v. City of Newport News*, 249 U.S. 540-44 (1918), quoted in Capper, Power, and Shivers, *Chesapeake Waters*, 95-96.

Chapter 11: Riparian Resurrection

1 George Chandler Whipple, "Shall Waterways Be Sewers Forever?" *American City* 35 (August 1926): 197, quoted in Martin V. Melosi, *The Sanitary City: Urban Infrastructure in America from Colonial Times to the Present. Creating the North American Landscape* (Baltimore: Johns Hopkins University Press, 2000), 257.
2 H.G. Maurice, "Come Ye to the Waters," *New Statesman*, 1 January 1944, 7.
3 W.L. Stevenson, "The Sanitary Conservation of Streams by Co-operation," *American Institute of Chemical Engineers, Transactions* 27 (1931): 9.
4 A.V. Delaporte, in *American Institute of Chemical Engineers, Transactions* 16 (1924): 215-16.
5 H.D. Turing, "River Pollution: Its Cause and Cure," *Illustrated London News*, 16 April 1949, 504.
6 GKM, "Passing of a River," *Blackwood's Magazine*, Janury 1947, 24.
7 Aldo Leopold, *A Sand County Almanac* (1949, repr., New York: Sierra Club and Ballantine Books, 1970), 124.
8 Fairfield Osborn, *Our Plundered Planet* (Boston and Toronto: Little, Brown, 1948), 194; see also William Vogt, *Road to Survival* (New York: William Sloane, 1948).
9 Leopold, *Sand County Almanac*, 240; Roderick Nash, *Wilderness and the American Mind*, 3rd ed. (New Haven: Yale University Press, 1982), 194-99.

10 Correspondence, Attorney General's Records, 1913, RG 4-32, no. 239, Archives of Ontario.
11 Dr. T.E. Kaiser to the Honourable J.B. Lucas, 23 April 1915, RG 4-32, no. 660, Archives of Ontario.
12 H.W. Streeter, "Disposal of Sewage in Inland Waters," in Langdon Pearse, ed., *Modern Sewage Disposal: Anniversary Book of the Federation of Sewage Works Associations* (New York: Federation of Sewage Works Associations, 1938), 200.
13 A.V. Delaporte to F.A. Dallyn, "Re Don River," 19 May 1925, RG 62, G-1, file 489.8, Archives of Ontario.
14 Craig E. Colten, "Creating a Toxic Landscape: Chemical Waste Disposal Policy and Practice, 1900-1960," *Environmental History Review* 85 (1994): 86-95.
15 George W. Fuller, "Administrative Problems in the Control of Pollution of Streams," *American Journal of Public Health* 16 (1926): 778.
16 Melosi, *The Sanitary City*, 226-28; Joel A. Tarr, "Industrial Wastes and Municipal Water Supplies in the 1920s," in Howard Rosen and Ann Durkin Keating, eds., *Water and the City: The Next Century* (Chicago: Public Works Historical Society, 1991).
17 John Emerson Monger, "Administrative Phases of Stream Pollution Control," *American Journal of Public Health* 16 (1926): 789.
18 Melosi, *The Sanitary City*, 216.
19 Monger, "Administrative Phases of Stream Pollution Control," 791.
20 E.B. Besselièvre, "Statutory Regulation of Stream Pollution and the Common Law," *American Institute of Chemical Engineers, Transactions* 16 (1924): 225.
21 William L. Andreen, "The Evolution of Water Pollution Control in the United States: State, Local and Federal Efforts, 1789-1972: Part I," *Stanford Environmental Law Journal* 22, 1 (January 2003): 171.
22 Dr. McNally to Dr. McCulloch, enclosing Canadian Milk Products letter by W.A. Rutter to Dr. George A. Shannon of St. Thomas, 21 August 1920, RG 62, B2A, Archives of Ontario.
23 *Water Pollution in the United States: Third Report of the Special Advisory Committee on Water Pollution* (Washington, DC: Government Printing Office, 1939), 68-72. A similar inventory of limitations is found in Abel Wolman, "State and Other Governmental Functions in the Control and Abatement of Water Pollution in the United States," in Langdon Pearse, ed., *Modern Sewage Disposal: Anniversary Book of the Federation of Sewage Works Associations* (New York: Federation of Sewage Works Associations, 1938), 291-93.
24 *Groat v. City of Edmonton*, [1928] S.C.R. 522 at 532-33.
25 Ibid., 528.
26 Keith Hawkins, *Environment and Enforcement: Regulation and the Social Definition of Pollution* (Oxford: Clarendon Press, 1982).
27 *Groat v. City of Edmonton*, [1928] S.C.R. 522 at 534.
28 Ibid., 540.
29 Ibid., 541.
30 Quoted in Roger Bate, *Saving Our Streams: The Role of the Anglers' Conservation Association in Protecting English and Welsh Rivers* (London: Institute of Economic Affairs, 2001), 36.
31 In 1994 the ACA merged with the Pure Rivers Society and adopted the name Anglers' Conservation Association.
32 Roger Bate, "English and Welsh Rivers: A Common Law Approach to Pollution Prevention" (MPhil diss., Cambridge University, 1993).
33 *Pride of Derby v. British Celanese*, [1952] 1 All E.R. 179 at 204.

34 Ibid., 203.
35 "A Halt to Pollution," *Times*, 16 December 1952.
36 Bate, "English and Welsh Rivers," 67, 69.
37 P. Boyer, *A Passion for Justice: The Legacy of James Chalmers McRuer* (Toronto: Osgoode Society for Canadian Legal History, 1994), 228-35.
38 *McKie et al. v. KVP Co. Ltd.*, [1948] 3. D.L.R. 201 at 214 (Ont. H.C.).
39 An Act to Amend the Lakes and Rivers Improvement Act, S.O. 1949, c. 48, s. 6, amending the *Lakes and Rivers Improvements Act*, R.S.O. 1937, c. 45, s. 30. Elizabeth Brubaker describes the measure as a device to "crush the property rights of those living downstream from pulp and paper mills." *Property Rights in the Defence of Nature* (London: Earthscan Publications, 1995), 74.
40 The Court of Appeal's deliberations and a judgment handed down on 22 November 1948 were unaffected by Ontario's legislative amendment, which did not go into effect until 1 April 1949.
41 Ontario, Legislative Assembly, *Debates* (24 February 1950), A-10 (Premier Leslie Frost).
42 *K.V.P. v. McKie*, [1949] S.C.R. 698 at 701.
43 *Stephens v. Richmond Hill*, [1955] 4 D.L.R. 572 at 575 (Ont. H. C.).
44 Ibid., 572.
45 Ibid., 578-79.
46 *Stephens v. The Village of Richmond Hill*, [1956] O.R. 88, [1956] 1 D.L.R. (2nd) 569.
47 The prohibition originated in the *Statute Law Amendment Act*, S.O. 1906, c. 19.
48 Ibid., 105.
49 *Borough of Westville v. Whitney Home Builders*, 40 N.J. Super. 62, 122 A.2d 233 (1956).
50 Arnold Reitze Jr., "Wastes, Water, and Wishful Thinking: The Battle of Lake Erie," *Case Western Reserve Law Review* 20 (1968): 65.
51 Quoted in Tom Lewis, *The Hudson: A History* (New Haven: Yale University Press, 2005), 270.
52 William H. Rodgers Jr., "Industrial Water Pollution and the Refuse Act: A Second Chance for Water Quality," *University of Pennsylvania Law Review* 119 (April 1971): 767; Arnold Reitze Jr., *Environmental Law*, 2 vols. (Washington, DC: North American International, 1972), vol. 1, pt. 4, 128-32.
53 Ibid., 130-32.
54 Rodgers, "Industrial Water Pollution and the Refuse Act," 766-69.
55 Reitze, *Environmental Law*, vol. 1, pt. 4, 118.
56 *United States v. Republic Steel*, 362 U.S. 482 (1960); *United States v. Standard Oil*, 384 U.S. 224 (1966).
57 W.O. Douglas, *Go East, Young Man: The Early Years* (New York: Random House, 1974), 207.
58 Justice Harlan in *United States v. Standard Oil*, 234.
59 The promise of early victories under the resurrected *Refuse Act* of 1899 was never fulfilled, in part as a consequence of administrative uncertainty and bottlenecks amounting to virtual paralysis in the permit process. Companies argued successfully that where permits were effectively unavailable because of administrative delays they should not be subject to conviction. Moreover, enforcement authorities determined, as a matter of policy, not to pursue prosecutions as widely as circumstances warranted. Reitze, *Environmental Law*, vol. 1, pt. 4, 125; Rodgers, "Industrial Water Pollution and the Refuse Act," 794-96.

60 Bate, *Saving Our Streams*, 15, 76, 90.
61 *KVP Company Limited Act*, S.O. 1950, c. 33.
62 *Toronto Globe and Mail*, 15 July 1952.
63 *Public Health Amendment Act*, S.O. 1956, c. 71, s. 6(1), amending the *Public Health Act*, R.S.O. 1950, c. 306, s. 106(22).
64 *Public Health Act*, R.S.O. 1950, c. 306, ss. 106(13)(d), (19)-(22).
65 *Toronto Globe and Mail*, 2 April 1956.
66 For general discussion, see Charles R. Wise, "Property Rights and Regulatory Takings," in Robert F. Durant, Daniel J. Fiorino, and Rosemary O'Leary, eds., *Environmental Governance Reconsidered: Challenges, Choices and Opportunities* (Cambridge, MA: MIT Press, 2004), 292-97.
67 C.A. Emerson Jr., Foreword, in Pearse, ed., *Modern Sewage Disposal*, viii.
68 Joel A. Tarr et al., "Water and Wastes: A Retrospective Assessment of Wastewater Technology in the United States, 1800-1932," *Technology and Culture* 25 (1984): 239; Joel A. Tarr, "The Separate vs. Combined Sewer Problem: A Case Study in Urban Technology Design Choice," *Journal of Urban History* 5 (1979): 329.
69 *Water Pollution in the United States: Third Report*, 1.
70 See "Sewage Disposal Plants Built in 1,200 Cities since 1935," *Public Management* 21 (April 1939): 117; "Sewage Disposal Construction at New Peak," *Public Management* 22 (June 1940): 178; "Census of Sewerage Systems Completed by Public Health Service," *Public Management* 24 (April 1942): 115-16.
71 "Sewage Disposal Construction at New Peak," *Public Management* 22 (1940): 178; "Sewage Disposal Plants Built in 1,200 Cities Since 1935," *Public Management* 21 (1939): 117.
72 "Census of Sewerage Systems Completed by Public Health Service," *Public Management* 24 (1942): 115; data current as of 1938 may be found in *Water Pollution in the United States: Third Report*, 6-7.
73 Melosi, *The Sanitary City*, 333.
74 Colten, "Creating a Toxic Landscape," 99.
75 On the Colorado, for example, see J.R. McNeill, *Something New under the Sun: An Environmental History of the Twentieth-Century World* (New York: W.W. Norton, 2000).
76 Kenneth A. Reid, "Pollution Control: A Post-war Public Works Opportunity for the States," *State Government* 18 (1945): 30.
77 James H. Duff, "The Effective Use of Water Resources," *State Government* 20 (1947): 219.
78 "Sewage Treatment Plants Needed to Reduce Water Pollution," *Public Management* 33 (1951): 134.
79 Reid, "Pollution Control: A Post-war Public Works Opportunity," 31.
80 Jamie Benidickson, *Water Supply and Sewage Infrastructure in Ontario, 1880-1990s: Legal and Institutional Aspects of Public Health and Environmental History*, Walkerton Inquiry, Commissioned Paper 1 (Toronto: Ontario Ministry of the Attorney General, 2002), 24-27, 35-36; J.R. Menzies, "Water Pollution in Canada by Drainage Basin," in *Resources for Tomorrow: Conference Background Papers*, 3 vols. (Ottawa: Queen's Printer, 1961), 1: 355.
81 Bate, "English and Welsh Rivers," 55.
82 "England Considers Water Problems: White Paper on Water Policy," *National Municipal*

Review, December 1944, 643; "Water Policy in Great Britain," *Nature,* 5 April 1944, 159-60; "Engineering the Rivers," *Illustrated London News,* 19 April 1944, 222; "Water Supply and Health," *Nature,* 25 November 1944, 660-62; "River Pollution: Its Cause and Cure," *Illustrated London News,* 16 April 1949, 504; "Water Pollution Research Board: Annual Report 1949," *Nature,* 3 March 1951, 351; Dirty Rivers, "Gardy Loo!" *Economist,* 5 December 1959, 193.

83 Ministry of Health, *Third Report of the Central Advisory Water Committee. River Boards* (London: HMSO, 1943); Ministry of Health, *Prevention of River Pollution* (London: HMSO, 1949).

84 It was nearly a quarter of a century before the last of these exemptions were brought within the scheme through the 1974 *Control of Pollution Act.*

85 David Kinnersley, *Troubled Water: Rivers, Politics and Pollution* (London: Hilary Shipman, 1988), 75-78.

86 F.T.K. Pentelow, *River Purification: A Legal and Scientific View of the Last 100 Years* (London: Edward Arnold, 1953), 18-19.

87 Louis Klein, *River Pollution* (London: Butterworths, 1959), 14-16.

88 Ibid., 14.

89 *Anglers' Cooperative Association Review,* 1978, 3, quoted in Bate, *Saving Our Streams,* 92.

90 Quoted in P.C. Yeager, "Industrial Water Pollution," in Michael Tonry and Albert J. Reiss Jr., eds., *Beyond the Law: Crime and Complex Organizations* (Chicago and London: University of Chicago Press, 1993), 119.

91 Frank J. Barry, "The Evolution of the Enforcement Provisions of the Federal Water Pollution Control Act: A Study of the Difficulty in Developing Effective Legislation," *Michigan Law Review* 68 (1970): 1111.

92 The Ohio board developed regulations to meet requirements of the federal *Water Pollution Control Act.*

93 Reitze, "The Battle of Lake Erie," 54-55.

94 Ibid., 54-55.

95 Peter Nemetz, "The Fisheries Act and Federal-Provincial Environmental Regulation: Duplication or Complementarity," *Canadian Public Administration* 29 (1986): 409-11.

96 Rodgers, "Industrial Water Pollution and the Refuse Act," 784.

97 Robert Andrews, Mary Biggs, and Michael Seidel, eds., *The Columbia World of Quotations* (New York: Columbia University Press, 1996), http://www.bartleby.com/66.

98 Aldous Huxley, "Hyperion to a Satyr," in *Adonis and the Alphabet and Other Essays* (London: Chatto and Windus, 1975), 147-48.

99 Ibid., 160, 156.

Chapter 12: Governing Water

1 Dirty Rivers, "Gardy Loo!" *Economist,* 5 December 1959, 193.
2 Ibid.
3 Ibid.
4 Leslie B. Wood, *The Restoration of the Tidal Thames* (Bristol: Adam Hilger, 1982), 60-61; Martin V. Melosi, *The Sanitary City: Urban Infrastructure in America from Colonial*

Times to the Present. Creating the North American Landscape (Baltimore: Johns Hopkins University Press, 2000), 333.
5 A.S. Davidsohn and B. Milwidsky, *Synthetic Detergents*, 7th ed. (New York: Wiley, 1987), 4-5; H.R. Jones, *Detergents and Pollution: Problems and Technological Solutions* (Park Ridge, NJ: Noyes Data Corporation, 1972), 6; Arnold Reitze Jr., *Environmental Law*, 2 vols. (Washington, DC: North American International, 1972), vol. 1, pt. 4, 26.
6 Russell L. Winget and Russell O. Blossor, "The Pulp and Paper Industry Pollution Abatement Program in the United States," in proceedings of the Ontario Water Resoures Commission, Fourth Industrial Waste Conference, 1957, 100.
7 *New York Times*, 25 July 1970, quoted in H.R. Jones, *Mercury Pollution Control* (New Jersey: Noyes Data Corporation, 1971), 1.
8 Jones, *Mercury Pollution Control*, 4; David H. Klein, "Sources and Present Status of the Mercury Problem," in Donald R. Buhler, ed., *Mercury in the Western Environment: Proceedings of a Workshop, Portland, Oregon, February 25-26, 1971* (Corvallis, OR: Continuing Education Publications, 1973), 5.
9 Katherine Montague and Peter Montague, *Mercury* (San Francisco and New York: Sierra Club, 1971), 76-77. See also William H. Rodgers Jr., "Industrial Water Pollution and the Refuse Act: A Second Chance for Water Quality," *University of Pennsylvania Law Review* 119 (April 1971): 793.
10 Frank M. D'Itri, *The Environmental Mercury Problem* (Cleveland: CRC Press, 1972), 96.
11 Klein, "Sources and Present Status," in Buhler, ed., *Mercury in the Western Environment*, 10.
12 National Research Council, Committee on Nitrate Accumulation, *Accumulation of Nitrate* (Washington, DC: National Academy of Sciences, 1972), 20.
13 In Texas in 2001, animal wastes amount to about 127 billion kilograms of manure annually – or 18 kilograms per Texan per day. (National Resources Defence Council report, cited in Molly Ivins, "Home, Home on the Latrine," *Time*, 6 August 2001, 29.) In Wisconsin, America's dairyland, livestock produce sufficient manure every day to fill the 76,000-seat football stadium in Madison. ("Manure Happens," *Economist*, 20 October 2001, 35.) In April 1993, bacterial contamination from manure triggered the closure of a Milwaukee water intake pipe, forcing 800 thousand people to boil tap water or buy bottled water. (G. Castle, "Agricultural Waste Management in Ontario, Wisconsin and British Columbia: A Comparison of Policy Approaches," *Canadian Water Resources Journal* 18 [1993]: 217.) Fish kills resulting from spilled animal wastes have in recent years severely damaged fish stocks in North Carolina, Iowa, Minnesota, and Missouri. (United States Senate Committee on Agriculture, Nutrition and Forestry, *Animal Waste Pollution in America: An Emerging National Problem – Environmental Risks of Livestock and Poultry Production*, December 1997, http://agriculture.senate.gov/animalw.htm.) Roughly 60 percent of US river and stream miles that are impaired have pollutants from agriculture, and it is estimated that 9 percent of impaired river and stream miles have been contaminated by animal feedlots. (Terence J. Centner and Jeffrey D. Mullen, "Enforce Existing Animal Feeding Operations Regulations to Reduce Pollutants," *Water Resources Management* 16 [2002]: 133-34.)
14 Ontario, Legislative Assembly, *Debates* (28 February 1956), 557.
15 Report of the Select Committee of the Ontario Legislature on Conservation, 1950, 103.

16 Jennifer Read, "Managing Water Quality in the Great Lakes Basin: Sewage Pollution Control, 1951-1960," in Lori Chambers and E.-A. Montigny, eds., *Ontario Since Confederation: A Reader* (Toronto: University of Toronto Press, 2000), 339-61.
17 *Toronto Globe and Mail*, 15 July 1952.
18 *Toronto Globe and Mail*, 2 April 1956.
19 Ontario, Pollution Control Board, "The Present Status of Stream Pollution in Ontario" (September 1955).
20 F.T.K. Pentelow, *River Purification: A Legal and Scientific Review of the Last 100 Years* (London: Edward Arnold, 1953), 12-15.
21 W.R.D. Sewell and L.R. Barr, "Evolution in the British Institutional Framework for Water Management," *Natural Resources Journal* 17 (1977): 401.
22 Ibid., 400.
23 Jon Tinker, "River Pollution: The Midlands Dirty Dozen," *New Scientist*, 6 March 1975, 552.
24 *Water Pollution in the United States: Third Report of the Special Advisory Committee on Water Pollution* (Washington, DC: Government Printing Office, 1939), 80-81.
25 James H. Duff, "The Effective Use of Water Resources," *State Government* 20 (1947): 219.
26 A.E. Berry, Address, Engineering Institute, 1955.
27 Arnold Reitze Jr., "Wastes, Water, and Wishful Thinking: The Battle of Lake Erie," *Case Western Reserve Law Review* 20 (1968): 7-11. For a critique of the Cuyahoga River's role in inspiring US environmental legislation, see Jonathan H. Adler, "Fables of the Cuyahoga: Reconstructing a History of Environmental Protection," *Fordham Environmental Law Journal* 14 (2002-3): 89.
28 *Costas v. City of Fond du Lac*, 24 Wisconsin 2d 409 (1963).
29 Donald M. Carmichael, "Forty Years of Water Pollution Control in Wisconsin: A Case Study," *Wisconsin Law Review* (1967): 358, 412-13, 417.
30 Andrew Halloran, "Water Pollution Control in New York," *Albany Law Review* 31 (1967): 50.
31 E. Donald Elliot, Bruce A. Ackerman, and John C. Millian, "Toward a Theory of Statutory Evolution: The Federalization of Environmental Law," *Journal of Law, Economics and Organization* 1 (1985): 313-40; S. Elworthy and J. Holder, *Environmental Protection: Text and Materials* (London: Butterworths, 1997); Jonathan R. Macey and Henry N. Butler, "Federalism and the Environment," in Roger E. Meiners and Andrew P. Morriss, eds., *The Common Law and the Environment: Rethinking the Statutory Basis for Modern Environmental Law* (London: Rowman and Littlefield, 2000), 158-81; R.E. Meiners and B. Yandle, "Constitutional Choice for the Control of Water Pollution," *Constitutional Political Economy* 3 (1992): 359-80; E.A. Parson, "Environmental Trends and Environmental Governance in Canada," *Canadian Public Policy* 26 (2000): S125-43; Anthony Scott, "Economists, Environmental Policies and Federalism," unpublished manuscript; Richard B. Stewart, "International Trade and the Environment: Lessons from the Federal Experience," *Washington and Lee Law Review* 49 (1992): 1329-71; P. Weiland, L.K. Caldwell, and R. O'Leary, "The Evolution, Operation and Future of Environmental Policy in the United States," in R. Baker, ed., *Environmental Law and Policy in the European Union and the United States* (Westport, CT: Praeger, 1997).
32 Leslie C. Frank, *American Journal of Public Health* 16 (1926): 795.

33 Ralph W. Johnson and Gardner M. Brown Jr., *Cleaning Up Europe's Waters: Economics, Management and Policies* (New York: Praeger, 1976), 199-201; Allen V. Kneese and Blair T. Bower, *Managing Water Quality: Economics, Technology, Institutions* (Johns Hopkins University Press, 1968), 262-65.
34 H.G. Maurice, "Come Ye to the Waters," *New Statesman*, 1 January 1944, 7.
35 C.M. Yonge, "Clean Rivers," *New Statesman*, 14 August 1948, 131.
36 David Kinnersley, *Troubled Water: Rivers, Politics and Pollution* (London: Hilary Shipman, 1988), 75-78.
37 *Spectator*, 27 June 1952, 856, and 4 July 1952, 21.
38 Louis Klein, *River Pollution* (London: Butterworths, 1959), 14.
39 Alexandre Kiss and Dinah Shelton, *Manual of European Environmental Law* (Cambridge: Cambridge University Press, 2004), 246; Kinnersley, *Troubled Water*, 92-93; Sewell and Barr, "Evolution in the British Institutional Framework for Water Management," 405.
40 Sewell and Barr, "Evolution in the British Institutional Framework for Water Management," 400-1.
41 Tinker, "River Pollution: The Midlands Dirty Dozen," 551.
42 *Control of Pollution Act* (U.K.), 1974, c. 40.
43 Kinnersley, *Troubled Water*, 121.
44 Ibid., 120-21.
45 Charles L. Reese, discussion following W.L. Stevenson, "The State vs. Industry or the State with Industry," *American Institute of Chemical Engineers, Transactions* 16 (1924): 208.
46 For a proposal to establish river-basin authorities under federal jurisdiction, see Marc J. Roberts, "River Basin Authorities: A National Solution to Water Pollution," *Harvard Law Review* 83 (1970): 1544-56.
47 Henry W. Toll, "The States: Co-operation or Obliteration?" *State Government* 9 (November 1936): 229.
48 Ellwood J. Turner, "Four States and a River," *State Government* 11 (1939): 210; *Water Pollution in the United States: Third Report*, 57-60, 76-77.
49 *Water Pollution in the United States: Third Report*, 77.
50 Turner, "Four States and a River," 210.
51 Ellwood J. Turner, "The Delaware," *State Government* 19 (1946): 220.
52 Russell F. Weigley, *Philadelphia: A 300-Year History* (New York: W.W. Norton and Company, 1982), 693; Roscoe C. Martin et al., *River Basin Administration and the Delaware* (Syracuse, NY: Syracuse University Press, 1960).
53 Seth G. Hess, "Interstate Action to Control Pollution," *State Government* 23 (1950): 204. In the US, citizen efforts dating from 1922 in the form of the National Coast Anti Pollution League encouraged legislative action against oil pollution. Corresponding measures were enacted in Britain at the same time.
54 Ibid.
55 "Ohio River Sanitation Compact," *State Government* 12 (1939): 202.
56 Joel A. Tarr, "Industrial Wastes and Municipal Water Supplies in the 1920s," in Howard Rosen and Ann Durkin Keating, eds., *Water and the City: The Next Century* (Chicago: Public Works Historical Society, 1991), 267.
57 Frank Raschig, "The Ohio," *State Government* 19 (1946): 230-31; US Public Health Service and US Army Corps of Engineers, *Ohio River Pollution Control*, pt. 1, Report

of the Ohio River Committee (H.R. Doc. No. 299) (Washington, DC: Government Printing Office, 1944).
58 W.L. Stevenson, "The Sanitary Conservation of Streams by Cooperation," *American Institute of Chemical Engineers, Transactions* 27 (1931): 21-24. The participating states were Illinois, Indiana, Kentucky, Maryland, New York, North Carolina, Ohio, Pennsylvania, Tennessee, Virginia, and West Virginia.
59 "State Action Approved," *State Government* 13 (1940): 192.
60 *Water Pollution in the United States: Third Report,* 75.
61 "New England Water Control Compact," *State Government* 21 (1948): 101.
62 Gilbert F. White, "A Perspective on River Basin Development," *Law and Contemporary Problems* 22 (1957): 158.
63 Frank Trelease, "A Modern State Code for River Basin Development," *Law and Contemporary Problems* 22 (1957): 303.
64 Raschig, "The Ohio," 231.
65 Kenneth A. Reid, "Pollution Control: A Post-war Public Works Opportunity for the States," *State Government* 18 (1945): 32.
66 Ibid.
67 Ibid.
68 "Congress Tackles River Pollution Control," *State Government* 13 (1940): 67.
69 Ibid., 73.
70 Craig E. Colten, "Creating a Toxic Landscape: Chemical Waste Disposal Policy and Practice, 1900-1960," *Environmental History Review* 85 (1994): 96-98. Some of the proposed legislation is listed in Abel Wolman, "State and Other Governmental Functions in the Control and Abatement of Water Pollution in the United States," in Pearse, ed., *Modern Sewage Disposal,* 287.
71 Cleveland M. Bailey, "Shall We 'Pass the Buck' or Pass Legislation?" *State Government* 19 (1946): 9.
72 "Congress Tackles River Pollution Control," 74.
73 Melosi, *The Sanitary City,* 315.
74 *Water Pollution Control Act,* 62 Stat. 1155, c. 758 (1948).
75 Colten, "Creating a Toxic Landscape," 107.
76 L.B. Dworsky, "Assessing North America's Management of Its Transboundary Waters," *Natural Resources Journal* 33 (1993): 424.
77 Frank E. Smith, *The Politics of Conservation* (New York: Harper Colophon, 1971), 290.
78 Reitze, *Environmental Law,* vol. 1, pt. 4, 34.
79 The most notable reports addressing water issues were those of the President's Water Resources Policy Commission (1950), the Missouri Basin Survey Commission (1953), the Commission on Intergovernmental Relations (1955), the Commission on Organization of the Executive Branch of Government (1955), and the Presidential Advisory Committee on Water Resources Policy (1955).
80 Ernest A. Englebert, "Federalism and Water Resources Development," *Law and Contemporary Problems* 22 (1957): 325.
81 James W. Fesler, "National Water Resources Administration," *Law and Contemporary Problems* 22 (1957): 444.
82 Melosi, *The Sanitary City,* 335.
83 Smith, *The Politics of Conservation,* 291.

84 Frank J. Barry, "The Evolution of the Enforcement Provisions of the *Federal Water Pollution Control Act*," *Michigan Law Review* 68 (1970): 1112.
85 Ibid., 1108.
86 Ibid., 1105.
87 Ibid., 1106.
88 Ibid., 1111.
89 Ibid., 1118-25.
90 Rodgers, "Industrial Water Pollution and the Refuse Act," 769-84.
91 More comprehensive efforts oriented around ecosystem functions and total maximum daily loads subsequently added to the complexity of the regulatory challenge. For an overview, see Michael M. Wenig, "How 'Total' Are 'Total Maximum Daily Loads'?" *Tulane Environmental Law Journal* 12 (1998): 87.
92 James Boyd, *The New Face of the Clean Water Act*, Discussion Paper 00-12 (Washington, DC: Resources for the Future, 2000), 3.
93 Bruce A. Ackerman, Susan Rose Ackerman, James W. Sawyer, and Dale W. Henderson, *The Uncertain Search for Environmental Quality* (New York: Free Press, 1974), 317-27.
94 William L. Andreen, "Water Quality Today – Has the Clean Water Act Been a Success?" *Alabama Law Review* 55 (2004): 537; A. Dan Tarlock, "Putting Rivers Back in the Landscape," *Hastings West-Northwest Journal of Environmental Law and Policy* 6 (1999-2000): 167.
95 Ontario, Legislative Assembly, *Debates* (28 March 1957), 1778 (Hon. Leslie M. Frost).
96 B. Grover and David Zussman, *Safeguarding Canadian Drinking Water* (Ottawa: Inquiry on Federal Water Resources, 1985).
97 The Resources for Tomorrow Conference, 23-28 October 1961.
98 J.R. Menzies, "Water Pollution in Canada by Drainage Basin," in *Resources for Tomorrow: Conference Background Papers*, 3 vols. (Ottawa: Queen's Printer, 1961), 1: 353.
99 Ibid.
100 Express provision for federal financial contributions made the new statute more attractive to the provinces than earlier, more restrictive, cost-sharing arrangements under the *Canada Water Conservation Assistance Act* of 1953. Initiatives under the *Canada Water Act* – some seventy agreements during the following quarter-century – fell overwhelmingly within the scope of these provisions. Several hundred million dollars went toward comprehensive planning and capital expenditures on costly structural works. L. Booth and F. Quinn, "Twenty-five Years of the Canada Water Act," *Canadian Water Resources Journal* 20 (1995): 72.
101 Ibid., 73.
102 Melville McMillan, "Perspectives on the Restructuring of Environmental Decision-Making Institutions: The Case of the Canada Water Act," *Canadian Water Resources Journal* 4 (1979): 60.
103 *Canada Water Act*, S.C. 1970, c. 52, ss. 9, 11.
104 *Canada Water Act*, s. 13(1). Subject to appropriate ministerial approval, including provincial approval for jointly constituted bodies, the agencies might proceed to implement the water quality plan by taking over the construction and operation of waste treatment facilities, the collection of fees, monitoring, inspection, and so on. *Canada Water Act*, s. 13(3).
105 "Pay-as-you-go Pollution Suggestion Turned Down," *Ottawa Citizen*, 4 November

1969, 23; "Water Act Won't Work in Ontario," *Toronto Globe and Mail*, 30 January 1970, 5; "Greene Defends Fees in Canada Water Act as Incentive to Firms," *Toronto Globe and Mail*, 3 February 1970, 4.
106 Hon. Jack Davis, quoted in J.W. Parlour, "The Politics of Water Pollution Control: A Case Study of the Canadian Fisheries Act Amendments and the Pulp and Paper Effluent Regulations, 1970," *Journal of Environmental Management* 13 (1981): 131.
107 Canada, *House of Commons Debates* (20 April 1970), quoted in Thomas Conway and G. Bruce Doern, *The Greening of Canada: Federal Institutions and Decisions* (Toronto: University of Toronto Press, 1994), 213.
108 Jack Davis, "The Fish and the Law That Guards against Pollution," *Toronto Globe and Mail*, 7 August 1969.
109 Kathryn Harrison, *Passing the Buck: Federalism and Canadian Environmental Policy* (Vancouver: UBC Press, 1996), 67-70.
110 Canada, *House of Commons Debates* (15 October 1970), quoted in Conway and Doern, *The Greening of Canada*, 214.
111 Harrison, *Passing the Buck*, 4.
112 Simon Ball and Stuart Bell, *Environmental Law* (London: Blackstone Press, 1991), 299. As measured with reference to biochemical oxygen demand, significant improvements were achieved during the 1940s. See Carolyn Abbot, "Environmental Command Regulation," in Benjamin J. Richardson and Stepan Wood, eds., *Environmental Law for Sustainability*, 61-96 (Oxford: Hart Publishing, 2006).
113 Roger Bate, *Saving Our Streams: The Role of the Anglers' Conservation Association in Protecting English and Welsh Rivers* (London: Institute of Economic Affairs, 2001), 91-98.
114 *Cambridge Water Co. Ltd v. Eastern Counties Leather Plc.*, [1994] 1 All E.R. 53; *Marcic v. Thames Water Utilities*, [2002] 2 All E.R. 55.
115 Lee Botts and Paul Muldoon, *The Great Lakes Water Quality Agreement: Its Past Successes and Uncertain Failure* (Hanover, NH: Dartmouth College, 1997).
116 John E. Carroll, *Environmental Diplomacy: An Examination and a Prospective of Canadian U.S. Transboundary Environmental Relations* (Ann Arbor: University of Michigan Press, 1983), 133-34.
117 The 1972 agreement was replaced in 1978 and further amended by a protocol in 1987.
118 Carroll, *Environmental Diplomacy*, 130.
119 *Great Lakes Water Quality Agreement*, 15 April 1972, art. 2.
120 The 1972 agreement had previously extended the scope of operation beyond the narrowly defined 1909 definition of boundary waters, to encompass tributary waters of the Great Lakes system.
121 Although the 1972 agreement made reference to toxic materials, it had really addressed problems of eutrophication and so-called "hazardous polluting substances," with the latter understood to be "any element or compound . . . which, when discharged in any quantity into or upon receiving waters or adjoining shorelines, presents an imminent or substantial danger to public health or welfare."
122 Carroll, *Environmental Diplomacy*, 134; Royal Commission on the Future of the Toronto Waterfront, *Regeneration: Toronto's Waterfront and the Sustainable City: Final Report* (Toronto: Minister of Supply and Services Canada, 1992), 104.
123 Carroll, *Environmental Diplomacy*, 142.
124 Quoted in Thomas W. Kierans, "The GRAND Canal Project – A Large Scale Water

Recycling Concept" (paper presented at the Futures in Water Conference, Toronto, 13 June 1984).
125 Ibid.
126 Frank Moss, *The Water Crisis* (New York: F.A. Praeger, 1967).
127 Frank Moss, "Toward a North American Water Policy," in Claude E. Dolman, ed., *Water Resources of Canada* (University of Toronto Press for the Royal Society of Canada, 1967), 4-7.
128 Anthony Scott, John Olynyk, and Steven Renzetti, "The Design of Water Export Policy," in John Whalley, ed., *Canada's Resource Industries and Water Export Policy* (Toronto: University of Toronto Press, 1986), 184.
129 Moss, *The Water Crisis*, 13.
130 D.B. Luten and Gordon Gould Jr., "NAWAPA: The Problem of Water" (ms., n.d.), 7-8.
131 Michael Keating, "Birthright Is Now under Attack," *Globe and Mail*, 24 January 1984.
132 Ludwig Kramer, *Focus on European Environmental Law* (London: Sweet and Maxwell, 1992); Kiss and Shelton, *Manual of European Environmental Law*, ch. 7 "Freshwaters." In 2000 the European Union adopted a new framework directive on water quality.
133 Geoffrey Lean, "Britain's Dirty Beaches Can Make You Sick," *Independent*, 20 January 1994.
134 Ball and Bell, *Environmental Law*, 304.
135 Ibid.; Julie Argent and Timothy O'Riordan, "The North Sea," in Jeanne X. Kasperson et al., eds., *Regions at Risk: Comparisons of Threatened Environments* (Tokyo: United Nations University Press), 405.
136 William Howarth, *Water Pollution Law* (London: Shaw and Sons, 1988), 298. Subsequent efforts to promote a more resolute response to land-based marine pollution have included a set of guidelines, a global program of action, and the *Washington Declaration on Protection of the Marine Environment from Land-Based Activities*, available on the UN Environmental Program website, http:// www.gpa.unep.org/document_lib/en/pdf/washington_declaration.pdf.
137 Nigel Haigh, *EEC Environmental Policy and Britain*, 2nd rev. ed. (London: Longman, 1989), 22. US and Canadian coastal communities engaged in foot-dragging exercises of their own to avoid or delay introducing or upgrading wastewater treatment facilities. See R.H. Burroughs, "Ocean Dumping: Information and Policy Development in the USA," *Marine Policy* 12 (1988): 96; Eric Jay Dolan, *Political Waters: The Long, Dirty, Contentious, Incredibly Expensive but Eventually Triumphant History of Boston Harbor – A Unique Environmental Success Story* (Amherst: University of Massachusetts Press, 2004).
138 In 1987 there had been agreement to end the disposal of industrial waste in the North Sea by 1990. It was not until mid-decade, however, that the UK finally brought the practice to an end. Argent and O'Riordan, "The North Sea," 399.
139 Ibid., 405.
140 Ibid., 406.

Conclusion: Water Quality and the Future of Flushing

1 *The United Nations World Water Development Report: Water for People, Water for Life* (Paris: UNESCO, 2003).
2 US EPA, *National Water Quality Inventory 2000* (Washington, DC: United States

Environmental Protection Agency) see also Reed F. Noss and Robert L. Peters, Endangered Ecosystems: A Status Report on America's Vanishing Habitat and Wildlife (Washington, DC: Defenders of Wildlife, 1995), cited in Oliver Houck, "Biodiversity for Documenting Ecosystem and Species Decline; The Great Lakes: A Report Card," Canada-U.S. Law Journal 28 (2002): 455-85.

3 "Executive Summary," in Water for People: Water for Life, 9; see also P. Gleick, "Dirty Water: Estimated Deaths from Water-Related Diseases 2000-2020," Pacific Institute Research Report 2002, Pacific Institute for Studies in Development, Environment, and Security, http://www.pacinst.org/reports/water_related_deaths/water_related_deaths_report.pdf.

4 Joint Group of Experts on the Scientific Aspects of Marine Environmental Protection (GESAMP) and Advisory Committee on Protection of the Sea, *A Sea of Troubles* (The Hague:UN Environmental Programme, 2001).

5 Lena Lencek and Gideon Bosker, *The Beach: The History of Paradise on Earth* (New York: Viking, 1998).

6 *Harper's*, July 2000, 13.

7 UN Convention on the Law of the Sea, Article 207; Global Programme of Action (GPA) for the Protection of the Marine Environment from Land-Based Activities (1995); Montreal Guidelines to Implement the GPA (2001).

8 Melissa Bank, *The Girls' Guide to Hunting and Fishing* (New York: Penguin, 1999), 8. The comment is an updated trans-Atlantic echo of James Joyce's 1922 statement from *Ulysses*: "Unwholesome sandflats waited to suck his treading soles, breathing upward sewage breath."

9 Yahoo! Internet Life, December 1999. Matt Groening's assessment may be closer to the mark: when the creator of *The Simpsons* envisaged Futurama circa 3000, he incorporated inadequate plumbing as a very likely prospect.

10 Peter Mayle, *French Lessons: Adventures with a Knife, Fork, and Corkscrew* (New York: Knopf, 2001), 14.

11 Dennis R. O'Connor, Commissioner, *Report of the Walkerton Inquiry, Part Two: A Strategy for Safe Drinking Water* ([Toronto]: Ministry of Attorney General, 2002), pt. 2, 8.

12 Environment Commissioner of Ontario, *Thinking beyond the Near and Now*, Annual Report 2002-3, 35.

13 Jane Taber, "Martin Staff Flush with Anger," *National Post*, 11 November 2000.

14 "High Tech Toilets Catch Fire," *Toronto Globe and Mail*, 8 July 1999.

15 Linda Nash, "Water Quality and Health," in Peter H. Gleick, ed., *Water in Crisis: A Guide to the World's Freshwater Resources* (New York: Oxford University Press, 1993), 35.

16 D.P. Emond, "The Greening of Environmental Law," *McGill Law Journal* 36 (1993): 742; C.D. Hunt, "The Changing Face of Environmental Law and Policy," *University of New Brunswick Law Journal* 41 (1992): 79; R.M. M'Gonigle, "Taking Uncertainty Seriously: From Permissive Regulation to Preventative Design in Environmental Decision Making," *Osgoode Hall Law Journal* 32 (1994): 99. The theme is summarized in Neil J. Brennan, "Impediments to Environmental Quality and New Brunswick's Clean Environment Act: An Argument for a New Statute," *Journal of Environmental Law and Practice* 7 (1996): 93. See also Arnold Reitze Jr., "A Century of Air Pollution Control Law," *Environmental Law* 21: 1549.

17 David Robinson, "Regulatory Evolution in Pollution Control," in Tim Jewell and Jenny Steele, eds., *Law in Environmental Decision-Making: National, European and*

International Perspectives (Oxford: Clarendon Press, 1998), 29; Reitze, "A Century of Air Pollution Control Law," 1549.

18 I am grateful to Tobias Halverson for the concept of proxy in this context. See his "Solid Waste Agency vs. US Army Corps of Engineers: The Failure of 'Navigability' as a Proxy Demarcating Federal Jurisdiction for Environmental Protection," *Ecology Law Quarterly* 29 (2002): 181.

19 Arthur F. McEvoy, *The Fisherman's Problem: Ecology and Law in the California Fisheries, 1850-1980* (Cambridge: Cambridge University Press, 1986), 72.

20 Oliver A. Houck, "On the Law of Biodiversity and Ecosystem Management," *Minnesota Law Review* 81 (1997): 952-53.

21 David Robinson, "Regulatory Evolution in Pollution Control," in Jewell and Steele, eds., *Law in Environmental Decision-Making*, 29.

22 Quoted in Houck, "On the Law of Biodiversity," 875n17.

23 Mississauga mayor Hazell McCallion, quoted in Environmental Commissioner of Ontario, *Thinking beyond the Near and Now*, Annual Report 2002-3, 48.

24 On valuation of nature's services, see Ted Mosquin, "The Roles of Biodiversity in Creating and Maintaining the Ecosphere," in Stephen Bocking, ed., *Biodiversity in Canada: Ecology, Ideas, and Action* (Peterborough: Broadview Press, 1999), 112-13 ; R. Costanza et al., "The Value of the World's Ecosystem Services and Natural Capital," *Nature*, 1997, 253-60.

25 Patricia Birnie and Alan Boyle, *International Law and the Environment* (Oxford: Oxford University Press, 1992), 483.

26 Norman Myers, "The Meaning of Biodiversity Loss," in *Nature and Human Society*, 67. See also Peter H. Raven, ed., *Nature and Human Society* (Washington, DC: National Academy Press, 2000), 291-300.

27 Robert B. Keiter, "Beyond the Boundary Line: Constructing a Law of Ecosystem Management," *University of Colorado Law Review* 65 (1994): 302-3.

28 Ibid., 301.

29 James M. Omernik and Robert G. Bailey, "Distinguishing between Watersheds and Ecoregions," *Journal of the American Water Resources Association* 33 (1997): 935-49; M. Lynne Corn, *Ecosystems, Biomes, and Watersheds: Definitions and Use*, CRS Report for Congress (Washington, DC: Congressional Research Service, Library of Congress, 1993).

30 Carol Rose, "Demystifying Ecosystem Management," *Ecology Law Quarterly* 24 (1997): 865; Scott C. Anderson, "Watershed Management and Non-point Source Pollution: The Massachusetts Approach," *B.C. Environmental Affairs Law Review* 26 (1999): 339; R.W. Adler, "Addressing Barriers to Watershed Protection," *Environmental Law* 25 (1995): 973-1106.

31 N. LeRoy Poff et al., "The Natural Flow Regime: A Paradigm for River Conservation and Restoration," Bioscience 47 (1997): 780.

32 For an example of the former, see the International Year of Freshwater 2003 website, http://www.wateryear2003.org.

33 Some signals are provided in the recent declaration by the European Parliament and Council that water is "a heritage which must be protected, defended and treated as such." Directive 2000/60/EC. California's Porter-Cologne Water Quality Control Act, Cal. Water Code, s. 13263(g) offered further guidance in relation to the future of flushing when it asserted that "all discharges of waste into waters of the state are privileges and not rights."

Suggested Reading

The references provided in the endnotes will guide readers to specific sources of information or quotations in the text. They include official reports, archival records, and primary documents.

This section serves two more general purposes. First, I would like to identify some of the publications I found most useful for their insights into particular aspects of a project that brings together questions concerning the histories of law, engineering, public health, medicine, and environment. Second, I would like to acknowledge a number of authors whose work contributed in a somewhat different manner to my own efforts. These publications I will simply describe as encouraging and inspiring. I am grateful that they came across my desk and thankful for the silent dialogue I have enjoyed with their authors.

The following discuss the history of waterworks systems, including aspects of their financial implications and engineering design:

Armstrong, Ellis L., ed. *History of Public Works in the United States, 1776-1976.* Chicago: American Public Works Association, 1976. Offers a comprehensive overview of water and sewerage.

Blake, Nelson M. *Water for the Cities: A History of the Urban Water Supply Problem in the United States.* Syracuse, NY: Syracuse University Press, 1956. Provides an excellent initial orientation.

Graham-Leigh, John. *London's Water Wars: The Competition for London's Water Supply in the Nineteenth Century.* London: Francis Boutle, 2000. Adds considerable local social and political detail to the more recent studies of the experience of individual cities.

Hassan, J.A. "The Growth and Impact of the British Water Industry in the Nineteenth Century." *Economic History Review,* 2nd ser., 38 (1985): 531-47. Collects and analyzes valuable background data.

Stanbridge, H.H. *History of Sewage Treatment in Britain.* Maidstone, Kent: Institute of Water Pollution Control, 1976. A valuable multivolume reference from the perspective of technological innovation.

Canadian researchers have devoted less attention to these issues, but the following are useful studies:

> Anderson, Letty. "Water and the Canadian City." In *Water and the City: The Next Century*, edited by Howard Rosen and Ann Durkin Keating, 40. Chicago: Public Works Historical Society, 1991. Assembles basic information.
> Brace, Catherine. "Public Works in the Canadian City: The Provision of Sewers in Toronto, 1870-1913." *Urban History Review* 23 (March 1995): 33-43. This and the next contribution are important case studies.
> Jones, Elwood, and Douglas McCalla. "Toronto Waterworks, 1840-77: Continuity and Change in Nineteenth Century Toronto Politics." *Canadian Historical Review* 60 (1979): 300-23.

Although waterworks and wastewater systems operate at an institutional and community level, the interconnections between these arrangements and the personal, private residential and individual circumstances of the users of water appliances equally require consideration. The two works below are illuminating accounts of the evolution and adoption of the household fixtures featured here:

> Ogle, Maureen. *All the Modern Conveniences: American Household Plumbing, 1840-1890*. Baltimore and London: Johns Hopkins University Press, 1996.
> Wright, Lawrence. *Clean and Decent: The Fascinating History of the Bathroom and the Water Closet*. Toronto: University of Toronto Press, 1967.

The history of public health contains several studies of the influence of Edwin Chadwick and his associates. For my purposes, however, a number of other works were particularly helpful in placing developments in public health within a wider social, political, professional, or institutional context:

> Evans, Richard J. *Death in Hamburg: Society and Politics in the Cholera Years, 1830-1910*. Oxford: Clarendon Press, 1987.
> Rosenkrantz, Barbara Gutmann. *Public Health and the State: Changing Views in Massachusetts, 1842-1936*. Cambridge, MA: Harvard University Press, 1972.
> Wohl, Anthony S. *Endangered Lives: Public Health in Victorian Britain*. London: J.M. Dent, 1983.

Environmental history, including studies of human impacts on water quality, was, of course, a central element of my own research. Several publications provided valuable points of reference or orientation:

> Bogue, Margaret Beattie. *Fishing the Great Lakes: An Environmental History, 1783-1933*. Madison: University of Wisconsin Press, 2000. Demonstrates the opportunity awaiting those willing to look beneath the surface.
> Capper, John, Garrett Power, and Frank R. Shivers Jr. *Chesapeake Waters: Pollution,*

Public Health, and Public Opinion, 1607-1972. Centreville, MD: Tidewater Publishers, 1983. An important regional study of an area that continues to present great environmental challenges.

MacLeod, Roy M. "Government and Resource Conservation: The Salmon Acts Administration, 1860-1886." *Journal of British Studies* 7 (1968): 114-50. MacLeod's work is another reminder of the early voices of environmental management and of the obstacles they encountered.

Rosen, Christine. "Noisome, Noxious, and Offensive Vapors, Fumes and Stenches in American Towns and Cities, 1840-1865." *Historical Geography* 25 (1997): 49-82. This author has produced a number of pioneering studies.

I benefited greatly from insights provided by several other early publications in environmental history, whose specific focus is water and water quality:

Allardyce, Gilbert. "'The Vexed Question of Sawdust': River Pollution in Nineteenth-Century New Brunswick." *Dalhousie Review* 52 (1972): 177-90.

Cowdrey, Albert E. "Pioneering Environmental Law: The Army Corps of Engineers and the Refuse Act." *Pacific Historical Review* 44 (1975): 331-49.

Gillis, R. Peter. "Rivers of Sawdust: The Battle over Industrial Pollution in Canada, 1865-1903." *Journal of Canadian Studies* 21 (1986): 84-103.

Pisani, Donald J. "Fish Culture and the Dawn of Concern over Water Pollution in the United States." *Environmental Review* 8 (1984): 117-31.

The addition of law to the environmental history mix thins out the contributors somewhat, but there are a number of rewarding publications offering overviews or model case studies:

McLaren, John P.S. "Nuisance Law and the Industrial Revolution – Some Lessons from Social History." *Oxford Journal of Legal Studies* 3 (1983): 155-221.

—. "The Tribulations of Antoine Ratté: A Case Study of the Environmental Regulation of the Canadian Lumber Industry in the Nineteenth Century." *University of New Brunswick Law Journal* 33 (1984): 203-59.

Murphy, Earl Finbar. *Water Purity: A Study in Legal Control of Natural Resources.* Madison: University of Wisconsin Press, 1961.

Several US legal scholars have made multi-part or multi-article additions to the law journals, where a vast range of case law, legislation, and literature has been assembled and synthesized:

Andreen, William L. "The Evolution of Water Pollution Control in the United States: State, Local and Federal Efforts, 1789-1972: Part I." *Stanford Environmental Law Journal* 22, 1 (January 2003): 145-200. This and the following work are impressive examples.

Hines, N. William. "Nor Any Drop to Drink: Public Regulation of Water Quality, Part I: State Pollution Control Programs." *Iowa Law Review* 52 (1966): 186-235.

Reitze, Arnold, Jr. "Wastes, Water, and Wishful Thinking: The Battle of Lake

Erie." *Case Western Reserve Law Review* 20 (1968): 65-86. This and the following study not only review past developments, but helped to influence policy direction.

Rodgers, William H., Jr. "Industrial Water Pollution and the Refuse Act: A Second Chance for Water Quality." *University of Pennsylvania Law Review* 119 (April 1971): 761-822.

The relationship of water quality to waste management and the environment more generally is a vast and complex subject. Studies either that survey the terrain in an orderly and authoritative way, or that offer detailed insights into elements of this evolving story are invaluable. Several of these were enormously helpful guides to my inquiries:

Corbin, Alain. *The Foul and the Fragrant: Odor and the French Social Imagination.* Cambridge, MA: Harvard University Press, 1986.

Goubert, Jean-Pierre. *The Conquest of Water: The Advent of Health in the Industrial Age.* Translated by Andrew Wilson. Cambridge: Polity Press, 1989.

Hamlin, Christopher. *A Science of Impurity: Water Analysis in Nineteenth Century Britain.* Bristol: Adam Hilger, 1990.

Kinnersley, David. *Troubled Water: Rivers, Politics and Pollution.* London: Hilary Shipman, 1988.

Luckin, Bill. *Pollution and Control: A Social History of the Thames in the Nineteenth Century.* Bristol: Adam Hilger, 1986.

Melosi, Martin V. *The Sanitary City: Urban Infrastructure in America from Colonial Times to the Present. Creating the North American Landscape.* Baltimore: Johns Hopkins University Press, 2000.

Schultz, Stanley K. *Constructing Urban Culture: American Cities and Urban Planning, 1800-1920.* Philadelphia: Temple University Press, 1989.

Steinberg, Theodore. *Nature Incorporated: Industrialization and the Waters of New England.* Cambridge: Cambridge University Press, 1991.

Tarr, Joel A. *The Search for the Ultimate Sink: Urban Pollution in Historical Perspective.* Akron, OH: University of Akron Press, 1996.

Illustration Credits

80 / Pan water closet. Source: Maureen Ogle, *All the Modern Conveniences: American Household Plumbing, 1840-1890* (Baltimore and London: Johns Hopkins University Press, 1996), 81.

81 / The "Unitas" water closet, 1880. Source: Maureen Ogle, *All the Modern Conveniences: American Household Plumbing, 1840-1890* (Baltimore and London: Johns Hopkins University Press, 1996), 139.

100 / "Henry Hastings, Nightman and Poleman." Source: Lawrence Wright, *Clean and Decent: the Fascinating History of the Bathroom and the Water Closet* (Toronto: University of Toronto Press, 1967), 145.

104 / Microscopist's findings. Source: Christopher Hamlin, *A Science of Impurity: Water Analysis in Nineteenth Century Britain* (Bristol: Adam Hilger, 1990), 103.

105 / "Death's Dispensary," actually titled *Fun*, 1860. Source: Stephen Halliday, *The Great Stink* (Sutton Publishing), 131.

111 / Sir Joseph Bazalgette. Source: Stephen Halliday, *The Great Stink* (Sutton Publishing), 78.

116 / "King Street Sewer, Midland, Ontario." Source: Archives of Ontario, Acc. 2375 S 5272.

123 / Sewage used as fertilizer. Source: *Harper's Weekly*, 1890. Source: Joel A. Tarr, *The Search for the Ultimate Sink: Urban Pollution in Historical Perspective* (Akron, OH: University of Akron Press, 1996), 288.

125 / Sewerage gardens at the London Asylum. Source: Archives of Ontario, RG 15-82-0-1-15.

161 / Sewer gas. Source: Maureen Ogle, *All the Modern Conveniences: American Household Plumbing, 1840-1890* (Baltimore and London: Johns Hopkins University Press, 1996), 123. From [Harriette M.] Plunkett, *Women, Plumbers, and Doctors; or, Household Sanitation* (New York, 1885; reprinted from T. Pridgin Teale, *Dangers to Health* [London, 1878]).

188 / Map of the Sanitary District of Chicago, 1923. Source: F. Garvin Davenport,

Illustration Credits

"The Sanitation Revolution in Illinois, 1870-1900," *Journal of the Illinois State Historical Society* 63 (Autumn 1973): 312.

211 / The Campbell plan for the "Sixth Great Lake." Source: Wallace J. Laut, *A Canadian's Plea to Restore the Water Levels of the Great Lakes – The Campbell Plan for the Restoration of the Great Lakes Water Levels: The Mighty Albany – St. Lawrence of the Northland* (Toronto: n.p., 1925).

230 / Interior filter of the Toronto Filtration Plant. Source: Leo G. Denis, *Water Works and Sewerage Systems of Canada* (Ottawa: Commission of Conservation, 1916), 97.

235 / The Toronto sewerage system. Source: Leo G. Denis, *Water Works and Sewerage Systems of Canada* (Ottawa: Commission of Conservation, 1916), 16.

319 / Surfers Against Sewage. Source: Surfers Against Sewage.

324 / Sewage Spill in Florida. Source: AP/Wide World Photos.

Index

Abbott, Chief Justice, 52-53
Abrams, Robert, 29
Activated Sludge Ltd., 242, 290. *See also* sewage: treatment
Adderley, Charles (Lord Norton), 129-32, 135-37, 142, 147, 197, 237-38
Advisory Committee on Water Pollution, 295
Allegheny River. *See* Pittsburgh
Allen, Dr. Norman, 96
American Institute of Chemical Engineers, 259-61
American Waterworks Association, 68, 271
Amyot, Dr. John A., 170
Anglers' Cooperative Association (ACA), 276, 283, 288
Anglers' Conservation Association, 276, 283, 288
Army Corps of Engineers, 49, 303
Ashbridge's Bay, 235-37
Attorney-General v. Birmingham, 129-32
Attorney-General v. Birmingham, Tame and Rea District Drainage Board, 238-42
Attorney-General v. Metropolitan Board of Works, 110

Bailey, Cleveland, 306-7
Baird, Nicol H. *See* Scugog River

balancing. *See* cost-benefit analysis
Bamford v. Turnley, 358
Barkley, Alben William, 306
Bazalgette, Joseph, 109-11, 120-21, 126
Bealey v. Shaw, 13-14
Beck, James M., 206-8
Belcourt, Senator Napoleon, 177-81
Bentham, Jeremy, 107
Berry, A.E., 296
Besselièvre, E.B., 259
best practicable means, 37, 142-46, 262
"Big Odor" case. *See Fieldhouse v. Toronto*
Biochemical Oxygen Demand (BOD) test, xvii, 220, 112
Birmingham: adoption of water closets in, 79; Sewage Inquiry, 132-33; sewage treatment, 135, 150, 238-41; water works, 135, 215-16
Birmingham, Tame, and New District Drainage Board, 238-42
Blackstone, Sir William, 11
Blundell v. Catterall, 51-56
Boonton Reservoir, 229
Booth, J.R., 47-48
Boston: adoption of plumbing fixtures in, 82, 93-94; sewerage, 117, 125, 168, 233; typhoid in, 159; water supply, 65, 71; water consumption, 73, 114-15

Boundary Waters Treaty. *See* Great Lakes
Bowell, Sir Mackenzie, 179
Boyd, Caldwell and Company, 38-39
Bramah, Joseph, 78
Bramwell, Baron, 140-41, 156
British Celanese Ltd., 277
British Field Sports Society, 269
British Institute of Water Engineers, 93
British Standards Institute, 93
British Waterworks Association, 93
Bronson, H. F., 43
Bryce, Dr. Peter H., 157, 163, 202
Buckland, Francis Trevelyan, 37-38
Buckley, Justice, 248
Burns, John, 214
Burr, Aaron, 63

Cal-Sag diversion, 195-96
Canadian Institute of Sewage and Sanitation, 284
Canada Water Act, 311-13, 385n100
Canada Water Conservation Assistance Act, 385n100
Canadian Iron Corporation, 256-57
Canadian Milk Products, 273
Canadian Steel Foundries Ltd., 270
Campbell, C. Lorne, 210-11
Campbell Plan, 210-11
Cartwright, Richard, 41-44
Centre for Pollution Studies, 175
Central Advisory Water Committee, 287
cesspools, 4, 79, 836, 91, 93, 96-97, 99-100, 108, 112, 115, 119, 128
Chadwick, Edwin, 61, 76, 107, 112, 120, 122, 124, 126
Chamberlain, Joseph, 132, 135-36, 152
Chelsea Water Works, 252
Chesbrough, Ellis Sylvester, 113, 184-85
Chicago River, 113, 183-95, 199, 202, 204-5
Chicago Sanitary District. *See* Sanitary District of Chicago
Chicago Steal, 186-95
chlorination, xiv, 176, 229-30, 234, 243, 249, 281, 286, 327
cholera, viii, ix, xii, 61, 103, 155, 162, 183-84, 214
Clean Water Act (US 1972), 309-10

Clean Water Restoration Act (US 1966), 309
Colne Valley Sewage Board, 283
Commission of Conservatio (Canada), 177-78, 202, 222, 228,
common law: criticism of rights holders for failing to assert their interests, 36, 137-42, 274-81; legislation restricting common law rights, 25-62, 283; limitations of private litigation, 137-142, 153; nuisance, 95, 141, 157, 163-66, 194, 231, 236-37, 247-49, 360n110; occupancy theory, 11, 13; prescription, 138, 246-49, 253-54; riparian rights, 12-30; reasonable use, 19, 21-23, 28-29, 40-41, 140, 337n38, 338n69. *See also* public rivers doctrine (Scotland)
conservancy principle, 99, 119-26, 130, 132
Control of Pollution Act, 314
Cooley, Lyman E., 189
Cooper v. Wandsworth, 87-89
cost-benefit analysis, 32-33, 45, 139-41, 160-62, 230-33, 254-59
Cozens-Hardy, Master of the Rolls, 240
Craigentinny Meadows, 124
Crapper, Thomas, 78
Cumberland, Fred, 66
Cumming, Hugh S., 303
Cuyahoga River, 296

Dallyn, Frederick A., 167
Darling, Frank W., 265
Darling v. City of Newport News, 265
Davyhulme Sewage Works, 219, 220
De Veber, Senator L. George, 178-80
Delaporte, Anthony V., 262-3
Delaware River, 168, 229, 301-2
Deneen, Governor Charles S., 198
Denman, Chief Justice Thomas, 33
Denman, Justice, 250
Denning, Lord Justice Alfred, 277
Detroit and Cleveland Navigation Company, 174
Detroit Sanitation Company, 118, 168
Dibdin, William Joseph, 156
dilution, xi; at Chicago, 183-212; economic advantages, 118, 221; reliance on, 168, 221, 226, 257, 263, 292
Dilution Project, 199, 188

Index

disease, water-borne, viii-ix, 154-62, 175-76. *See also* cholera
Don River, 271, 279, 294
Douglas, Justice William O., 282
Drainage of Trade Premises Act (1937), 262
Duff, Governor James H., 296

Earl of Harrington, 247-49
earth closet, 121, 126, 164
East London Water Works Company, 59, 344n24
Eastwood, John, 276, 283
Eckhart, B.A., 191
Ecological Society of America, 269
Edinburgh: adoption of plumbing fixtures in, 79; fire insurance in, 75; sewage irrigation, 99, 124, 246
Ellenborough, Lord, 13-14
enforcement of legislation and regulations, 37, 43, 44, 46-48, 145-46, 147-52, 164-67, 262-64, 272-74, 295
engineers and engineering, 65, 222; Bazalgette, Joseph, 109-11, 120-21, 126; Chesbrough, Ellis Sylvester, 113, 184-85; Hazen, Allen 159, 168, 175-76, 231-34, 271; Hering, Rudolph, 186, 199-200, 215, 229; Jervis, John B., 64, 65-66; Keefer, Thomas Coltrin, 65-66, 70-77, 201; Kierans, Tom, 317-18; McAlpine, William J., 43-44, 64-65; Mansergh, James, 74, 172, 216-17; Murray, T. Aird, 169, 178, 223
environmental values, 8, 9, 154, 268-70, 330-32. *See also* cost-benefit analysis; willingness to pay
environmental organizations. *See* Anglers' Cooperative Association; British Field Sports Society; Ecological Society of America; Izaak Walton League of America; National Coast Anti-Pollution League
Environmental Protection Agency, 309
Equitable Gas Company, 32-34
Europe: conservancy system, 122-24; influence on North American sewerage, 113, 150; sewerage, 110-2, 114, 353n63;

water-borne disease in, 169, 171, 228; water consumption, 73-74; water quality, 318-23; water treatment, 229; eutrophication, 292, 315, 323
excrement: impact on water quality, 142, 148, 151, 155, 171, 217; nature of 241; price of 119-26

Farwell, Lord Justice, 241
Federal Water Pollution Advisory Committee, 261
federalism and inter-governmental water management arrangements 7, 175-82, 294-98, 305-10. *See also* river basin management
Ffennell, William, 36
Fieldhouse, Samuel E., 235-37
Fieldhouse v. Toronto, 235-237
filtration. *See* Lawrence Experimental Station; sewage: treatment
finance, 113, 150, 306-9. *See also* cost-benefit analysis; willingness to pay
fire: Chicago losses 185; fire insurance and water supply, 75-77; influence on water supply systems, 57, 60-69, 74, 177; London losses, 75; New York losses, 63
fish and fisheries: affected by water pollution, 31-36, 48-49, 142, 146, 271, 328-39; British salmon protection legislation 36-38; *Fisheries Act* and water quality, 312-13; litigation by fishing interests against pollution, 35, 129, 264-66, 270, 328; mercury contamination, 292-93; relationship to water quality, 299. *See also* Anglers' Cooperative Association
Fisher, Justice Harry M., 210
Fisheries Act, 47, 312-13
Fleming, Sandford, 103-4, 106
Fletcher Moulton, L.J., 240
Flower, Major Lamorock, 147
Folsom, C.D., 150
Fowler, Dr. Gilbert John, 219-20
Frank, Leslie C., 242, 298
Frankland, Dr. Edward, 142-43, 155
Franklin, Benjamin, 62
Frost, Leslie, 279, 294, 310
Fuller, George Warren, 271, 284

Gaskell, Charles Milnes, 148-49
George Legge & Son, 252
Goodell, Edwin B., 164-65, 254
Grand Canal, 317-18
Great Lakes: decline of fisheries in, 34; impact of Chicago diversion, 186-212; municipal contamination of, 118, 168-69; restoration of, 310, 315-18; water quality in, 106, 170-73
Great Lakes Water Quality Agreement, 315
Great Stink, 107-10
Great Recycling and Northern Development Canal, 317-18
Griscom, John H., 136
Groat, Malcolm Forbes and Groat, Walter S., 274-76
Groat v. Edmonton, 274-76
Gzowski, Casimir Stanislaus, 66

Halsbury, Lord, 251
Hamburg, 76, 110-11, 113, 228
Hamlin, Christopher, 101-2
Hampton v. Watson, 264-65
Harington, John, 78
Hassall, Arthur Hill, 103-4
Hazen, Allen 159, 168, 175-76, 231-34, 271
Hemenway, Dr. Henry Bixby, 91, 98
Hering, Rudolph, 186, 199-200, 215, 229
Hertfordshire County Council, 283
Hind, Henry Youle, 103-4, 106
Hole v. Barlow, 140-41
Holmes, Justice Oliver Wendell, 194-95, 205-6, 244-45, 265, 275
Hudson River: contamination of, 40, 173-74, 226; as a municipal water supply source, 158, 215; Storm King development, 281
Hughes, Charles Evans, 169, 206
Humphrey, Hubert, 289
Hurst, Willard, 25
Huxley, Aldous, xxi-ii, 290
Hyperion Activated Sludge Plant, 290

Illinois and Michigan Canal, 184-87
Imhoff tank, 357n34
Institute of Sanitary Engineers, 284

Institute of Sewage Purification, 284
institutional design and reform for water management, 7-8. *See also* river basin management
insurance. *See* fire
international aspects of water quality: Great Lakes, 315-18; North Sea, 318-21
International Joint Commission, 169-73, 181, 315-16
International Waterways Commission, 191, 195, 199
Irwell, x, 12-15, 148-49
Izaak Walton League of America, 268, 285, 304-5

Jephson, Dr. Henry, 83, 97,
Jersey City Water Company, 229
Jervis, John B., 64-66
Jessel, Sir George, 238
Johnson, Cuthbert, 119, 126
Jones and Attwood Ltd., 219, 242

Kalamazoo Vegetable and Parchment Company, 278-79
Keefer, Thomas Coltrin, 65-66, 70-77, 201
Kekewich, Justice, 239
Kent, Chancellor James, 21-22
Kierans, Tom, 317-18
Killaly, Hon. William Hamilton, 44-46
Kinebuhler, Carl, 190
Kinnersley, David, 287, 300

Lachine Rapids, 65, 71
Landis, Judge Kenesaw Mountain, 205
Latrobe, Benjamin Henry, 62, 74
Lauder, Albert E., 119, 122
Lavoisier, Antoine-Laurent, 11-12
Lawrence Experimental Station, 159, 219, 231, 242, 271
Le Roy, Emmanuel Ladurie, 12
Leith, 150, 245-46
Leopold, Aldo, 269-70, 282
Lindley, Lord Justice, 149, 251
Lindley, William, 112
Local Government Amendment Act. *See Attorney-General v. Birmingham, Tame and Rea District Drainage Board*

Local Government Board, 93, 145, 148-49, 152, 218, 222
London: disease in, viii-xi, 103; fire 75-76; Metropolitan Board of Works, 108-10; sewerage, 83-87, 107-12, 117, 155-56; water consumption, 74, 115; water supply, 58-61, 101, 214, 228. *See also* Thames River
London Bridge Water Works, 58
Los Angeles, 118, 126, 290
Lougheed, Senator James, 180
Lowell, Mass., 73, 158-59
lumber industry: impact on water quality (sawdust) 38-49

McAlpine, William J., 43-44, 64-65
McCullough, Dr. John W.S., 166, 170
McKie v. KVP, 278-79
McLaren v. Caldwell, 38-39
McLaren, J.S.P., 141, 153
McLaughlin, Dr. Allan J., 170, 172-74
MacLeod, Roy, 36-37
Macmillan, Lord, 253
Macnaughton, Lord, 253-54
McRuer, Chief Justice James Chalmers, 278-79
McSweeney, Senator, 178
Manchester (UK), 12, 14, 16, 61, 79; sewerage, 219; water consumption, 73; water storage, 215;
Mansergh, James, 74, 172, 216-17
Marsh, George Perkins, 48
Massachusetts State Board of Health, 115, 119, 150-51, 159
Maurice, H.G., 268, 299
Melosi, Martin V., 101, 230
Menzies, J.R., 311
Merrimack River, 118, 151, 245; typhoid along, 158-59
Metropolis Local Management Act (1855), 84-89
Metropolitan Commission of Sewers, 84
Metropolitan Sewerage Commission, 227
Middelkerke (Belgium), 229
mill dams: environmental impact of, 16-17, 26; legislation interfering with riparian rights, 25-26; and water power, 15-17, 23-28

Miner v. Gilmour, 27-28
Missouri v. Illinois, 194-95, 244-45
Monger, Dr. John Emerson, 271-72
Montagu, Lord Robert, 142
Montreal: nineteenth-century water supply, 65-66, 70-74; sewage arrangements, xvii, 179-80
Moule, the Reverend Henry, 121, 164
Mulock, Chief Justice, 235-37
Mundt Bill, 306
municipal government, xii, 61, 113-14, 294-97. *See also* sewage: treatment
Murray, T. Aird, 169, 178, 223
Myddleton, Hugh, 58, 187

Native Guano Company, 134
National Coast Anti Pollution League, 383n53
National Resources Committee, 304
National Resources Planning Board, 304
National Rivers Authority, 314
navigation and water quality, 46-51, 177-82, 199-205. See also *Refuse Act*
Navigable Waters Protection Act, 48, 177
Nepisiguit River, 256-57
New York (City): consumption of water in, 73, 215; condition of harbour, 49, 151, 219, 224-26, 301-2; regulation of plumbing fixtures, 82, 93-95; sewerage 113, 116, 226; water supply, 63-64, 71-72, 214-15;
New York (State): water pollution control, 165, 169, 206, 223, 226-27, 297
New York Fish Commission, 48-49
Nichol, John, 117

ocean pollution, 117-18, 125, 178-79, 264-65, 318-20, 323
Ogle, Maureen, 82
O'Hanly, J.L.P., 200-1
Ohio, 271-72, 288-89, 296, 302
Ohio River Valley Water Sanitation Commission, 303
Ohio Water Pollution Control Board, 288
Ontario: water quality in, 262-63. *See also* Don River; Ottawa River; Toronto
Ontario Board of Public Health, 166-67

Ottawa: sewage arrangements, 43, 121; water supply, 176-77
Ottawa River, 38, 41-45, 47, 175-76, 178
Oysters, 264-66

Parker, Alderman Lawley, 136
Passaic River, 213-14, 224-26
Pennsylvania: mill development, 25; response to water pollution 165, 214, 259-64, 301-3, 375n60; sewage irrigation, 126; water quality, 158-61, 223, 245, 249, 296. See also *Sanderson v. Pennsylvania Coal*
Pennsylvania Administrative Code, 260
Pennsylvania Coal Company, 254. *See also Sanderson v. Pennsylvania Coal*
Pennsylvania State Sanitary Water Board, 302
Philadelphia: adoption of plumbing fixtures, 82, 93; nineteenth-century water supply, 62-63, 67, 229; sewage arrangements, 118, 302; typhoid in, 158; 363n71; water consumption, 73;
Pinchot, Gifford, 259
Pittsburgh, 73, 117, 233-34
plumbing standards, 91-93
pollution. *See* water pollution
Pooh, vii, xxi, 325
President's Water Resources Policy Commission, 304
Pride of Derby and Derbyshire Angling Association, 276-77
Pride of Derby v. British Celanese, 276-77
Prince of Wales, 110
privy pit, 4, 15, 84-91, 95-99, 112, 114, 121-22
Public Health: theories of disease, 99-107, 112-14, 159-62
Public Health Act (1848), 61
Public Health Act, 1875, 132, 136, 149, 162, 238-39, 250, 252
Public Health Act (Ont. 1912), 165
Public Health Act (Ont. 1950), 283
Public interest in water, 37, 39
public rights in water, 52-56. *See also* fish and fisheries; navigation and water quality
public rivers doctrine (Scotland), 245-46, 254, 265
Purdy, William. *See* Scugog River

Quincy, Josiah 71

R.C. Harris Filtration Plant (Toronto), xvi, xx
R. v. Medley, 339n2
Ramapo Water Company, 215
Ramsay, Sir William, 241
Randolph, Isham, 187, 195-99, 201, 209
Ratté, Antoine, 45
Ratté v. Booth, 45
reasonable use. *See* common law
Red River Roller Mills v. Wright, 338n68
Refuse Act (1899), 50-51, 191, 227, 281-82, 289, 293, 378n59
regulation of plumbing and household fixtures, 86-96, 112-14
regulation of water quality: Canada, 165-68, 310-13; Great Britain, 142-47, 298-300, 313-14; US, 150-51, 157-58, 164-65, 300-5, 305-10
Reitze, Arnold, 288
Ribble Joint Committee, 148
Rinfret, Justice Thibodeau, 274-76
risk, xviii
river basin management, 298-305, 311
River Irwell. *See* Irwell
River Boards Act (1948), 298
Rivers and Harbors Act (1899), 50-51, 191, 227, 281-82, 289, 293, 378n59
Rivers Pollution Prevention Act (1876), 144, 146, 149, 152, 248, 252, 295, 298
Rivers Pollution Commission, 135-43, 155
Rivers (Prevention of Pollution) Act (1951), 287
Rivers (Prevention of Pollution) Act (1961), 300
Robinson, Hiram, 47-48
Rose, Carol, 19, 40
Rosen, Christine, 255-58
Royal Commission on Sanitary Laws, 132, 135
Royal Commission on Sewage Disposal, 218-22, 262

Sanderson v. Pennsylvania Coal, 254-55, 258
Sanitary District of Chicago, 187-92, 195-210

Index 403

Sanitary District of Chicago v. United States, 205-6
Sax, Joseph, 3
Schuylkill River, 62, 168, 229
Scotland: public rivers doctrine, 245-46, 254; water pollution, 145, 152, 251; water treatment, 228. *See also* Edinburgh; Leith
Scugog River, 23-25
sea bathing, 51-56, 209, 347
Sedgwick, William Thompson, 159, 190, 222, 228, 232
Seeger, Pete, 281
septic tank, 357n34
Severn-Trent regional water authority, 300
sewage: sewage farming, 122-26, 238; treatment, 132-36, 152-53, 159-60, 218-23, 242-43, 325; used as fertilizer, 119-26; value, 119-26; at World's Columbian Exposition, 190
sewerage systems: extent and growth of, xvi, 114-15, 126-27; 222, 284-85, 286, 294, 353n59; technology of, 344n27
Shanley, Walter, 66
S—t. *See* excrement
Sludge Acid Act (1886), 227
Smith, Adam, 11
Smith, J. Toulmin, 84-85
source protection. *See* water supply
Spinks, William, 94
Steinberg, Ted, 19
Stephens, Annie, 279-81
Stephens v. Richmond Hill, 279-81
Stevenson, W.L., 260-61
Stimson, Henry L., 202-4
Story, Joseph, 20, 177

Taft, William Howard, 200, 209
Thames River: restoration of, 108-10; sewage discharge into, xi, 31-33, 60-61, 108, 155-56; as a source of water supply, ix, 58-59, 101, 103, 214
Thirlmere Dam, 215
Thom, Brigadier General, 43, 48
Tigger. *See* Pooh
Tinkler v. Wandsworth, 86-87
toilets: adoption of, 79-85; 89-96; adoption of, through elimination of privies, cesspools, 86-96, 112; early water closets, 78-81; modern design developments, 325-26
Toronto: 163, 165; growth of wastewater treatment system, xvi-xvii; nineteenth-century water supply, 66-69, 215-17, 369n10; nuisance abatement, 95-96; sewerage and sewage treatment, 235-37, 294; water consumption, 73, 74; water treatment, 229-30; waterworks competition 103-6. *See also* R.C. Harris Filtration Plant (Toronto)
Toronto Gas, Light & Water Company, xiii, 66
Tyler v. Wilkinson, 20-22
typhoid. *See* disease, water-borne

United States v. Republic Steel, 282
United States v. Standard Oil, 282
urinals, 89-90
US Army Corps of Engineers, 49, 303
US Public Health Service, 170, 177, 285, 303, 306-7

Van Cortland, Dr. E., 43
Vancouver, xii, xvii-xviii
Vettabia Canal (Milan), 124
von Liebig, Justus, 120
Vyrnwy Dam, 215

Walkerton, xxi, 5, 325
Walpole, Spencer, 37
Walton, Izaak 268. *See also* Izaak Walton League of America
Waring, Col. George Edwin Jr., 95, 121,
waste: assumptions about acceptability of, 4-5, 99, 115-29, 231-34, 259-262, 291, 327
water closet: adoption of, 78-82; legal requirements for, 83-89, 90, 93-96
water companies: Canada, 65-68; Great Britain, 58-61, 344n21, 348n49; US, 62-65, 350n90
water law, 3-4, 12, 18-19, 20-23, 129-32, 244-54, 274-83, 328. *See also* common law; enforcement of legislation and regulations

water pollution, 289; agricultural sources of, 34, 293, 381n13; at Chicago, 193; drug residues and personal care products, 323; industrial sources of, 32, 34, 35, 271-73; mills and, 15-17; municipal sources of, 142, 148, 151, 155, 171, 217; in New York Harbor, 49, 151, 219, 224-26, 301-2; along Passaic Valley, 224-25; from pulp and paper, 34, 292-93; synthetic detergents, 292; textiles, 14, 26-27, 141; at Toronto, 217

Water Pollution Control Act (US 1948, amended 1956 and 1961), 307-9

water quality: 5-6, 69, 102-3, 169, 322; classification of rivers according to, 259-61, 262-4; measurement of in Boundary Waters, 168-75; public health impacts of, 99-107, 112-14, 154-64, 186; self-purification of rivers, xi, 115-9, 154, 155, 158, 353n68; standards for, 142-46, 221-22, 238-42, 259-62, 287-89; *Water Quality Act* (US 1965), 309

water rights. *See* common law

water supply: 97, 99; protection of sources, 213-15, 223, 234; treatment, 228-34. *See also* chlorination; filtration; water companies

water uses: consumption of water, 57-58, 60, 64-67, 72-75, 114-15; hierarchy of, 9, 31, 208, 326; history of 7, 19-20

watershed management: by private agreement, 26-27, 147-50, 298-305. *See also* river basin management

waterworks: extent and growth of, 64, 65, 67-68, 97

West Riding of Yorkshire Joint Committee (Rivers Board), 148-49, 253

Whipple, George Chandler, 169, 232-34, 267

Wicksteed, Thomas, 157

willingness to pay for water, sewerage, and environmental services xviii, xx, 115, 146-47, 150-51, 223, 225, 267-68, 295-97, 312, 330, 349n67

Wisconsin, 25, 114, 296; opposition to Chicago diversion, 206-9

Wisconsin v. Illinois, 206-9

Wood v. Waud, 246

World's Columbian Exposition: sewerage arrangements, 189-90, 231

Wright v. Howard, 22

Yates, Richard, 193
Yonge, C.M., 299

Printed and bound in Canada by Friesens

Set in Adobe Garamond by Robert and Shirley Kroeger,
 Kroeger Enterprises

Copy editor: Deborah Kerr

Proofreader: Anna Eberhard Friedlander